Many brilliant resources for o[illegible]

Learning musculoskeletal anatomy and palpation can be a daun[illegible] [illegible] we're here to help. Explore this collection of study tools that can make it easier for you to master your classes, and pass your exams!

Trail Guide to the Body Textbook (5th Edition)

A fun and easy-to-understand guide to locating muscles, bones, and more. Before you can assess or treat a muscle, you must first be able to locate it. This acclaimed book delivers beautifully illustrated information for learning palpation and the musculoskeletal system, making mastering essential manual therapy skills interesting, memorable, and easy. With 504 pages and 1,400 illustrations covering more than 162 muscles, 206 bones, 33 ligaments and 110 bony landmarks, this text provides an invaluable map of the body. Our newest edition also includes an extensive appendix that describes the common trigger-point locations and pain patterns of 100 muscles.

Get the textbook and gain FREE ACCESS to amazing study tools. When you purchase the *Trail Guide to the Body* textbook — the parent to your Student Workbook — you'll have free access to an amazing repository of learning resources designed exclusively for students. ***Click into the FOR STUDENTS section at booksofdiscovery.com***, where you'll find palpation videos, overlay images, and an audio guide — all complimentary!

AnatomyMapp®

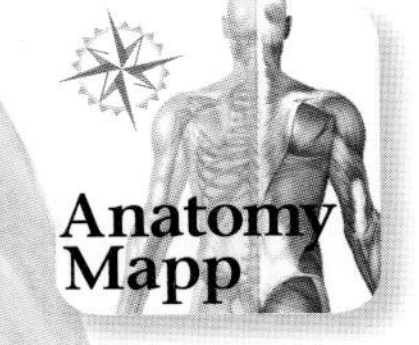

The best interactive app for learning musculoskeletal anatomy is now available on your Apple or Android mobile devices. AnatomyMapp includes information from all 364 *Trail Guide to the Body* flashcards; questions from the *Trail Guide to the Body* Student Workbook; and easy search and bookmarking capabilities. This app reinforces your learning and is a perfect companion tool for today's students. Available at Google Play and the App Store.

PalpationMapp®

PalpationMapp is a fun and explorative palpatory anatomy app. This useful visual resource is based on content from *Trail Guide to the Body* and led by author Andrew Biel. Watch detailed, high-definition videos on how to identify and palpate 91 essential muscles, including tips for isolating bony landmarks, attachment sites, and muscle borders. In addition, access more than 200 muscle overlay images for visualizing underlying anatomy. Available only at the App Store.

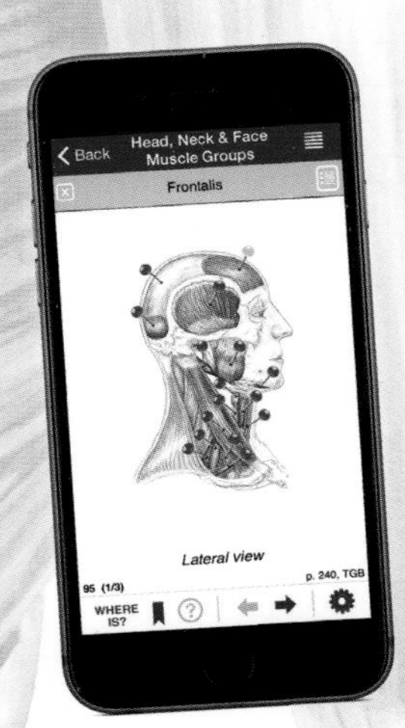

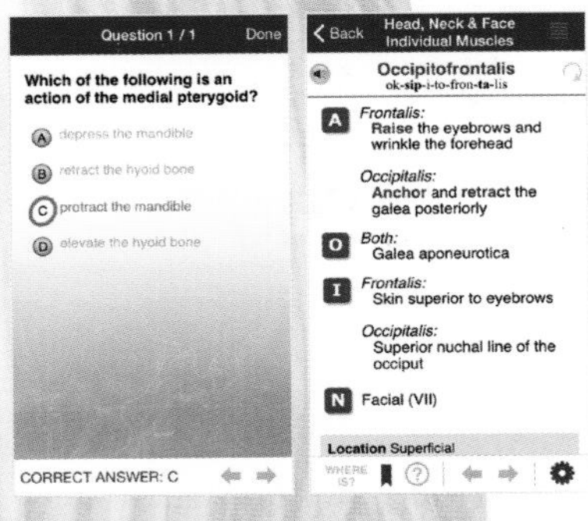

BodyMapp App Bundle

Get two apps in one! This app brings together **AnatomyMapp** and **PalpationMapp** in an all-in-one bundle, combining the best of musculoskeletal anatomy and palpation information. All content is from the acclaimed *Trail Guide to the Body* textbook and palpation videos. Available only at the App Store.

Trail Guide to the Body

A hands-on guide to locating muscles, bones and more

Student Workbook

Fifth Edition

Andrew Biel, LMP

Licensed Massage Practitioner

Illustrations by Robin Dorn, LMP

Licensed Massage Practitioner

Fifth Edition

Published by Books of Discovery
2539 Spruce St., Boulder, CO 80302 USA
www.booksofdiscovery.com
800.775.9227

Associate Editors
Shelly Barnard
Dana Ecklund

Special thanks to Aaron Adams, Cara Barbee, Ashley Bechel, Miranda Legge, Christine Malles, Gene Martinez, Mindy Morton, Lori Olcott and Alicia Pouarz.

Designer: Jessica Xavier of Planet X Design

Printed in South Korea by Four Colour Print Group, Louisville, Kentucky

Library of Congress Cataloging-in-Publication Data

Biel, Andrew R.
Trail Guide to the Body: Student Workbook
Fifth Edition

ISBN: 978-0-9829786-6-5

15 14 13 12 11 10 9 8 7

Disclaimer
The purpose of Books of Discovery's products is to provide information for hands-on therapists on the subject of palpatory anatomy. This book does not offer medical advice to the reader and is not intended as a replacement for appropriate healthcare and treatment. For such advice, readers should consult a licensed physician.

Table of Contents

How to Use This Workbook

Welcome to the *Trail Guide to the Body* Student Workbook. We designed this tool to help you develop and refine the techniques you have become familiar with in *Trail Guide* (5th Edition).

You can find the answers at the back of the workbook, and each answer is followed by the page number that corresponds to information in *Trail Guide*.

The **red oval** at the top of each page also provides the page number that you can refer to in *Trail Guide* for more information.

In the **CHOICES** box, a number after the choice indicates how many locations on the illustration it can be used.

Many of the anatomy illustrations have been left **uncolored** so you can learn their structures by highlighting them with colored markers.

The **Shorten or Lengthen?** section is designed to help you learn the span of a muscle when a joint is in a specific position. If you become confused, try standing up and performing the movement on your own body. (There's nothing like a little kinesthetic exploration to help you find the answer.)

The **Let's Palpate!** section can get you started as you create your own palpatory journal (see p. 9 in *Trail Guide*). Find three people—roommate, classmate, friend—and explore the specified bony landmark or muscle. Remember: There are no wrong answers, so you can describe what you feel in any way you prefer.

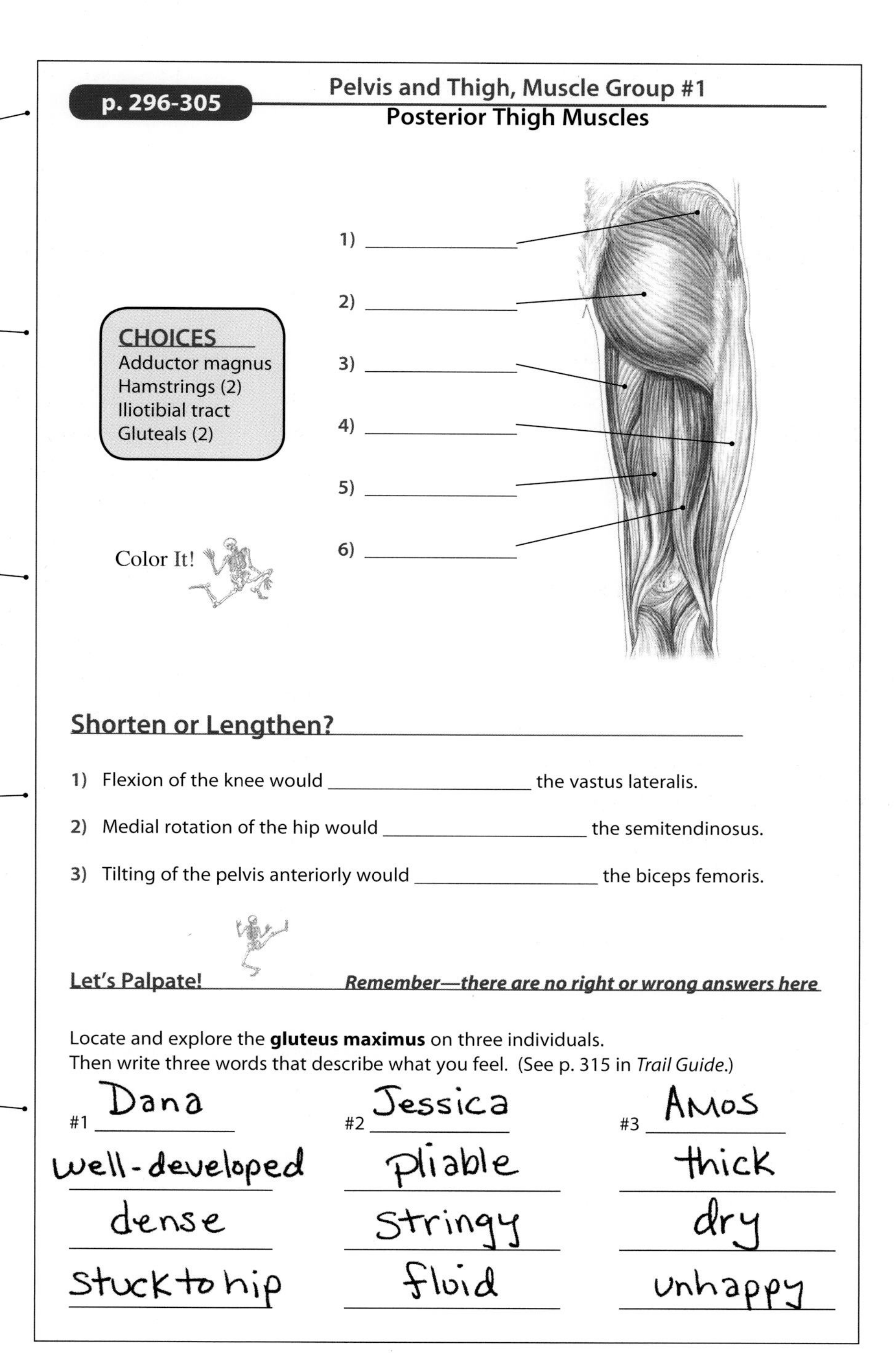

Hey, before you go: Those answer pages might be tempting, but try to use them only after you have completed the page or when you are really stuck. Happy trails!

Introduction

Tour Guide Tips #1

Please answer the following questions.

1) What are the "trail markers" that will help you locate muscles and tendons? ____________________

2) Since there are different body types and the terrain is never the same, explain how one "body map" could apply to all of them.

____

3) What does "palpation" mean? ____________________

4) Palpation is an art and skill that involves ____________ a structure, becoming ____________ of its characteristics and ____________ its quality or condition.

5) Laying one hand on the other allows the bottom hand to stay receptive while the top hand ____________ and ____________.

6) Name three ways palpation can be made easier by "working smart." ____________ ____________ ____________

7) When outlining the shape or edge of a bone, it is best to roll your fingers ____________ rather than ____________ its surface.

8) If the structure you are palpating is moving, your hands should ____________.

9) When a movement is performed by your partner it is called ____________ and when your partner relaxes and allows you to do the movement it is called ____________.

10) In *Trail Guide to the Body,* resisted movements are used to distinguish the ____________, ____________ and ____________ of different muscles and tendons.

11) As you improve your palpatory skills, what are three qualities or principles you will want to practice?

____________ ____________ ____________

12) Skeletal muscle is composed of nerves, blood vessels, ____________ and ____________.

13) A muscle's connective tissue layers merge at either end of the muscle to form a strong ____________.

Tour Guide Tips #2

p. 2-17

Please answer the following questions.

1) The muscle that carries out an action is called the ______________________, while the muscle that resists this action is the ______________________.

2) Please name the three physical characteristics which help to distinguish muscle from other tissues.

   ______________________ ______________________ ______________________

3) To distinguish a tendon from a ligament, explore its ______________________ and ______________________.

4) Name three types of connective tissue that are palpable.

   ______________________ ______________________ ______________________

5) Fascia is a continuous sheet of ______________________ located beneath the ______________________ and around muscles and organs.

6) A sharp, shooting sensation felt locally or down the corresponding appendage during palpation may be caused by ______________________.

Matching

Please match the term to the best definition.

1) ______	adipose	a)	a voluntary contractile tissue that moves the skeleton
2) ______	aponeurosis	b)	two types—superficial and deep
3) ______	artery	c)	a vessel easily seen on the dorsal surface of the hand
4) ______	bone	d)	a vessel in which a pulse can be felt
5) ______	bursa	e)	a small, fluid-filled sac that reduces friction between two structures
6) ______	fascia	f)	a broad, flat tendon
7) ______	ligament	g)	a structure connecting bones together at a joint
8) ______	lymph node	h)	easy to distinguish by its solid feel
9) ______	muscle	i)	bean-shaped, ranging in size from pea- to almond-sized
10) ______	nerve	j)	a tube-shaped vessel that becomes tender when compressed
11) ______	retinaculum	k)	a transverse thickening of deep fascia, strapping down tendons
12) ______	skin	l)	the largest organ in the body
13) ______	tendon	m)	attaches muscle to bone
14) ______	vein	n)	tissue with a gelatinous consistency

Exploring the Textural Differences of Structures #1

Please identify the following structures.

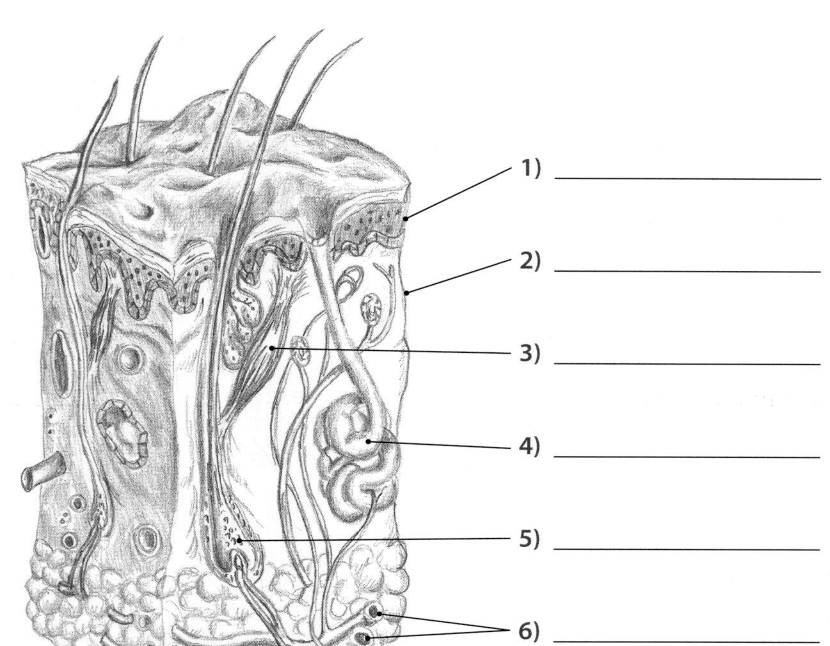

CHOICES
- Arrector pili muscle
- Blood vessels
- Dermis
- Epidermis
- Hair follicle
- Sweat gland

Cross section of the skin

Color Them!

7) MUSCLE FIBERS

8) ENDOMYSIUM

9) PERIMYSIUM

10) EPIMYSIUM

11) BONE

12) BLOOD VESSELS

13) NEUROVASCULAR BUNDLE

14) TENDON

15) PERIOSTEUM

Cross section of a typical skeletal muscle

CHOICES

Blood vessels	Epimysium	Perimysium
Bone	Muscle fibers	Periosteum
Endomysium	Neurovascular bundle	Tendon

Types of Muscle Bellies and Joints

p. 12 & 34

Please identify the types of muscle bellies and joints.

Color Them!

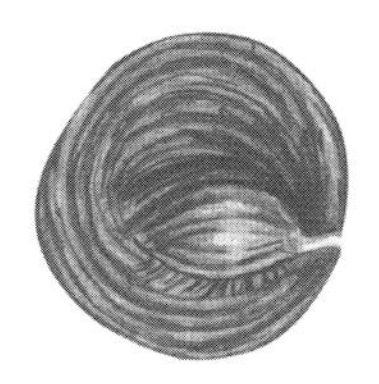

1) FUSIFORM 2) MULTIPENNATE 3) SPHINCTER

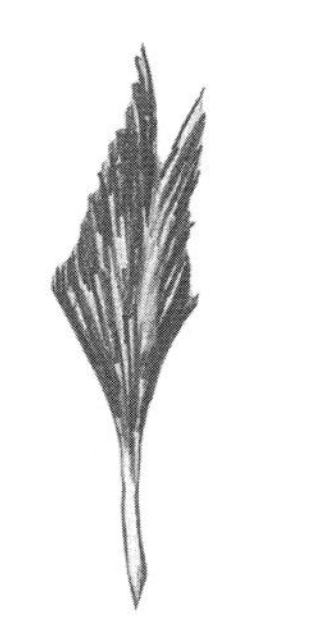

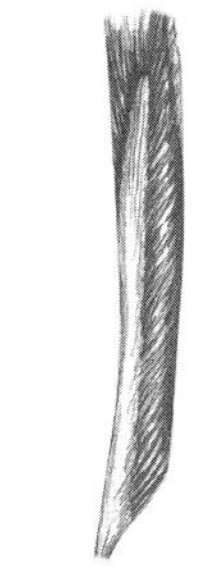

4) BIPENNATE 5) TRIANGULAR 6) UNIPENNATE

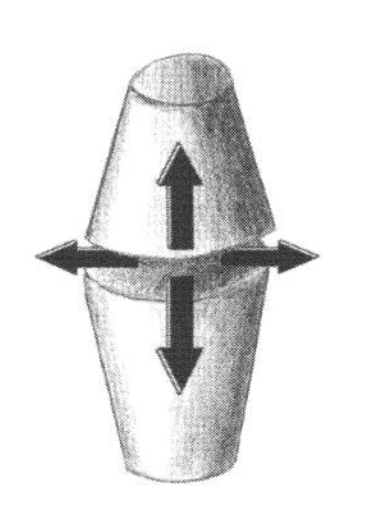

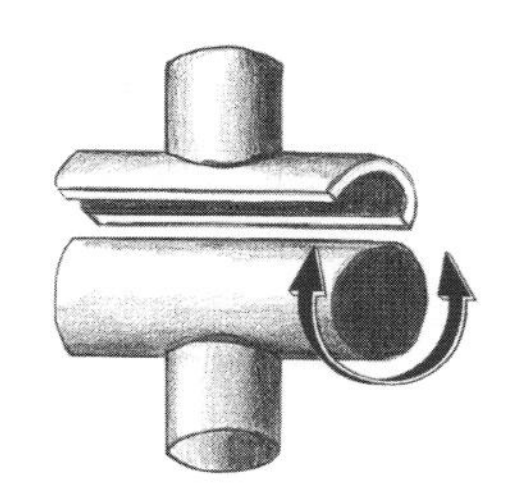

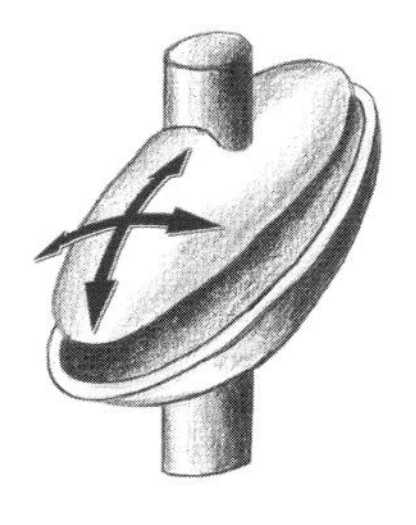

7) ____________ 8) ____________ 9) ____________

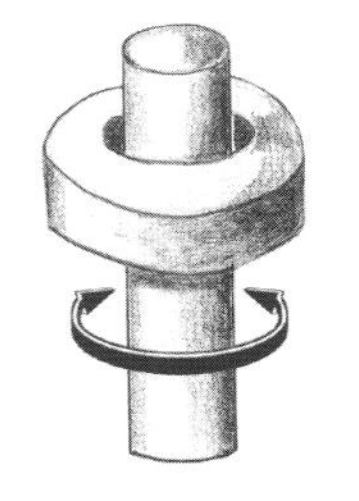

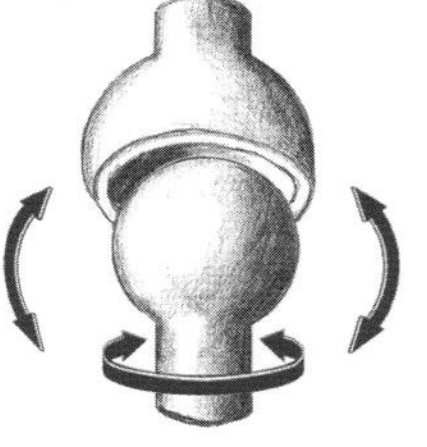

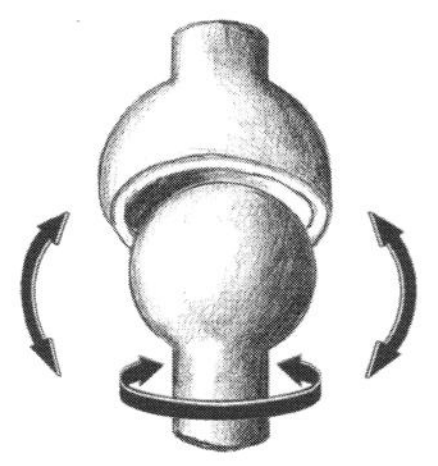

10) ____________ 11) ____________ 12) ____________

CHOICES

Ball-and-socket	Bipennate	Ellipsoid	Fusiform
Gliding	Hinge	Multipennate	Pivot
Saddle	Sphincter	Triangular	Unipennate

Exploring the Textural Differences of Structures #2

Please identify the following structures.

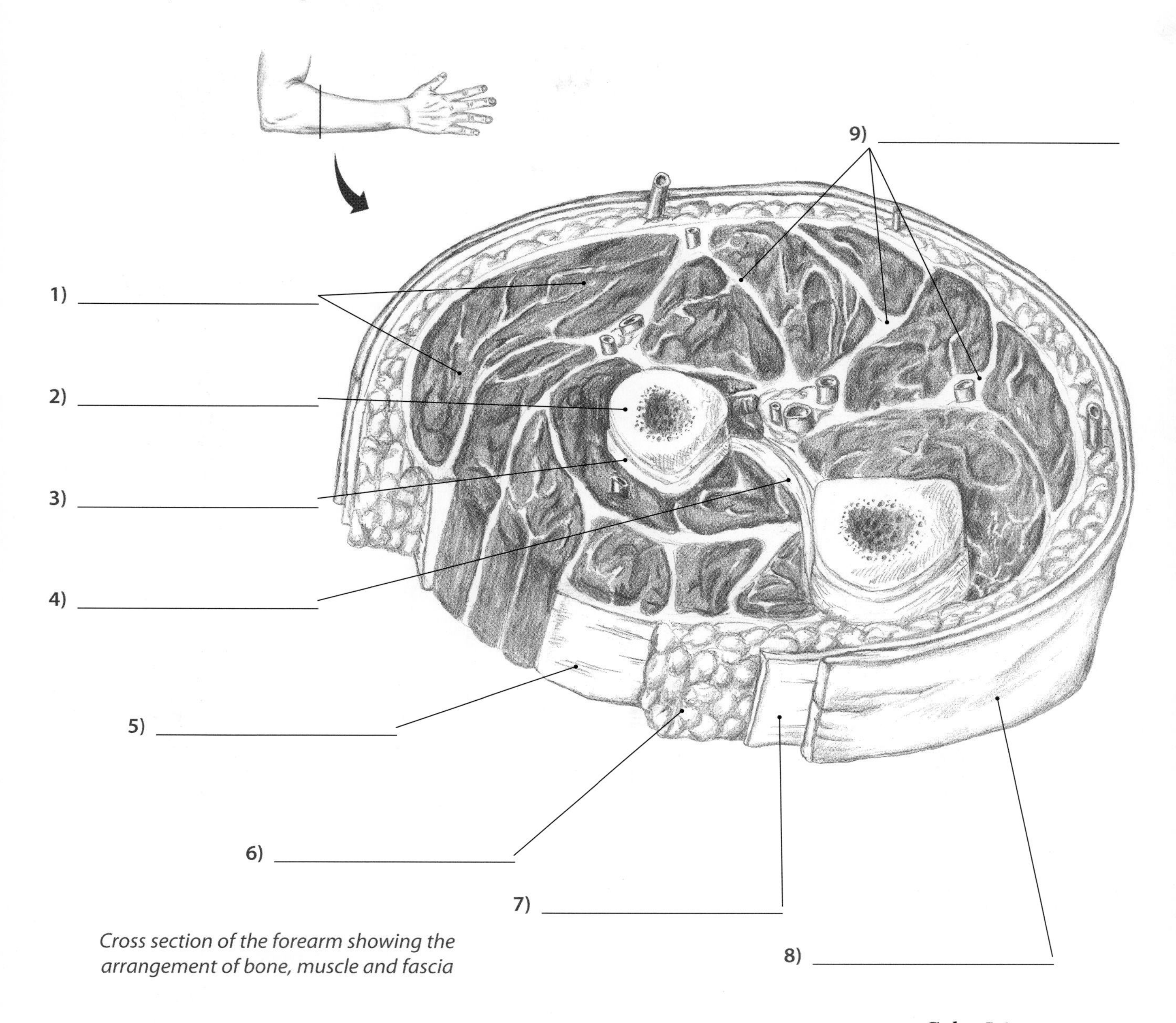

Cross section of the forearm showing the arrangement of bone, muscle and fascia

Color It!

CHOICES		
Adipose (fatty) tissue	Interosseous membrane	Skin
Bone	Muscle tissue	Superficial fascia
Deep fascia (2)	Periosteum	

Regions of the Body

Please identify the following regions.

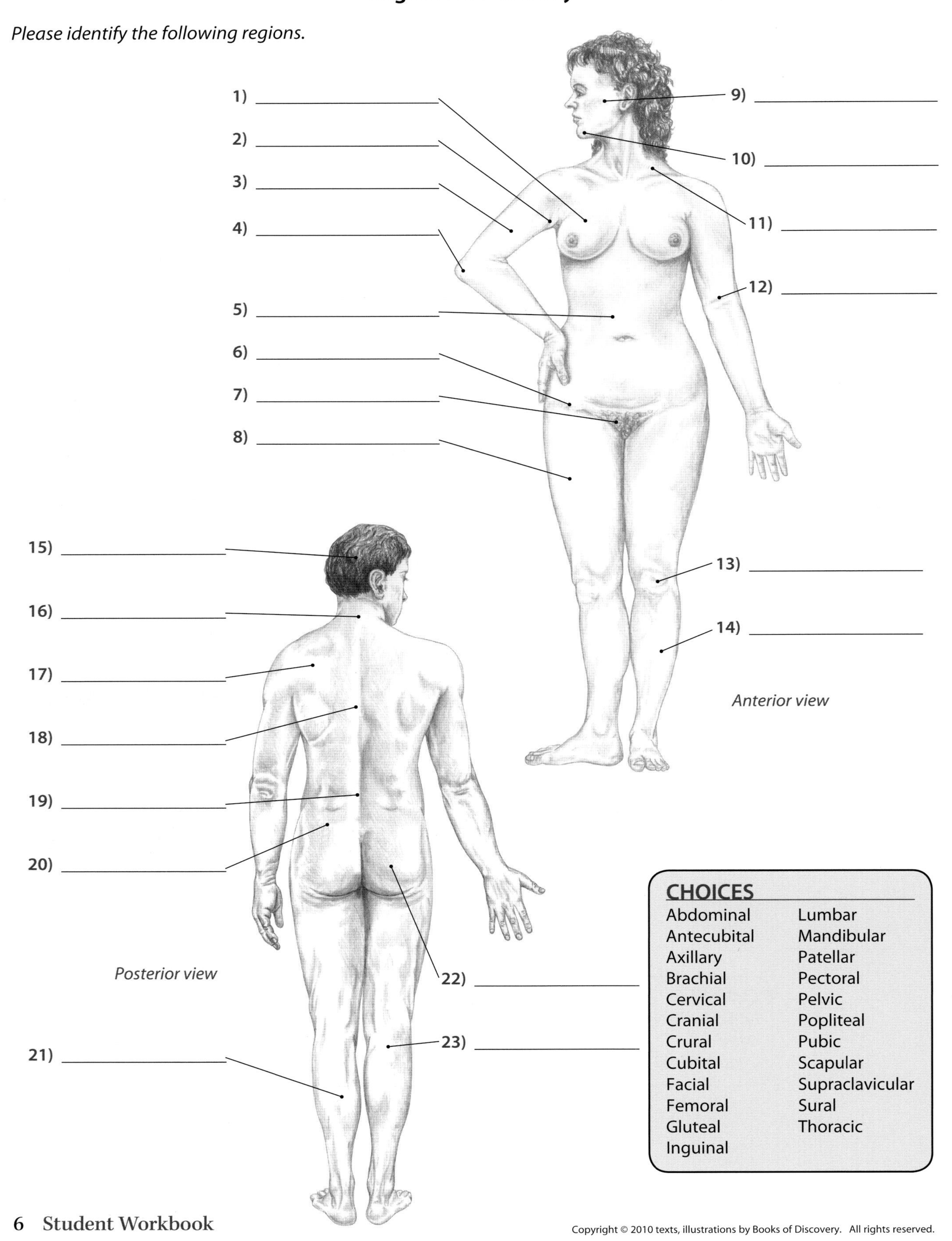

CHOICES

Abdominal	Lumbar
Antecubital	Mandibular
Axillary	Patellar
Brachial	Pectoral
Cervical	Pelvic
Cranial	Popliteal
Crural	Pubic
Cubital	Scapular
Facial	Supraclavicular
Femoral	Sural
Gluteal	Thoracic
Inguinal	

Navigating the Body

Planes, Directions, Positions and Movements #1

Please identify the following planes, directions and positions.

1) ______________

2) ______________

3) ______________

The head is

4) ______________

to the abdomen.

The abdomen is

5) ______________

to the head.

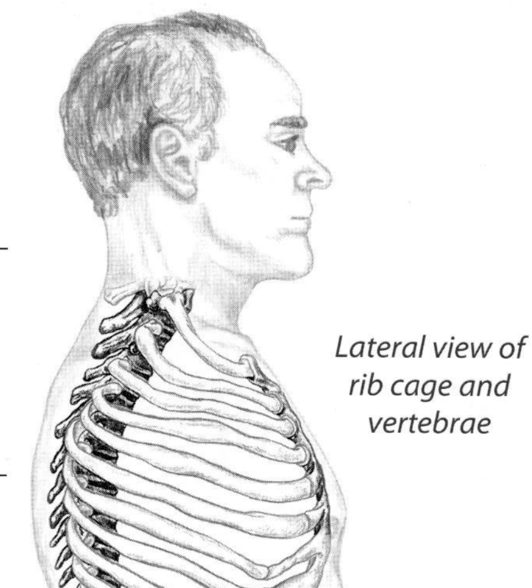

Lateral view of rib cage and vertebrae

The spine is **6)** ______________ to the sternum.

The sternum is **7)** ______________ to the spine.

The elbow is

8) ______________

to the wrist.

Right arm and forearm

The wrist is

9) ______________

to the elbow.

Right leg and foot

The big toe is

10) ______________

to the last (pinkie) toe.

The last (pinkie) toe is

11) ______________

to the big toe.

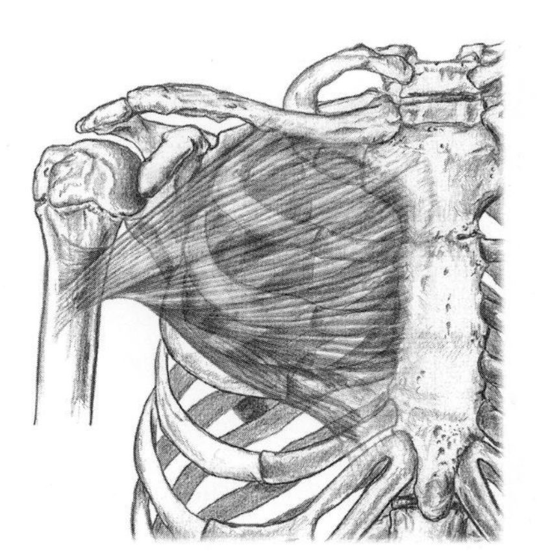

The pectoralis major muscle is

12) ______________

to the ribs. The ribs are

13) ______________

to the pectoralis major muscle.

CHOICES

Anterior	Posterior
Deep	Proximal
Distal	Sagittal
Frontal	Superficial
Inferior	Superior
Lateral	Transverse
Medial	

Planes, Directions, Positions and Movements #2

Please match the word to the appropriate definition.

Directions and Positions

1) ______	anterior	a)	further toward the back of the body
2) ______	deep	b)	a structure of the arm or leg that is further away from the trunk
3) ______	distal	c)	further toward the front of the body
4) ______	inferior	d)	a structure closer to the head
5) ______	lateral	e)	a structure closer to the feet
6) ______	medial	f)	a structure closer to the body's surface
7) ______	posterior	g)	a structure of the arm or leg that is closer to the trunk
8) ______	proximal	h)	closer to the midline of the body
9) ______	superficial	i)	further away from the midline of the body
10) ______	superior	j)	a structure deeper in the body

Movements of the Body

11) ______	abduction	k)	a movement that moves a limb laterally away from the midline
12) ______	adduction	l)	a limb at the shoulder or hip turns in toward the midline
13) ______	circumduction	m)	a limb at the shoulder or hip swings away from the midline
14) ______	dorsiflexion	n)	a movement bringing the radius and ulna parallel to one another
15) ______	extension	o)	ankle movement stepping on the car's gas pedal
16) ______	flexion	p)	a combination of flexion, extension, adduction and abduction
17) ______	lateral flexion	q)	when the head or vertebral column bend laterally to the side
18) ______	lateral rotation	r)	a movement of the head and vertebral column along the transverse plane
19) ______	medial rotation	s)	a movement that bends a joint or brings the bones closer together
20) ______	plantar flexion	t)	ankle movement letting off the car's gas pedal
21) ______	pronation	u)	a movement that straightens or opens a joint
22) ______	rotation	v)	a movement that brings a limb medially toward the body's midline
23) ______	supination	w)	a movement when the radius crosses over the ulna

Movements of the Body #1

Please identify the movement and its location.

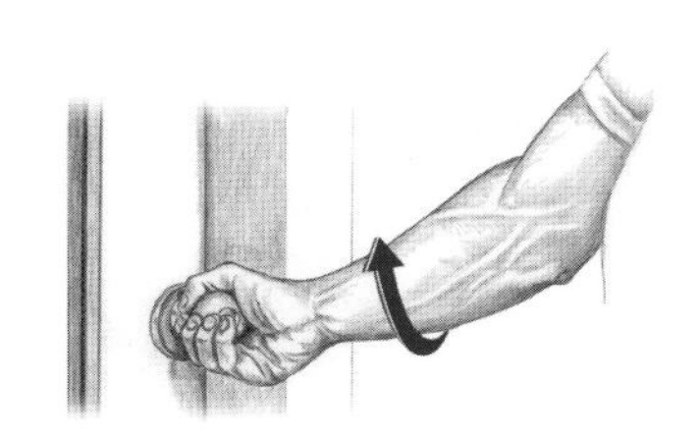

1) Supination of the forearm

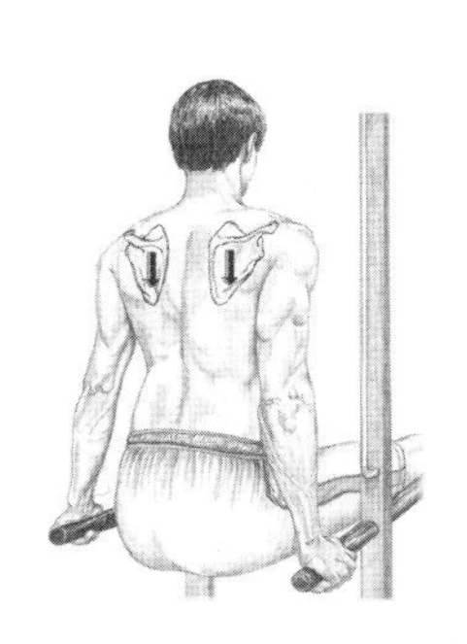

2) ____________________

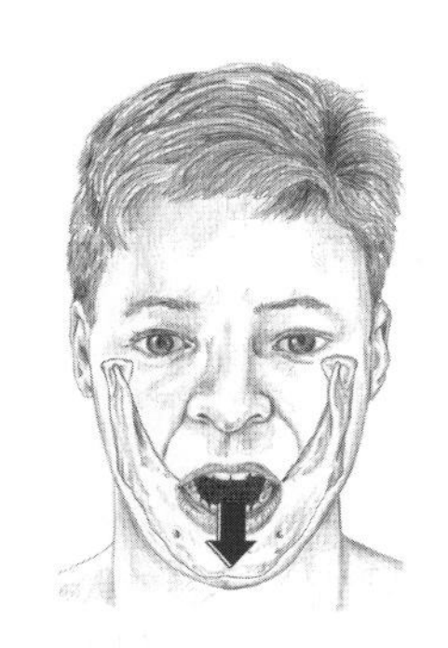

3) ____________________

4) ____________________

5) ____________________

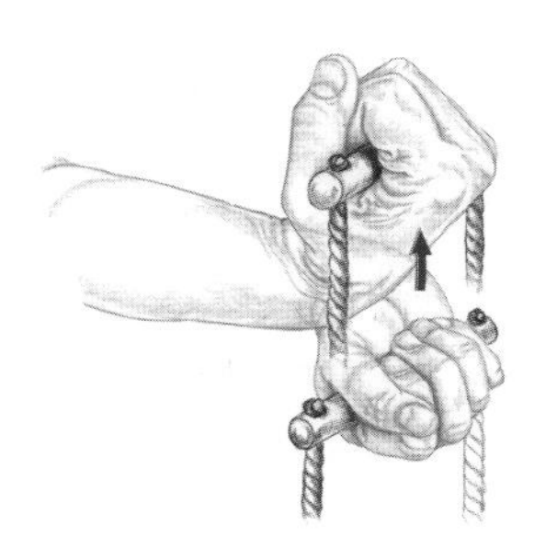

6) ____________________

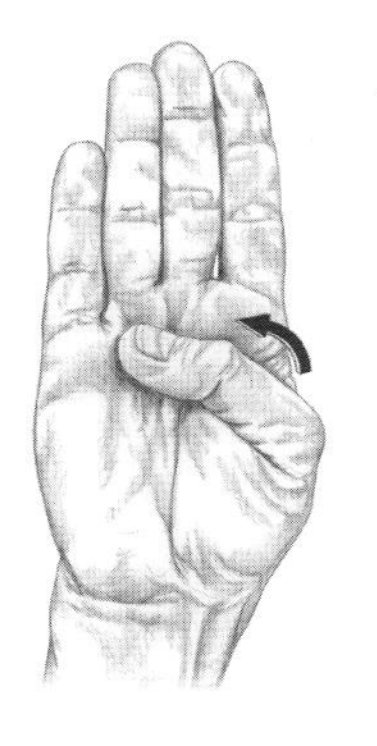

7) ____________________

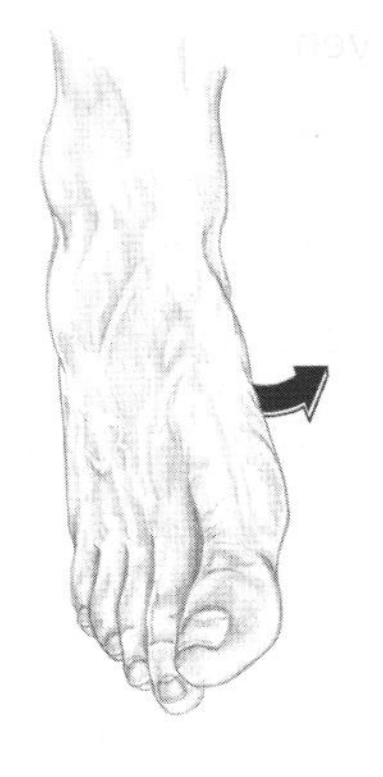

8) ____________________

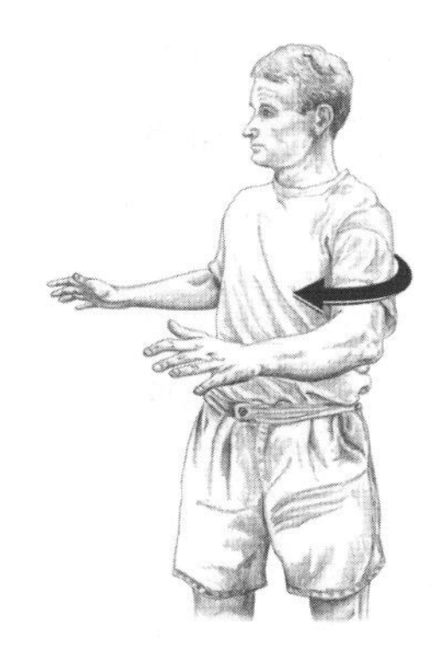

9) ____________________

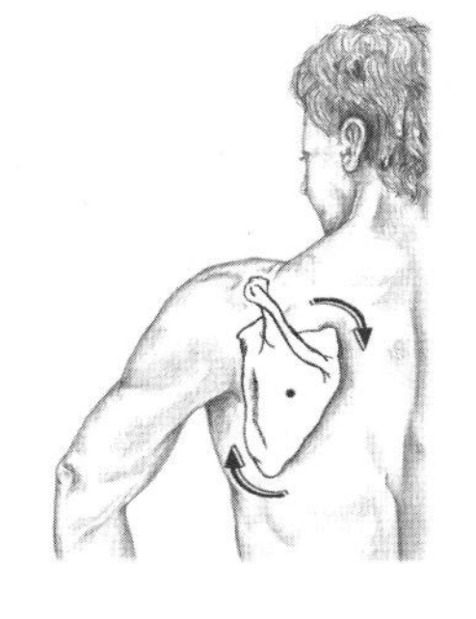

10) ____________________

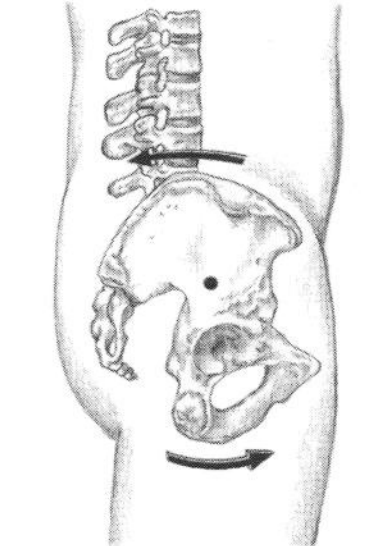

11) ____________________

12) ____________________

Movements of the Body #2

Please identify the movement and its location.

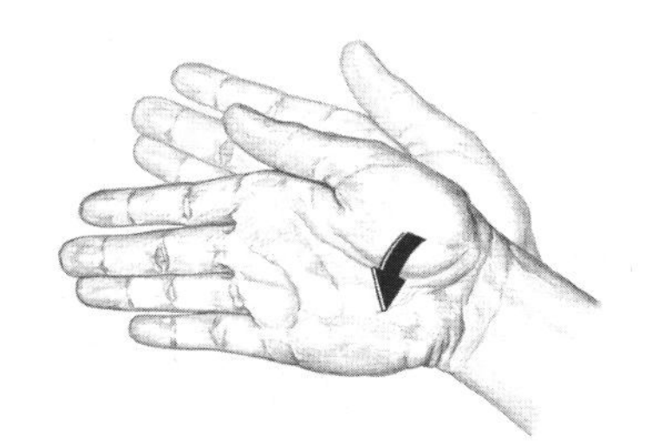

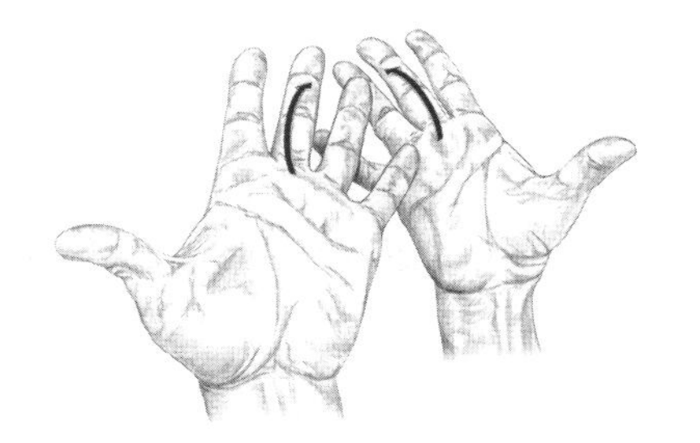

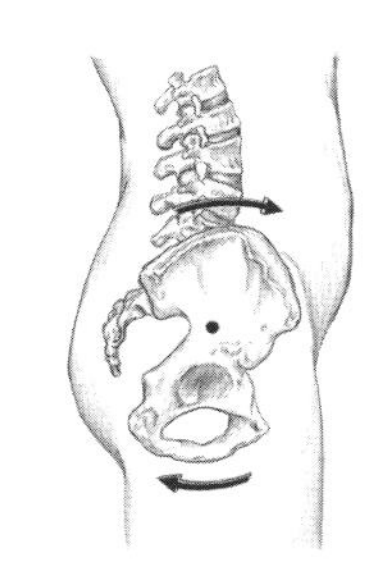

13) ______________________________ 14) ______________________________ 15) ______________________________

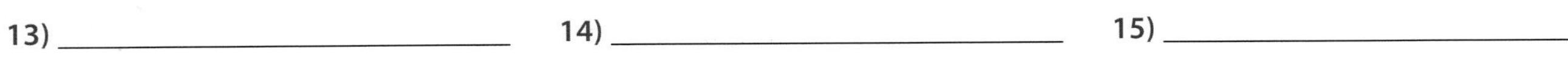

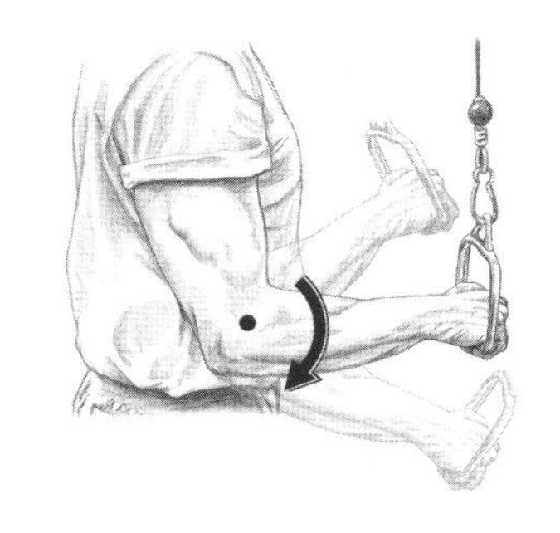

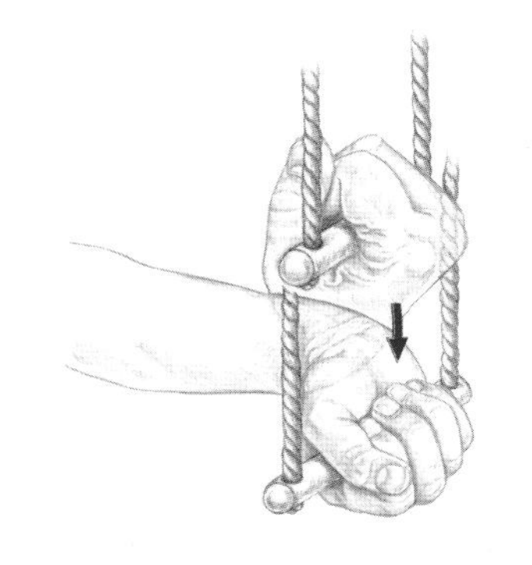

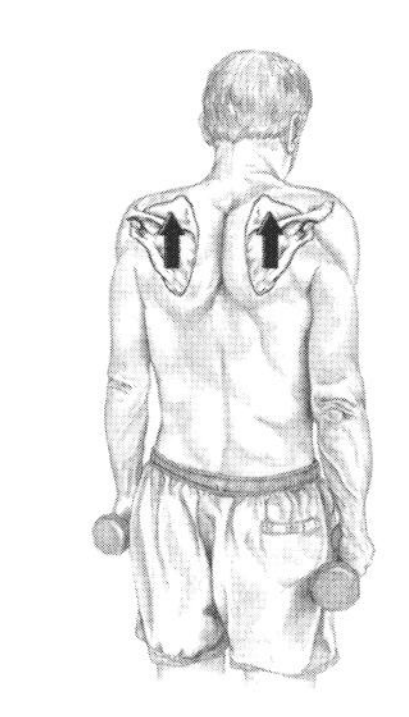

16) ______________________________ 17) ______________________________ 18) ______________________________

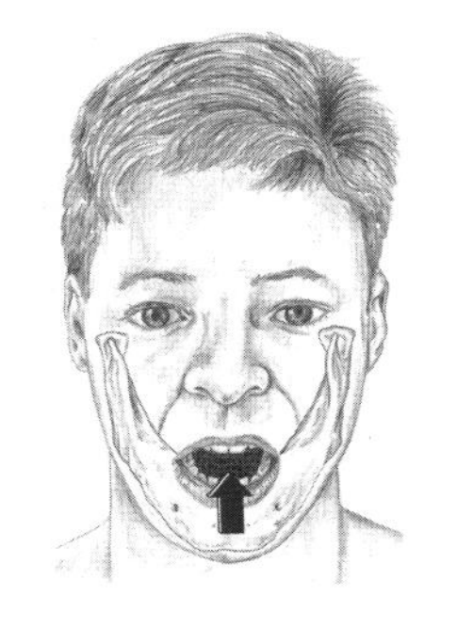

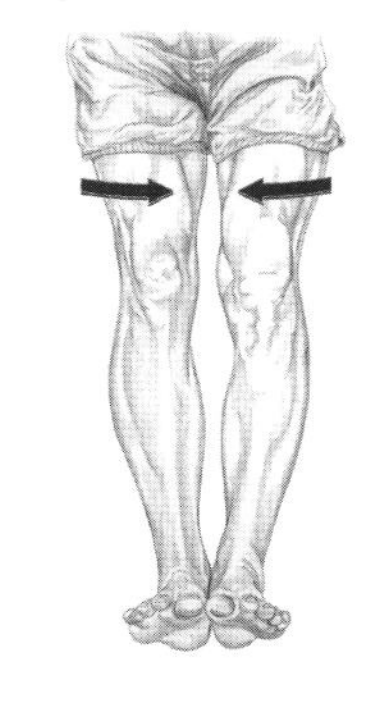

19) ______________________________ 20) ______________________________ 21) ______________________________

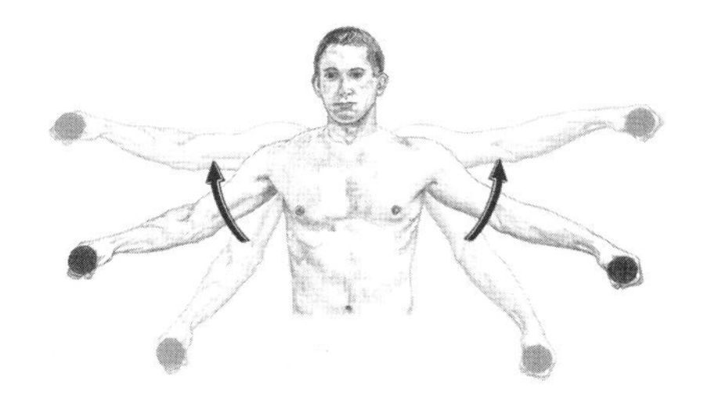

22) ______________________________ 23) ______________________________ 24) ______________________________

Movements of the Body #3

Please identify the movement and its location.

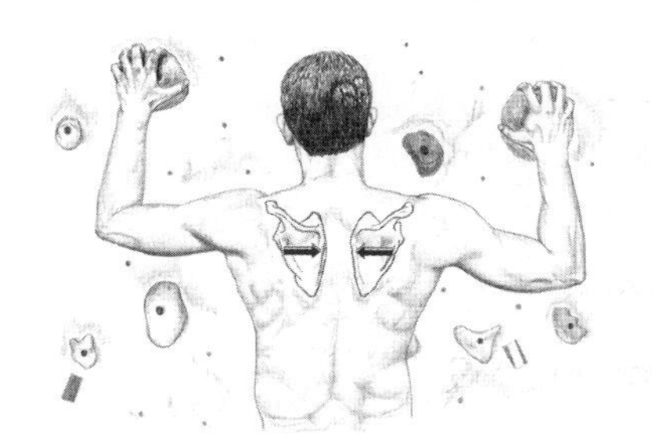

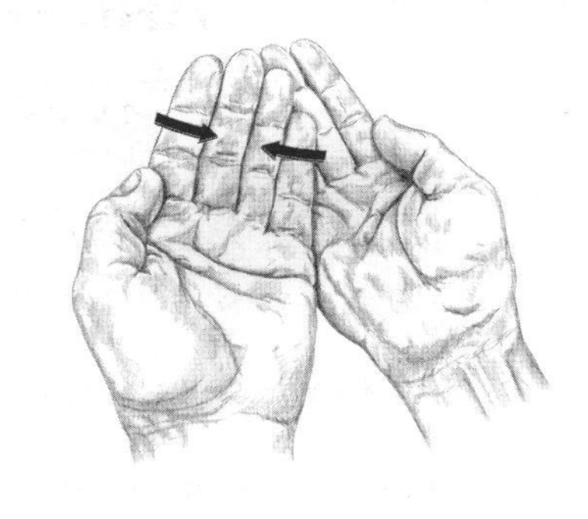

25) ______________________________

26) ______________________________

27) ______________________________

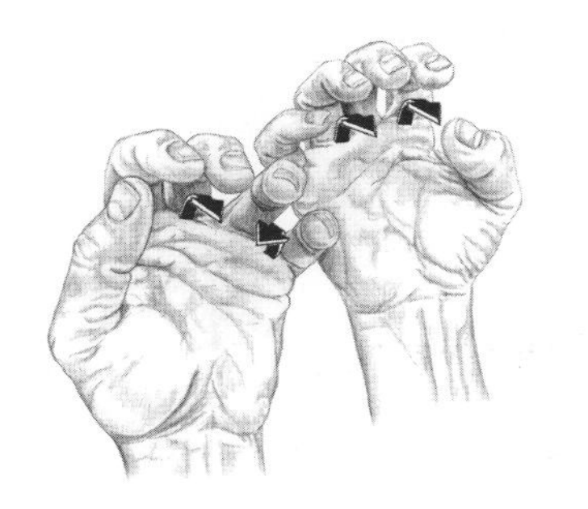

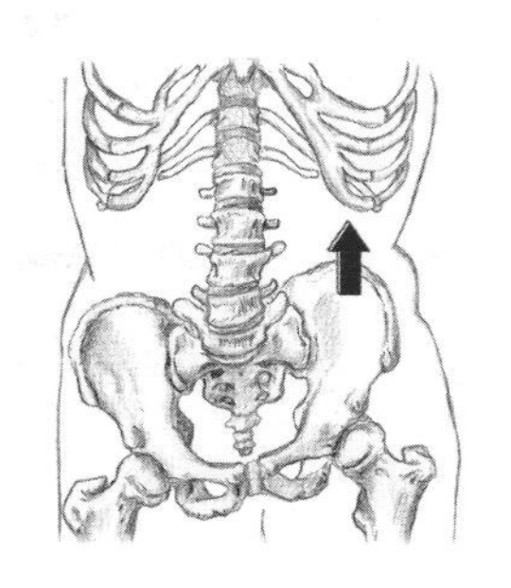

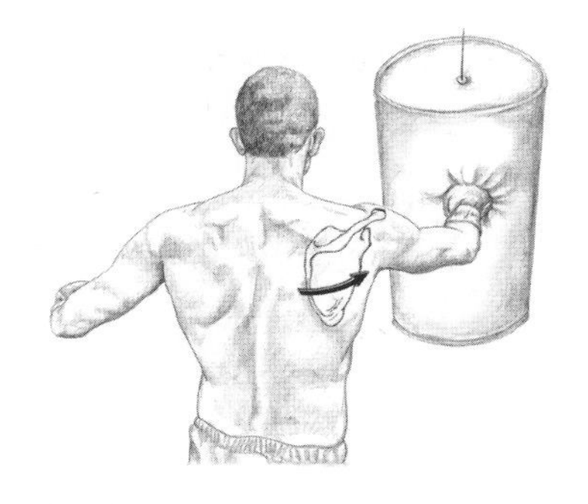

28) ______________________________

29) ______________________________

30) ______________________________

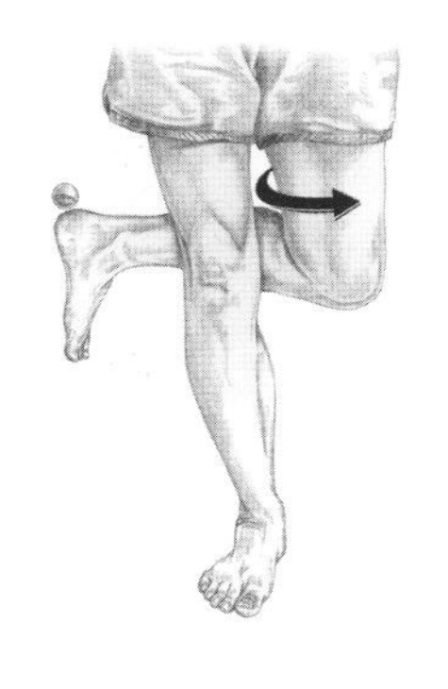

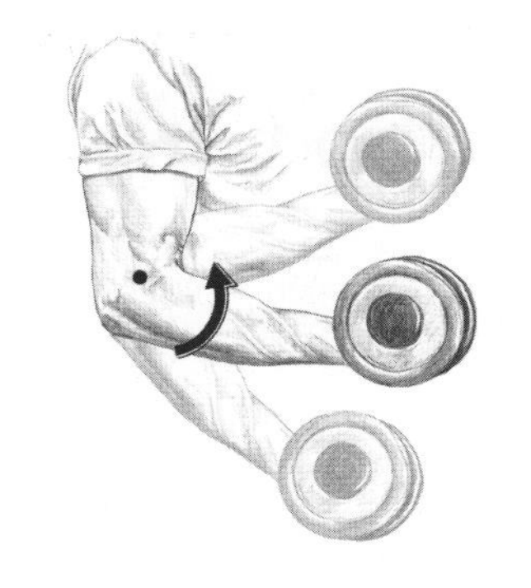

31) ______________________________

32) ______________________________

33) ______________________________

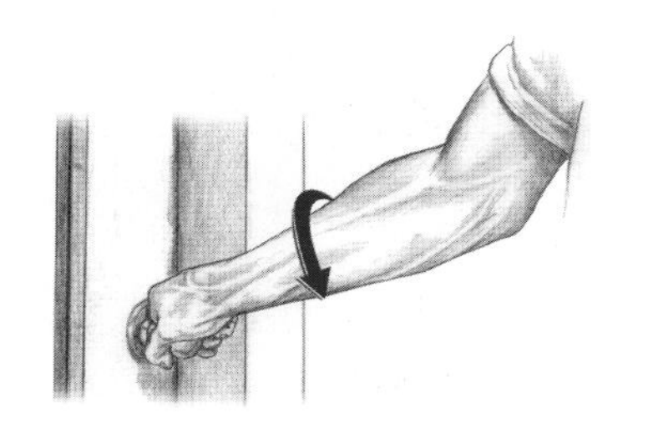

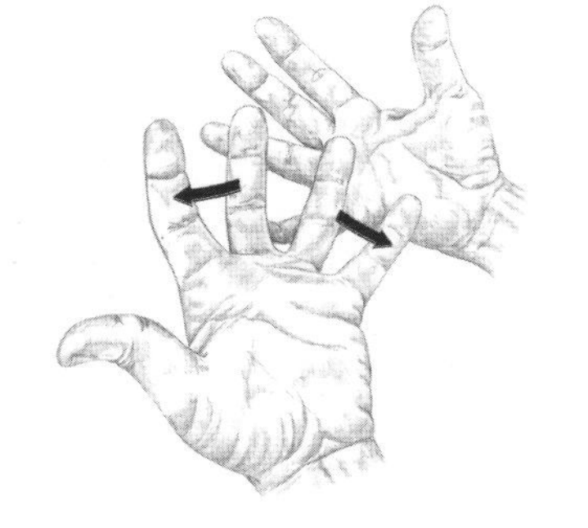

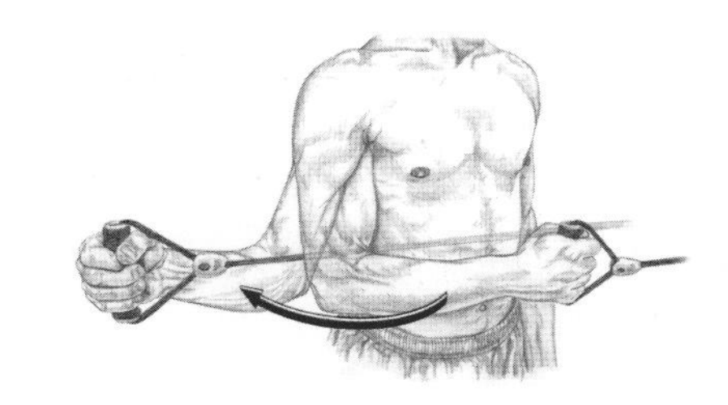

34) ______________________________

35) ______________________________

36) ______________________________

Movements of the Body #4

Please identify the movement and its location.

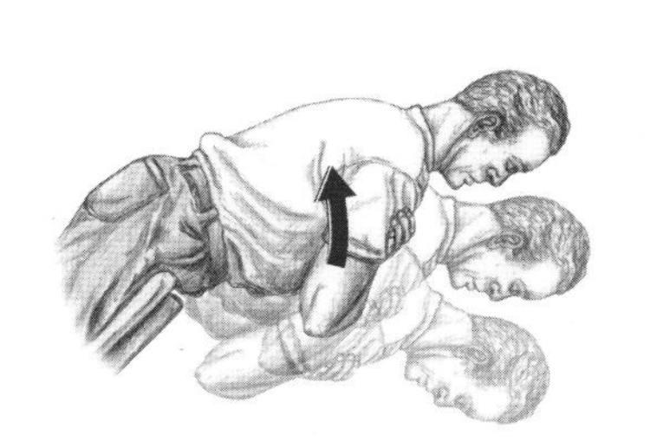

37) ______________________________

38) ______________________________

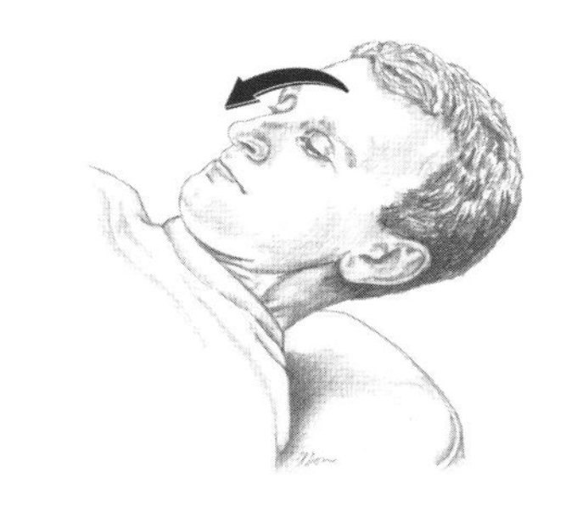

39) ______________________________

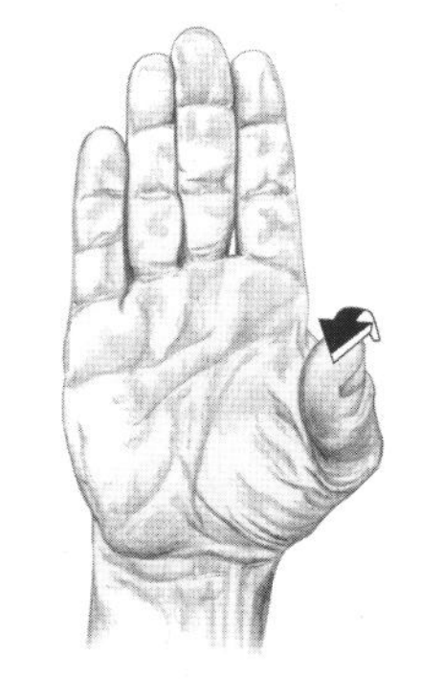

40) ______________________________

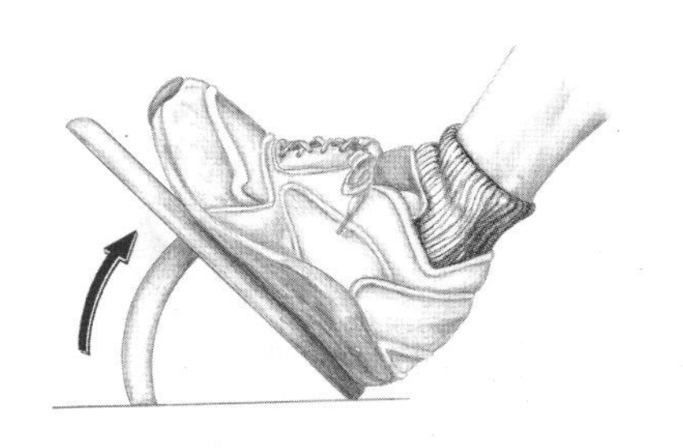

41) ______________________________

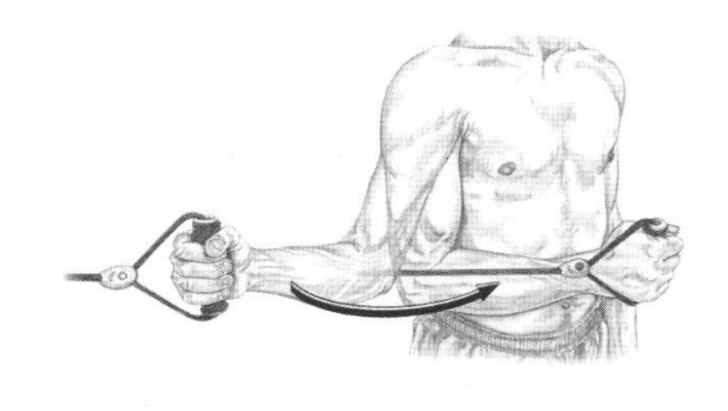

42) ______________________________

43) ______________________________

44) ______________________________

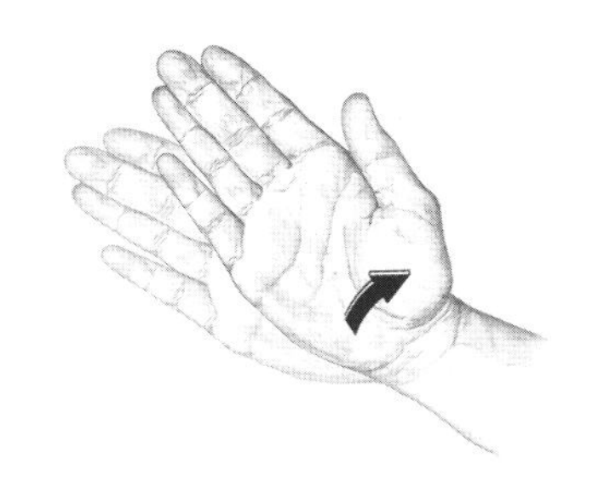

45) ______________________________

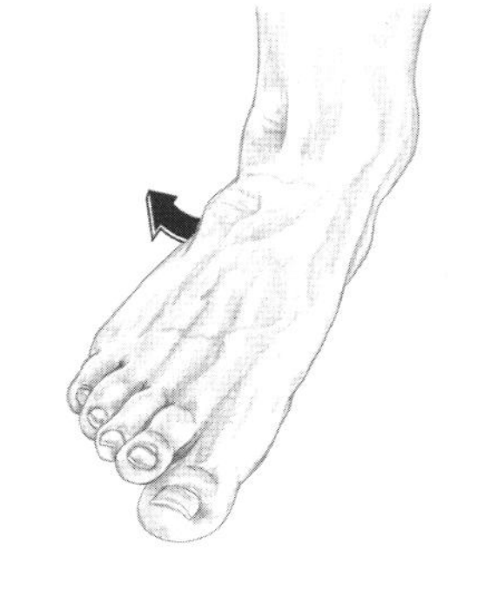

46) ______________________________

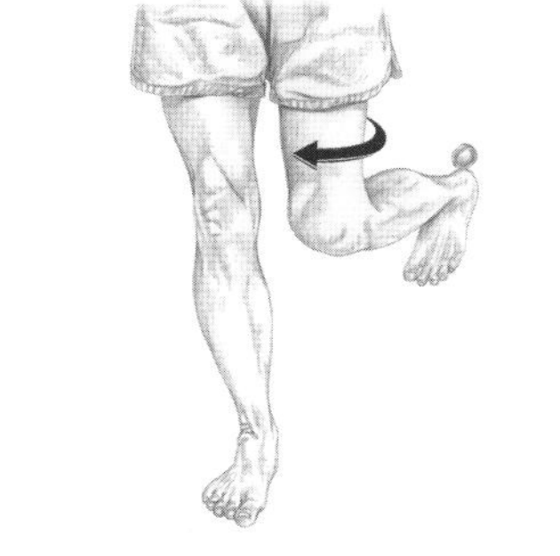

47) ______________________________

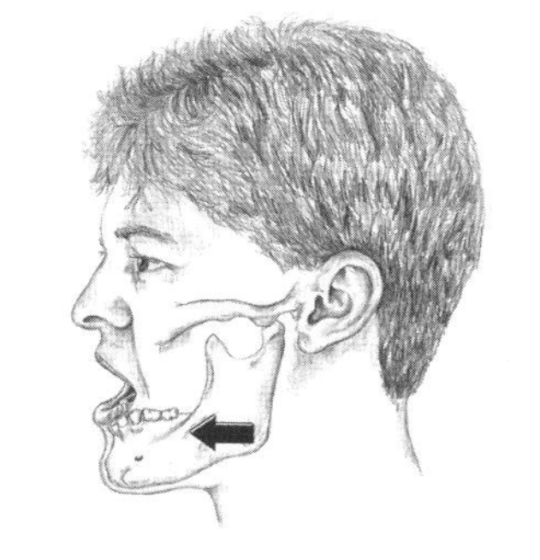

48) ______________________________

p. 26-31

Navigating the Body

Movements of the Body #5

Please identify the movement and its location.

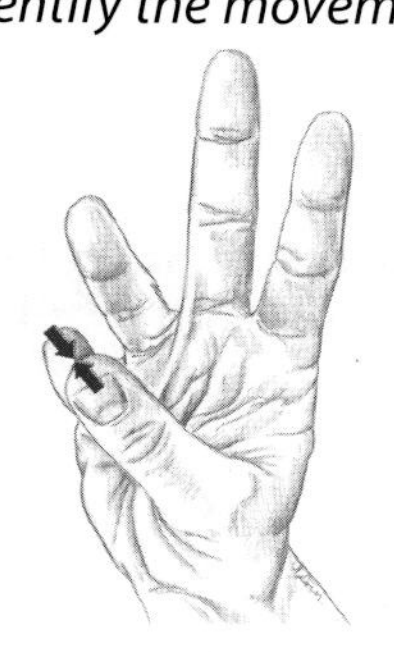

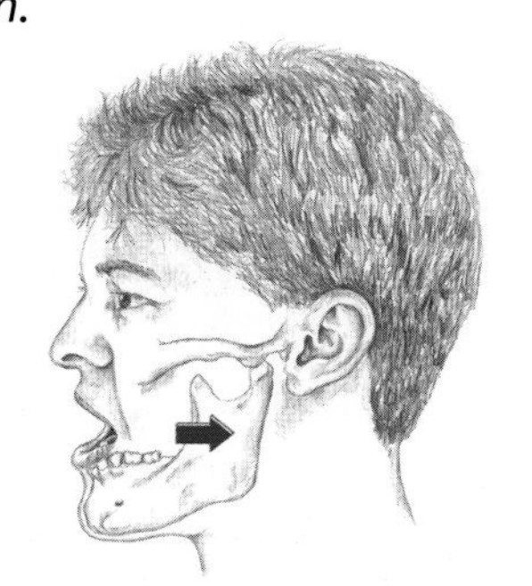

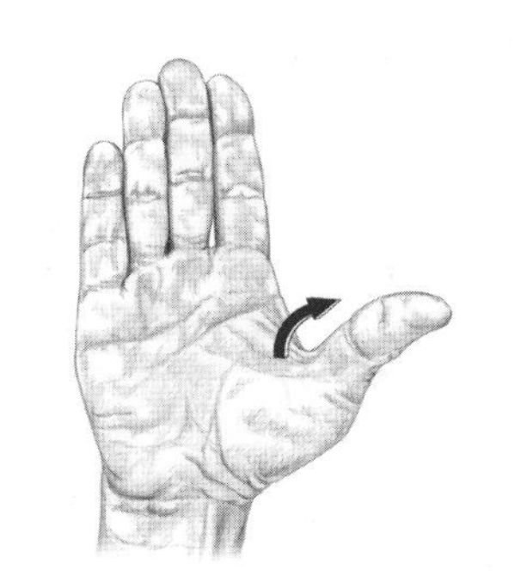

49) ______________________ 50) ______________________ 51) ______________________

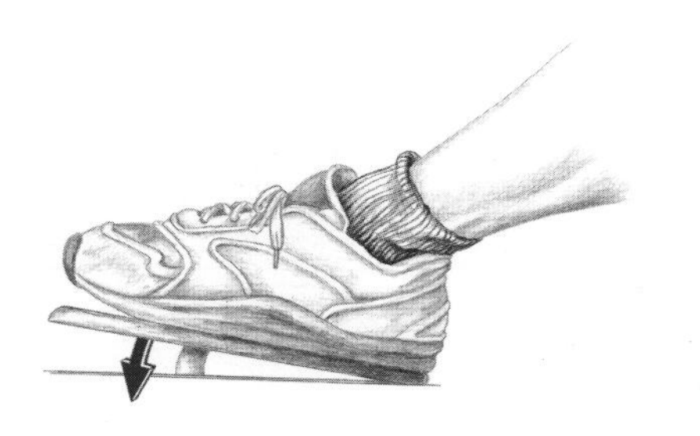

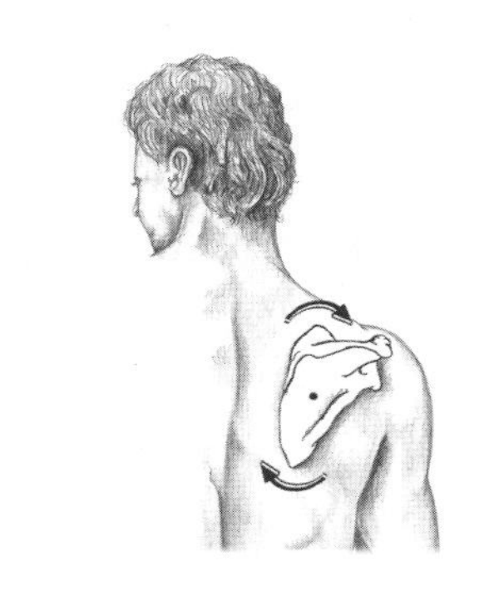

52) ______________________ 53) ______________________ 54) ______________________

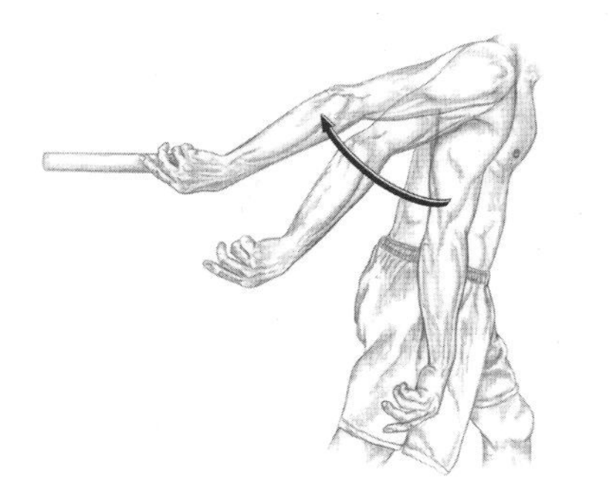

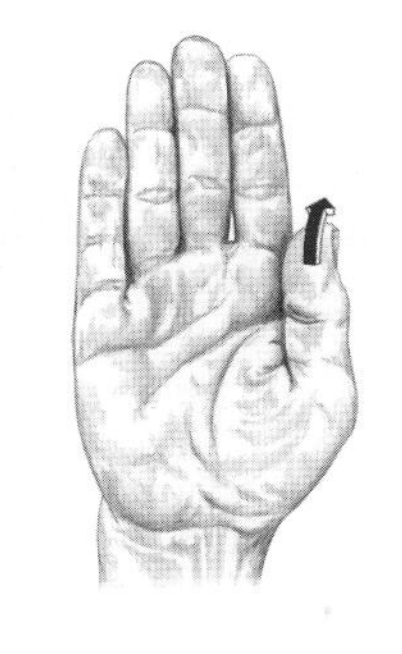

55) ______________________ 56) ______________________ 57) ______________________

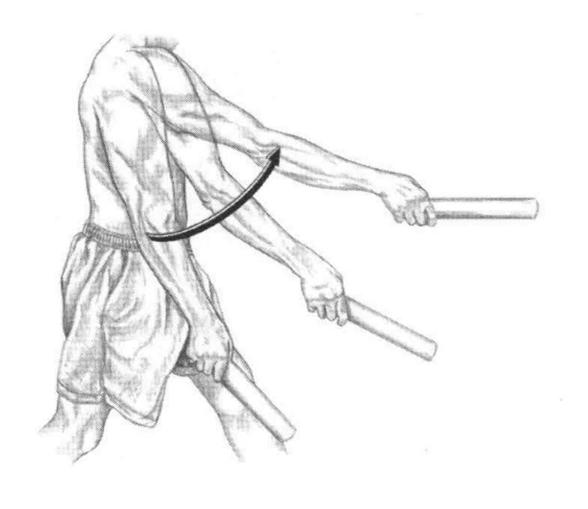

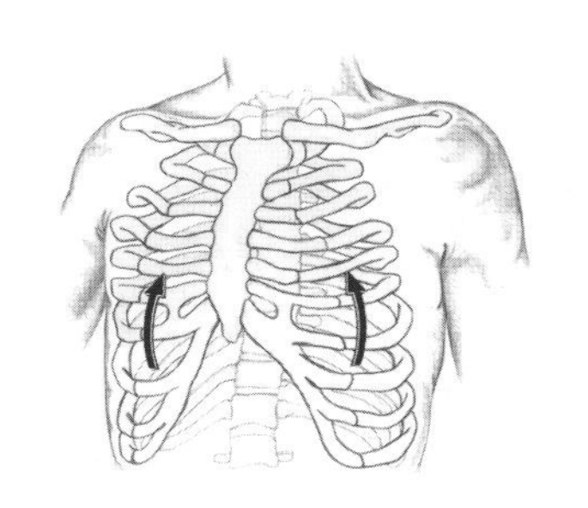

58) ______________________ 59) ______________________ 60) ______________________

Skeletal System #1

Please identify the following structures.

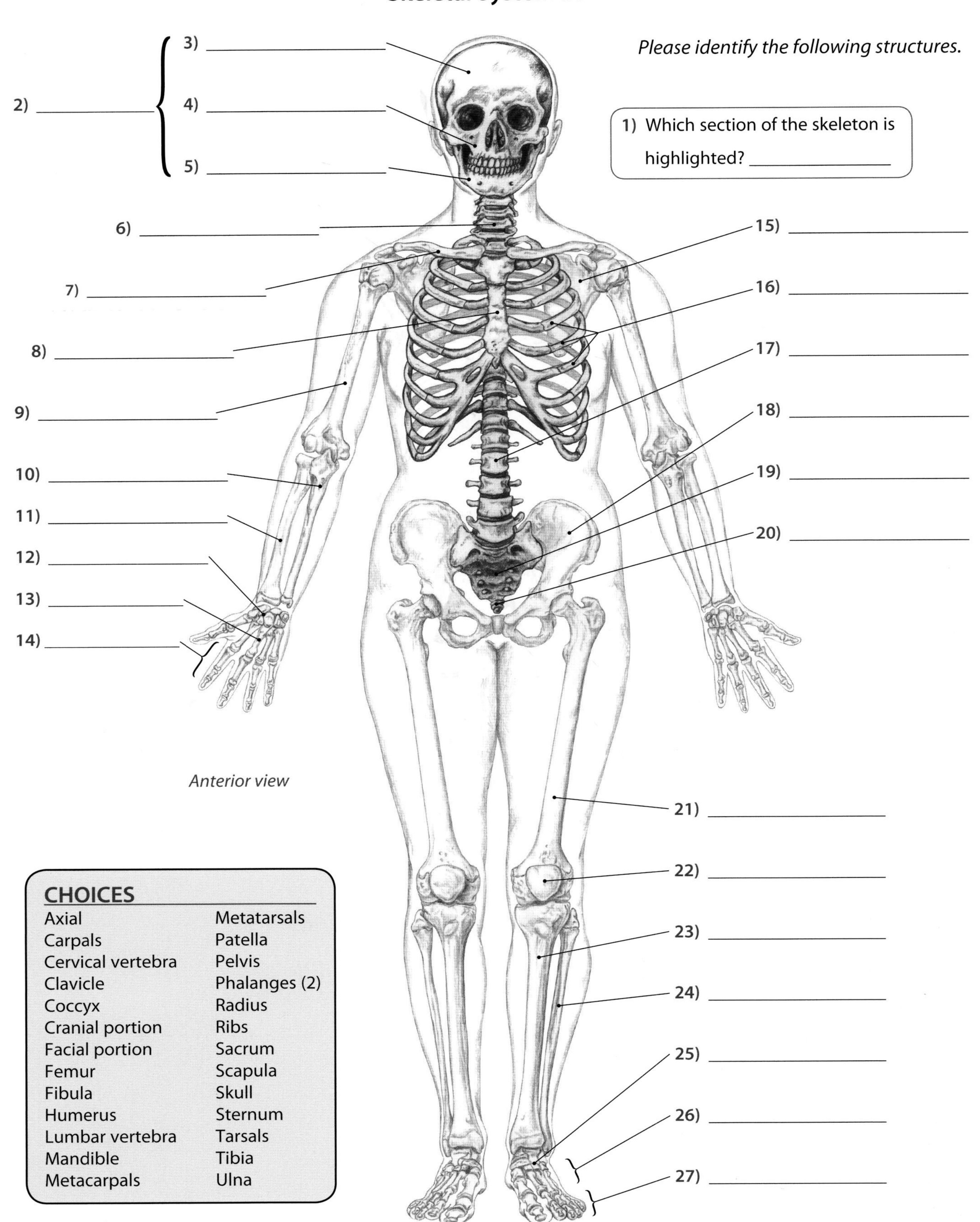

CHOICES

Axial	Metatarsals
Carpals	Patella
Cervical vertebra	Pelvis
Clavicle	Phalanges (2)
Coccyx	Radius
Cranial portion	Ribs
Facial portion	Sacrum
Femur	Scapula
Fibula	Skull
Humerus	Sternum
Lumbar vertebra	Tarsals
Mandible	Tibia
Metacarpals	Ulna

Navigating the Body

Skeletal System #2

1) Which sections of the skeleton are highlighted? ____________

Please identify the following structures.

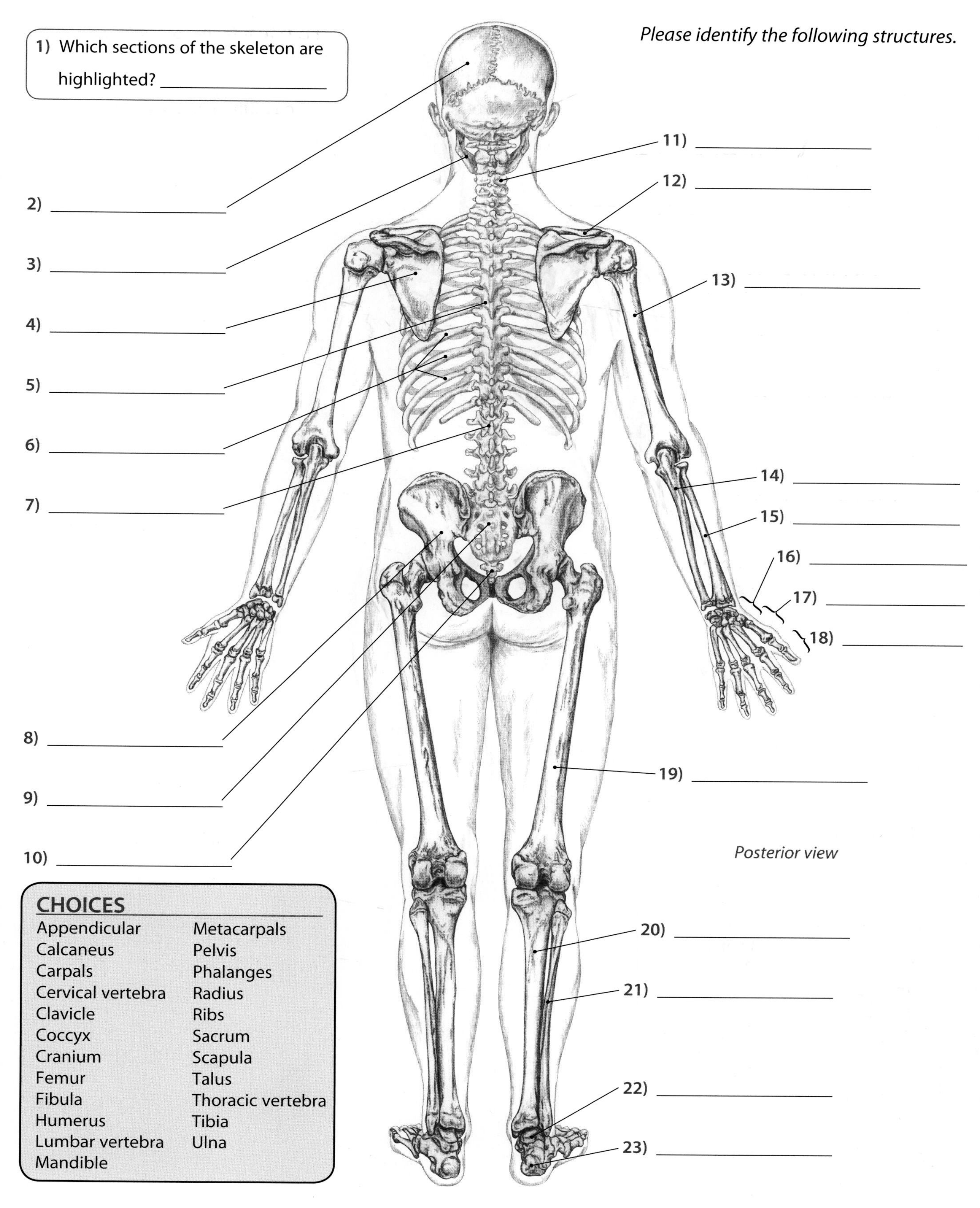

CHOICES

Appendicular	Metacarpals
Calcaneus	Pelvis
Carpals	Phalanges
Cervical vertebra	Radius
Clavicle	Ribs
Coccyx	Sacrum
Cranium	Scapula
Femur	Talus
Fibula	Thoracic vertebra
Humerus	Tibia
Lumbar vertebra	Ulna
Mandible	

Muscular System #1

Please identify the following structures.

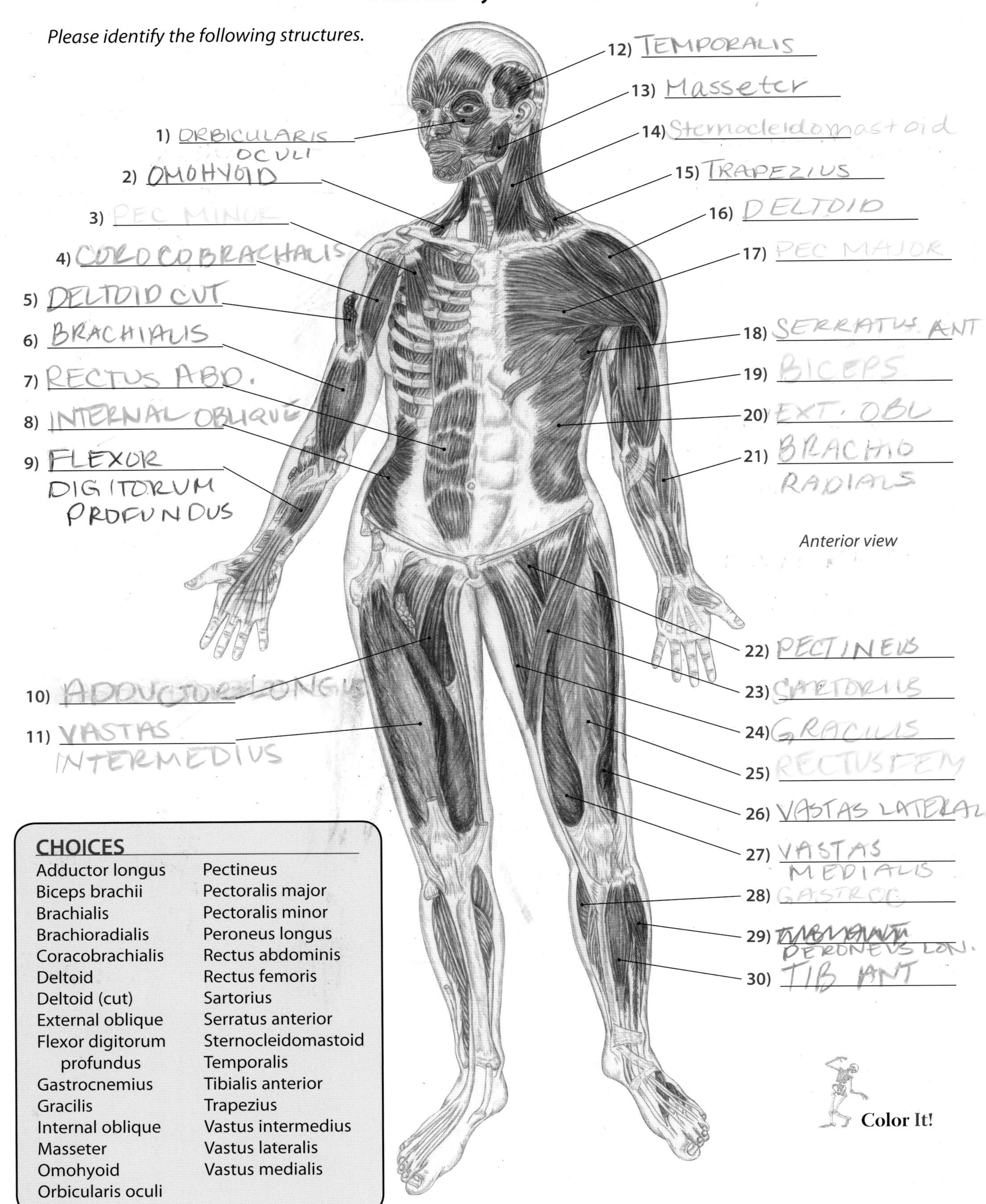

CHOICES

Adductor longus
Biceps brachii
Brachialis
Brachioradialis
Coracobrachialis
Deltoid
Deltoid (cut)
External oblique
Flexor digitorum profundus
Gastrocnemius
Gracilis
Internal oblique
Masseter
Omohyoid
Orbicularis oculi
Pectineus
Pectoralis major
Pectoralis minor
Peroneus longus
Rectus abdominis
Rectus femoris
Sartorius
Serratus anterior
Sternocleidomastoid
Temporalis
Tibialis anterior
Trapezius
Vastus intermedius
Vastus lateralis
Vastus medialis

Color It!

Navigating the Body
Muscular System #2

Please identify the following structures.

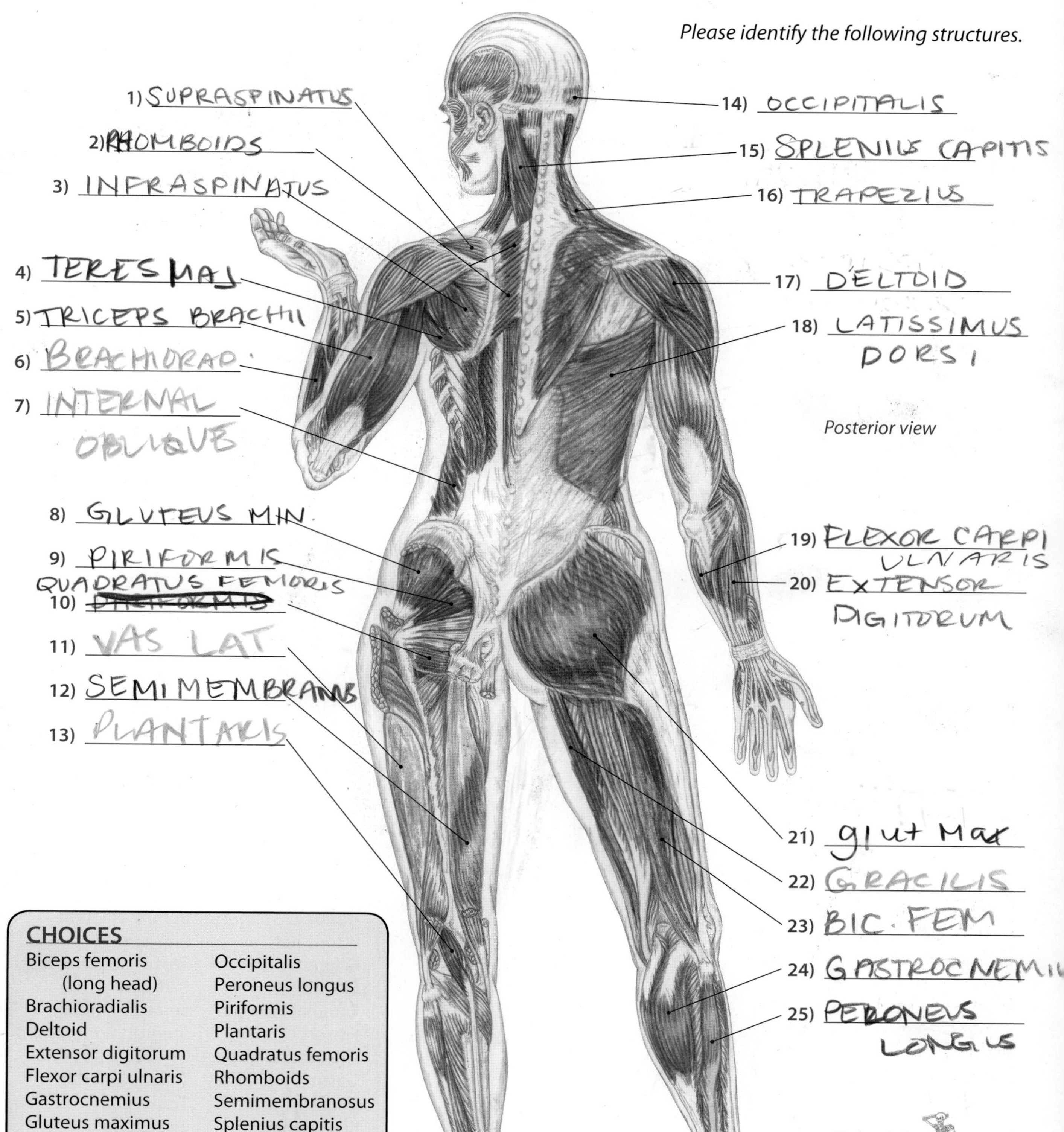

CHOICES

Biceps femoris (long head)	Occipitalis
Brachioradialis	Peroneus longus
Deltoid	Piriformis
Extensor digitorum	Plantaris
Flexor carpi ulnaris	Quadratus femoris
Gastrocnemius	Rhomboids
Gluteus maximus	Semimembranosus
Gluteus minimus	Splenius capitis
Gracilis	Supraspinatus
Infraspinatus	Teres major
Internal oblique	Trapezius
Latissimus dorsi	Triceps brachii
	Vastus lateralis

Color It!

Please identify the following structures.

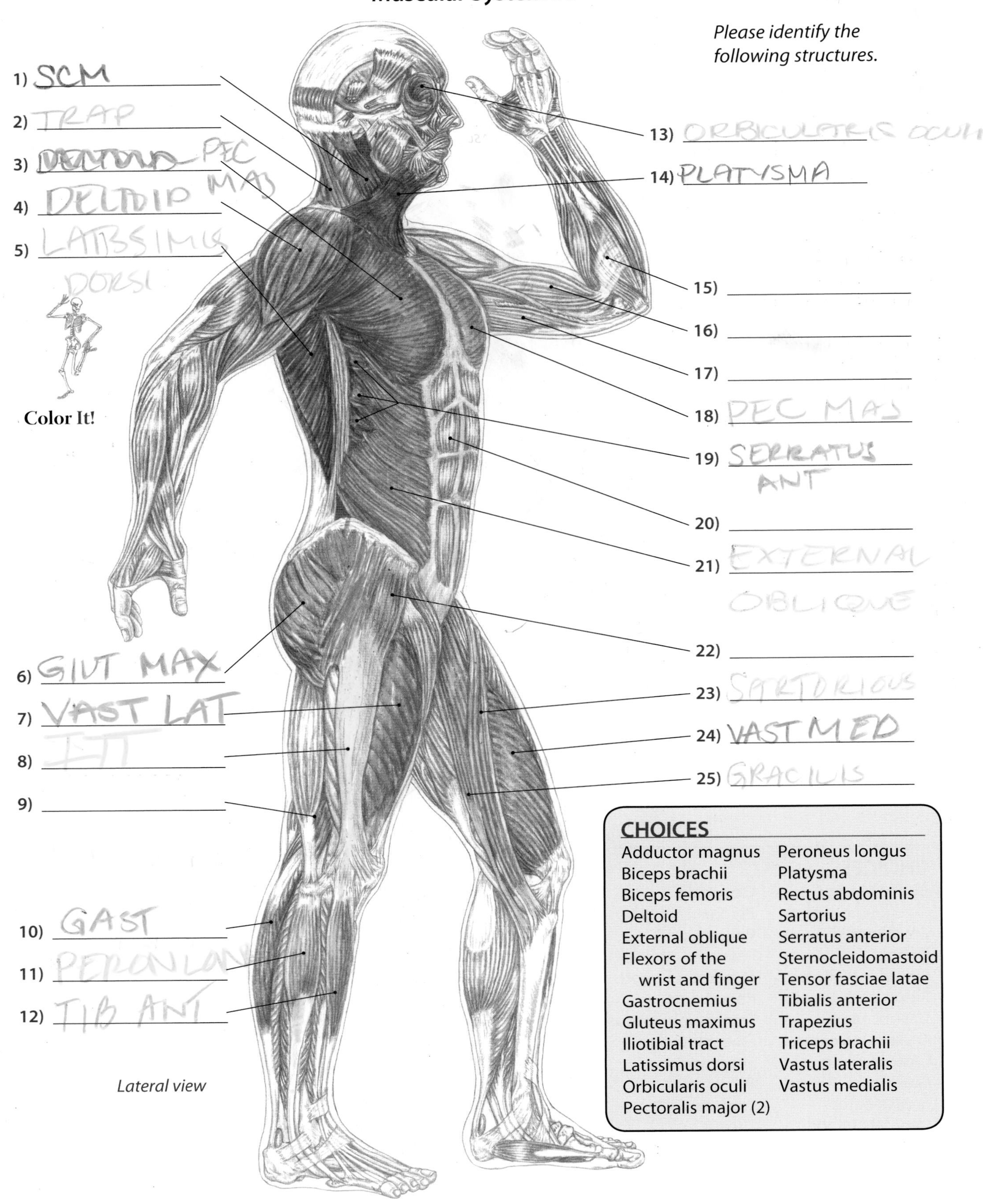

CHOICES

Adductor magnus	Peroneus longus
Biceps brachii	Platysma
Biceps femoris	Rectus abdominis
Deltoid	Sartorius
External oblique	Serratus anterior
Flexors of the wrist and finger	Sternocleidomastoid
Gastrocnemius	Tensor fasciae latae
Gluteus maximus	Tibialis anterior
Iliotibial tract	Trapezius
Latissimus dorsi	Triceps brachii
Orbicularis oculi	Vastus lateralis
Pectoralis major (2)	Vastus medialis

Navigating the Body
Fascial System #1

Please identify the following structures.

CHOICES
- Antebrachial fascia
- Biceps brachii
- Brachial fascia
- Extensor muscles
- Flexor muscles
- Humerus
- Interosseous membrane
- Lateral intermuscular septum
- Medial intermuscular septum
- Radius
- Triceps brachii
- Ulna

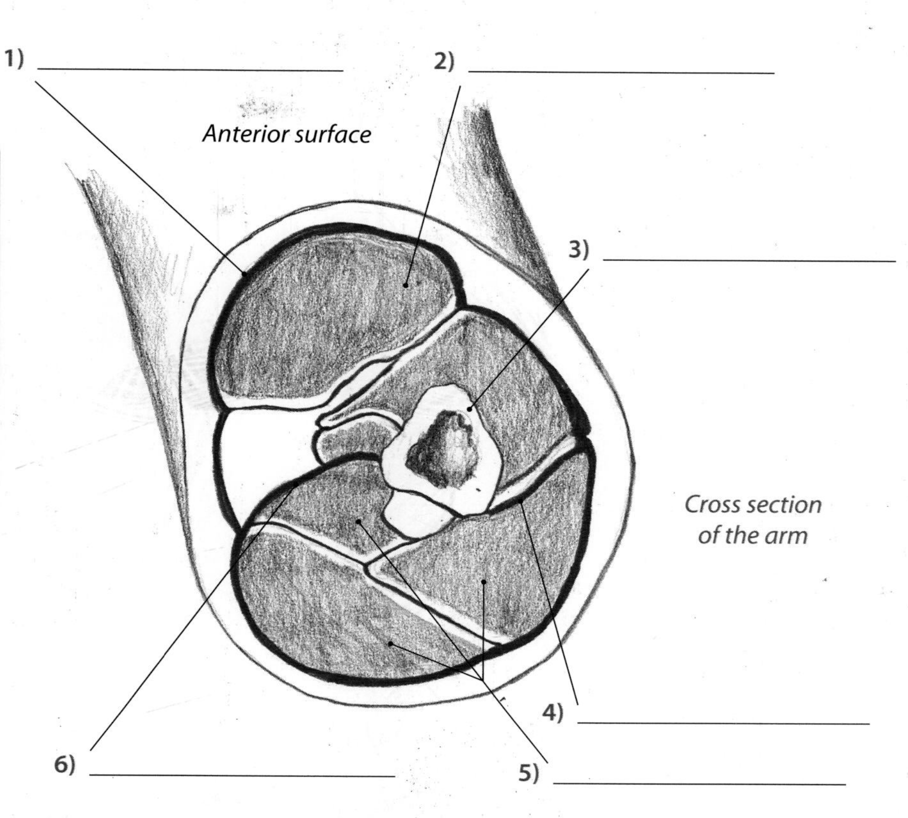

Cross section of the arm

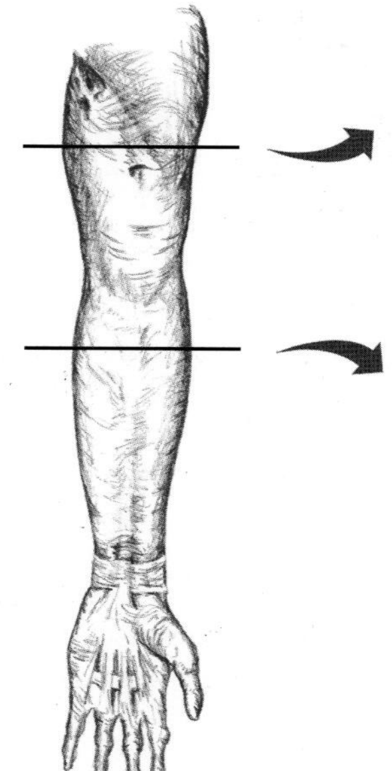

Anterior view of the left arm and forearm, skin removed

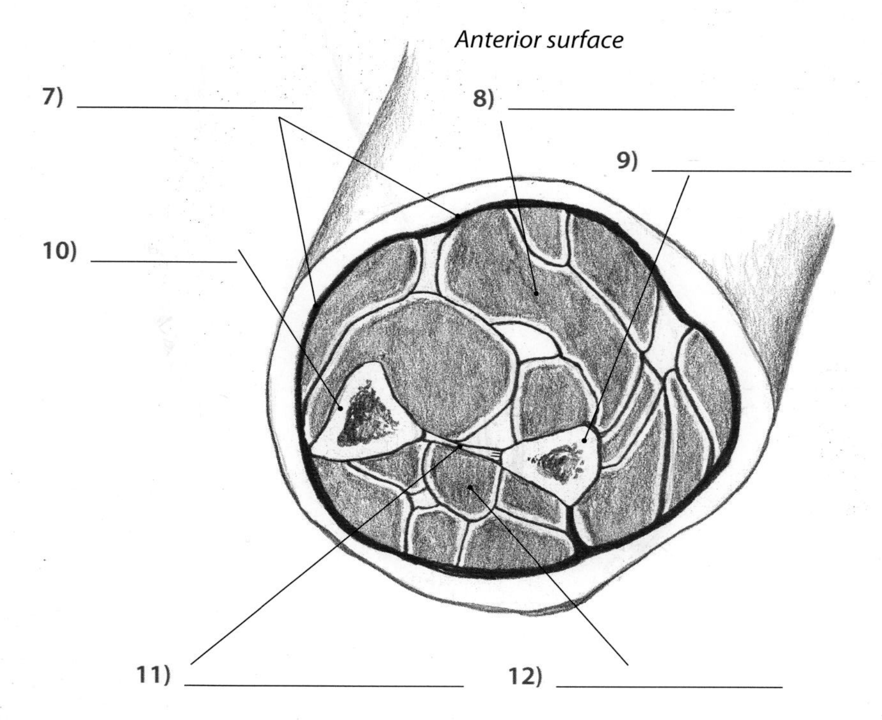

Cross section of the forearm

Fascial System #2

Please identify the following structures.

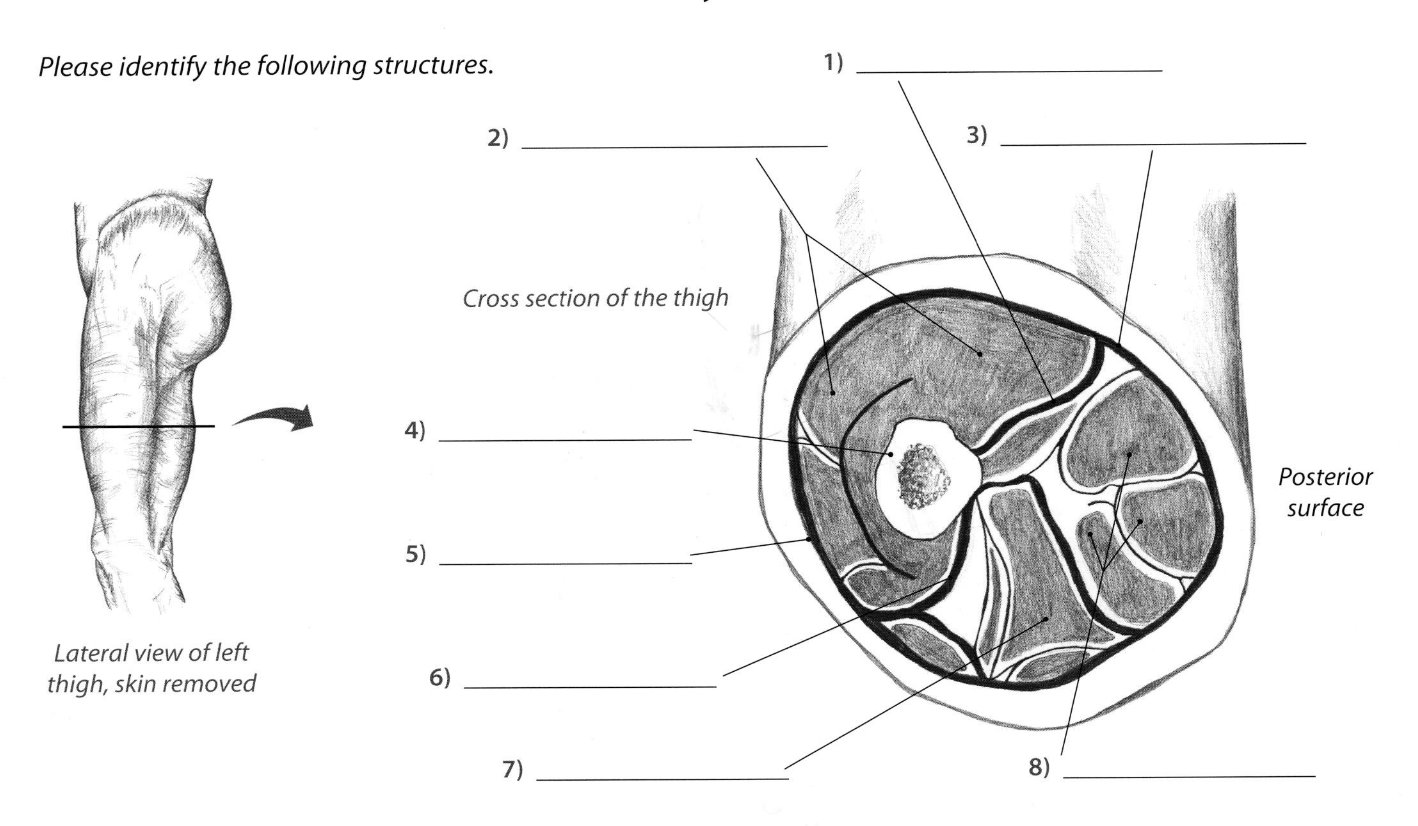

Lateral view of left thigh, skin removed

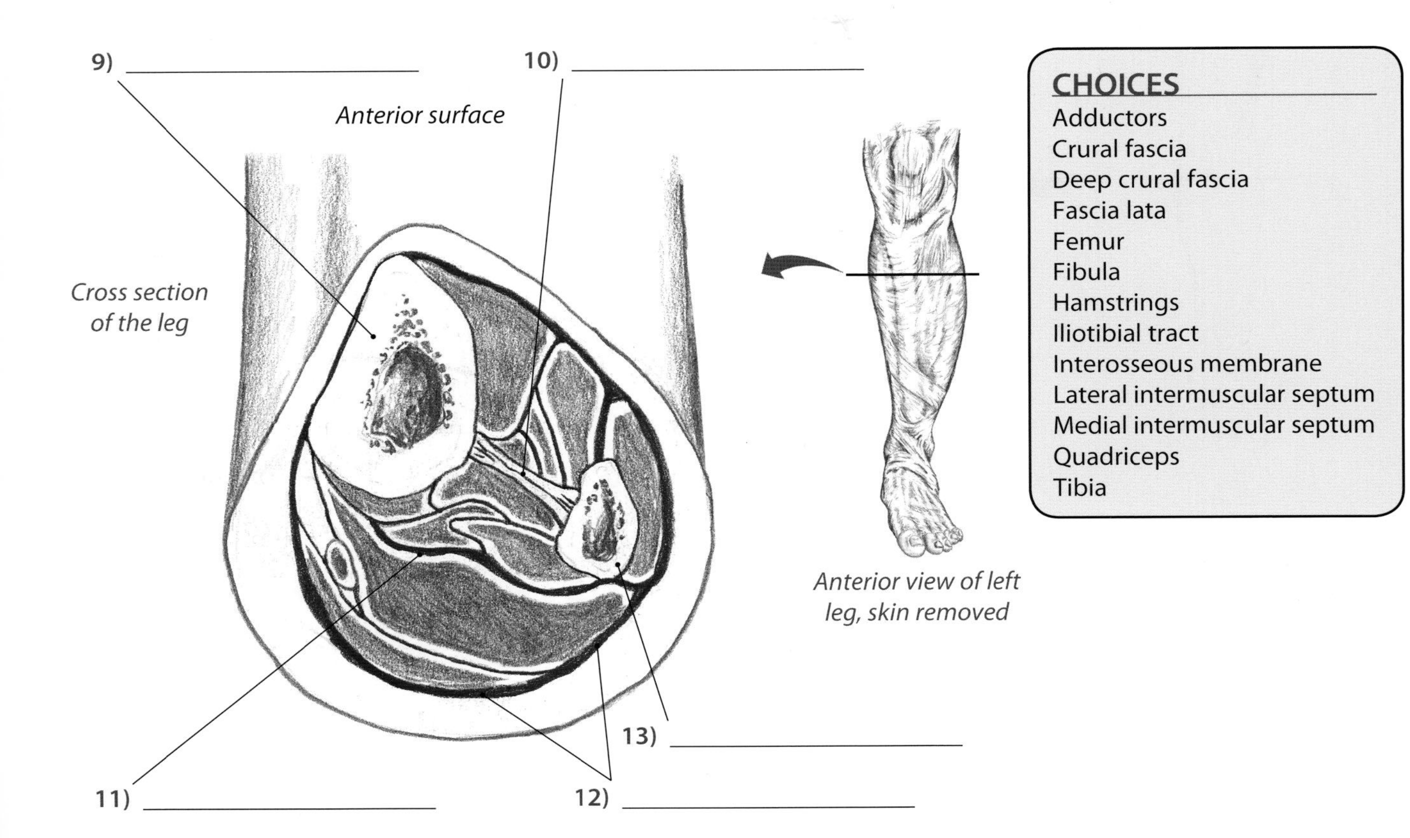

Cross section of the leg

Anterior view of left leg, skin removed

CHOICES

Adductors
Crural fascia
Deep crural fascia
Fascia lata
Femur
Fibula
Hamstrings
Iliotibial tract
Interosseous membrane
Lateral intermuscular septum
Medial intermuscular septum
Quadriceps
Tibia

Cardiovascular System—Arteries

Please identify the following structures.

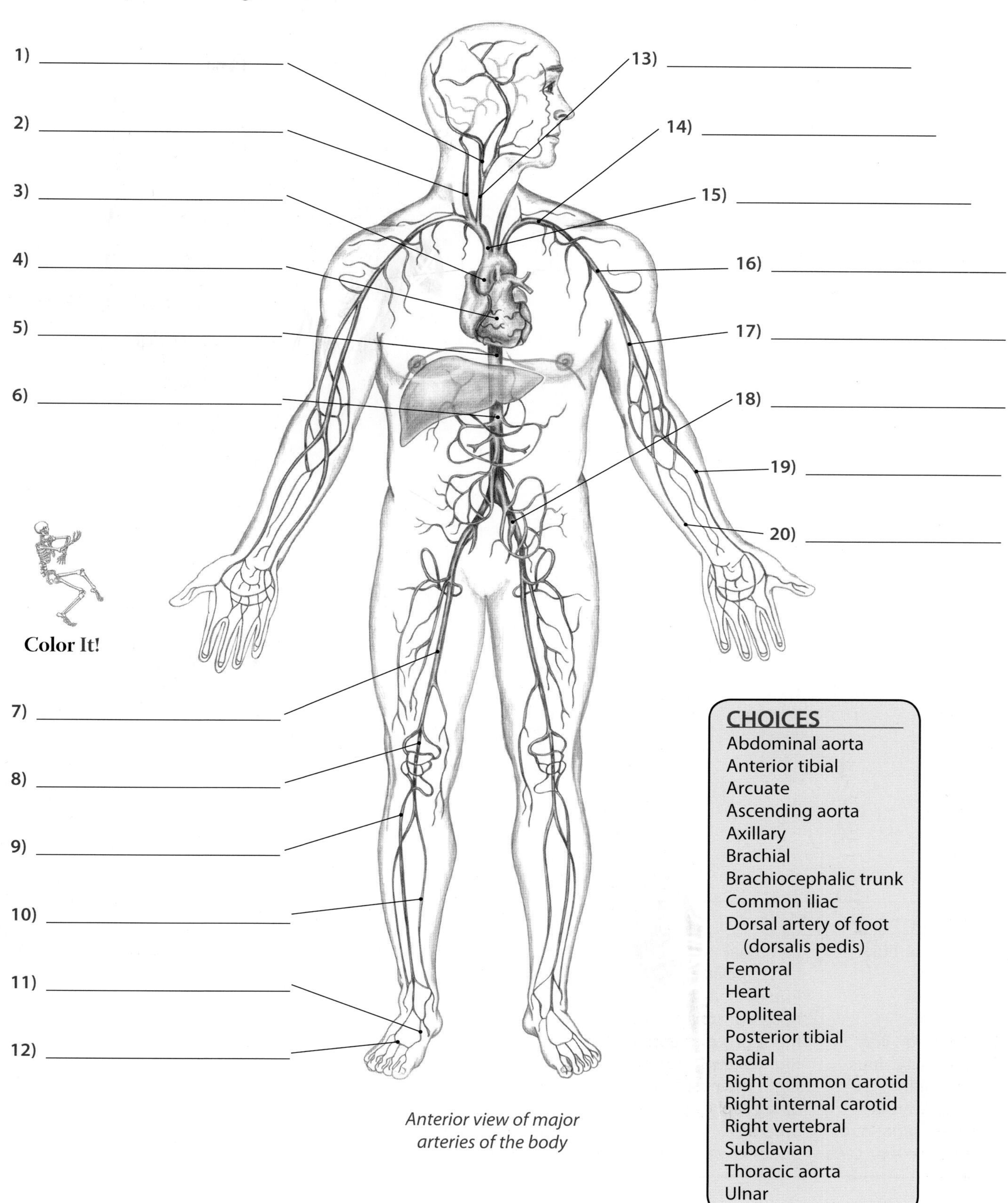

Anterior view of major arteries of the body

Cardiovascular System—Veins

p. 41

Please identify the following structures.

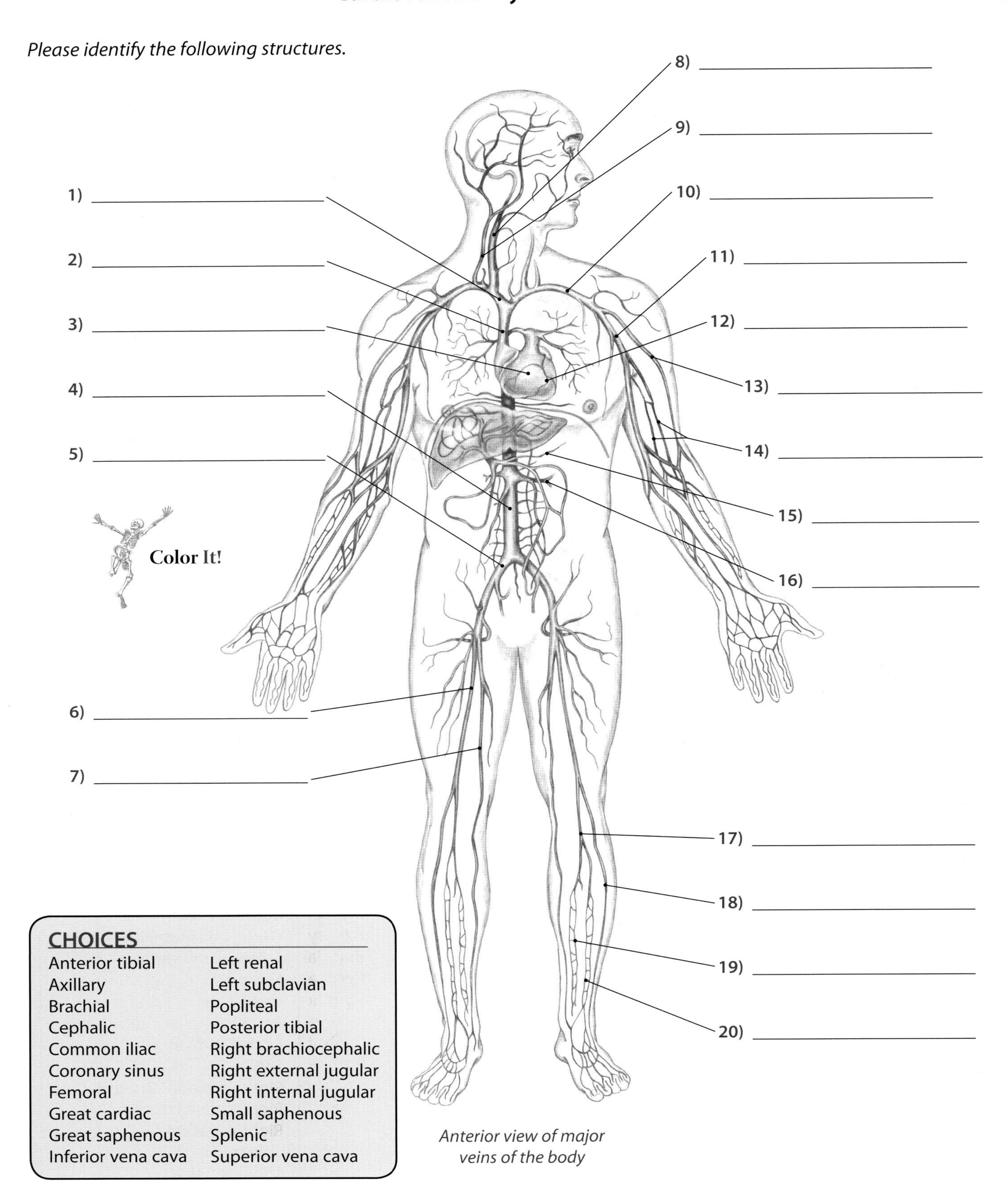

Anterior view of major veins of the body

CHOICES

Anterior tibial	Left renal
Axillary	Left subclavian
Brachial	Popliteal
Cephalic	Posterior tibial
Common iliac	Right brachiocephalic
Coronary sinus	Right external jugular
Femoral	Right internal jugular
Great cardiac	Small saphenous
Great saphenous	Splenic
Inferior vena cava	Superior vena cava

Navigating the Body

Nervous System

Please identify the following structures.

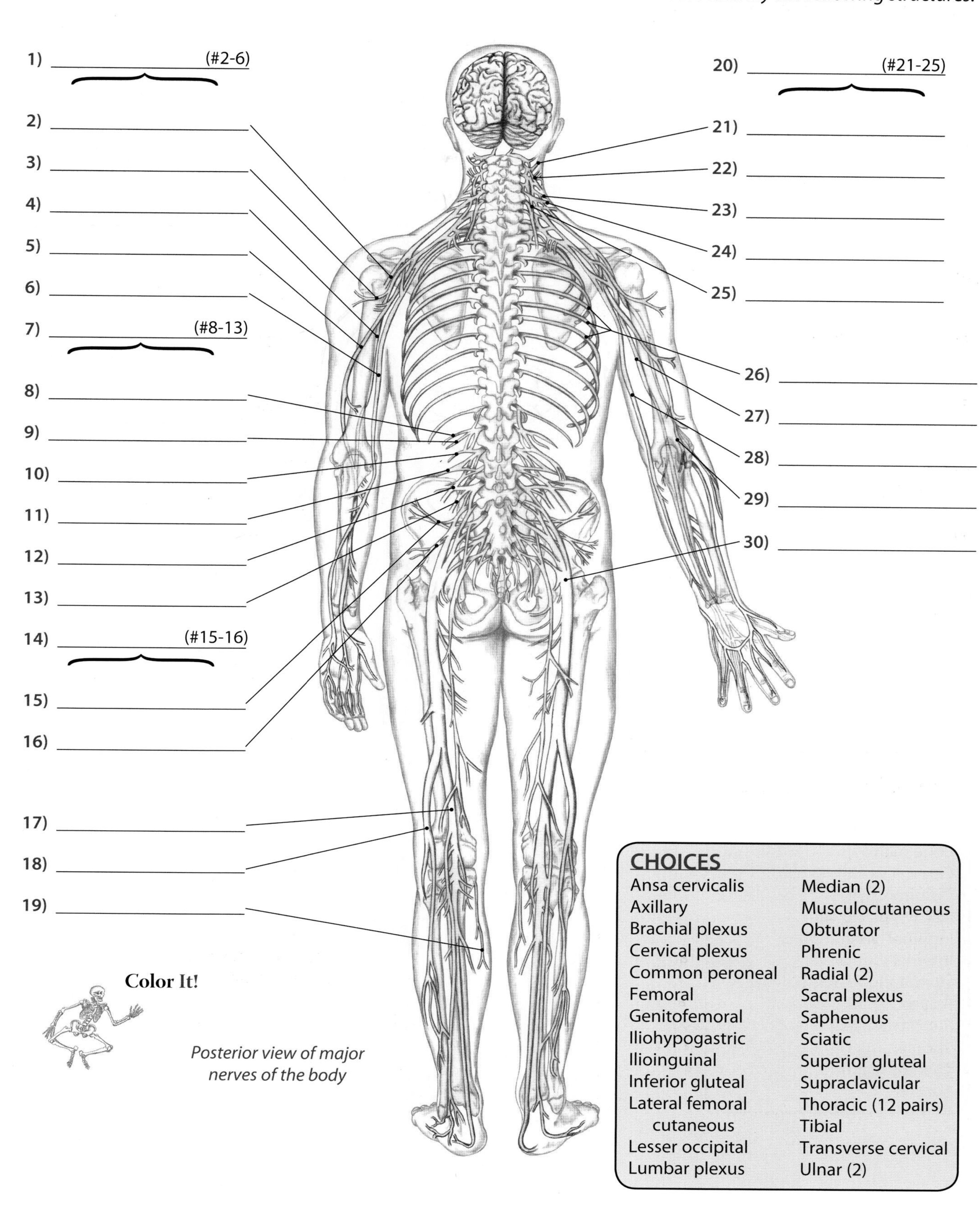

Posterior view of major nerves of the body

CHOICES

Ansa cervicalis	Median (2)
Axillary	Musculocutaneous
Brachial plexus	Obturator
Cervical plexus	Phrenic
Common peroneal	Radial (2)
Femoral	Sacral plexus
Genitofemoral	Saphenous
Iliohypogastric	Sciatic
Ilioinguinal	Superior gluteal
Inferior gluteal	Supraclavicular
Lateral femoral cutaneous	Thoracic (12 pairs)
Lesser occipital	Tibial
Lumbar plexus	Transverse cervical
	Ulnar (2)

Lymphatic System

Please identify the following structures.

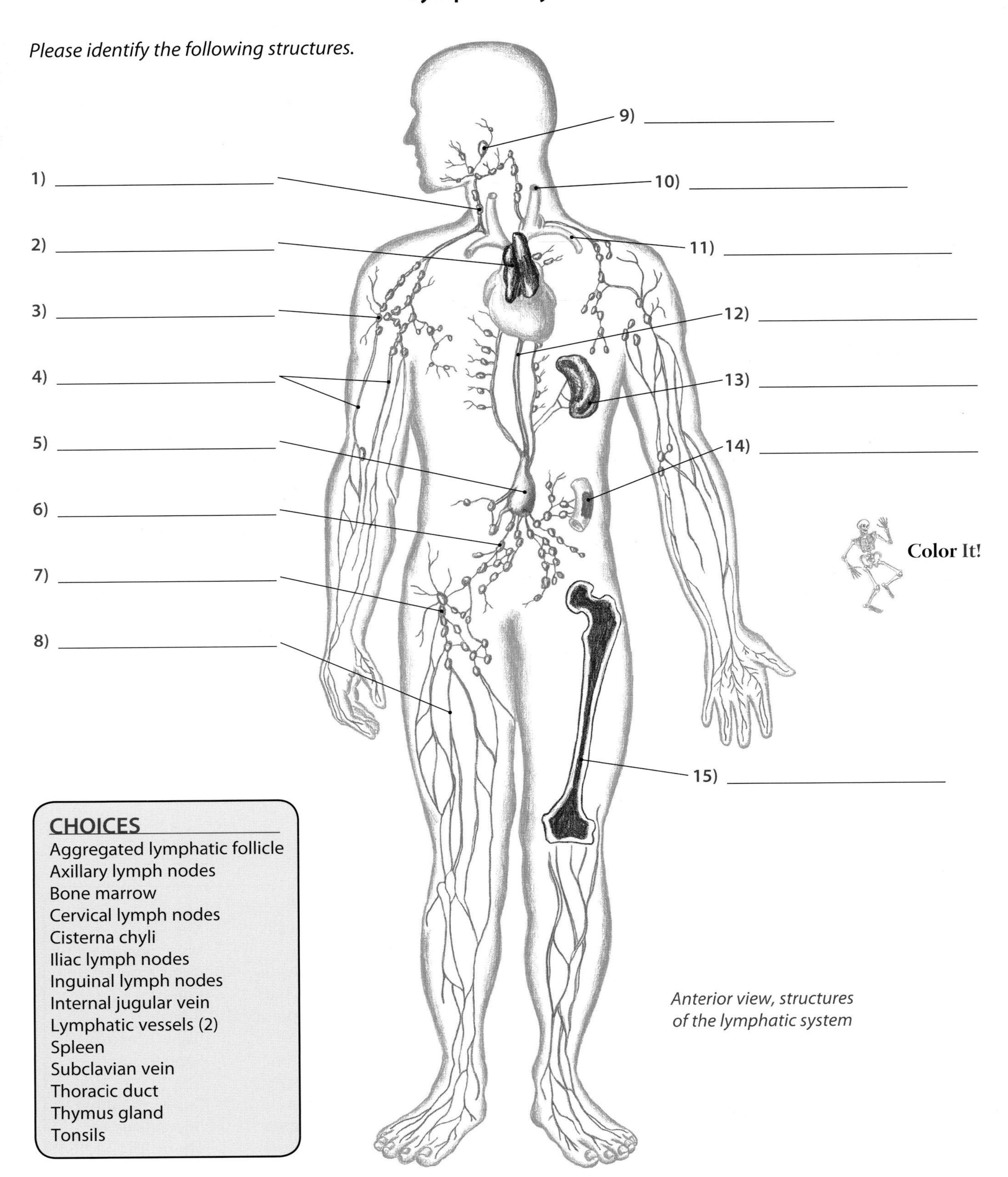

Anterior view, structures of the lymphatic system

CHOICES

- Aggregated lymphatic follicle
- Axillary lymph nodes
- Bone marrow
- Cervical lymph nodes
- Cisterna chyli
- Iliac lymph nodes
- Inguinal lymph nodes
- Internal jugular vein
- Lymphatic vessels (2)
- Spleen
- Subclavian vein
- Thoracic duct
- Thymus gland
- Tonsils

Shoulder and Arm
Topographical Views

Please identify the following structures.

CHOICES
Acromion
Axilla
Biceps brachii
Clavicle
Deltoid (2)
Inferior angle of the scapula
Latissimus dorsi (2)
Pectoralis major
Serratus anterior
Spine of the scapula
Superior nuchal line of the occiput
Trapezius (2)
Triceps brachii (2)

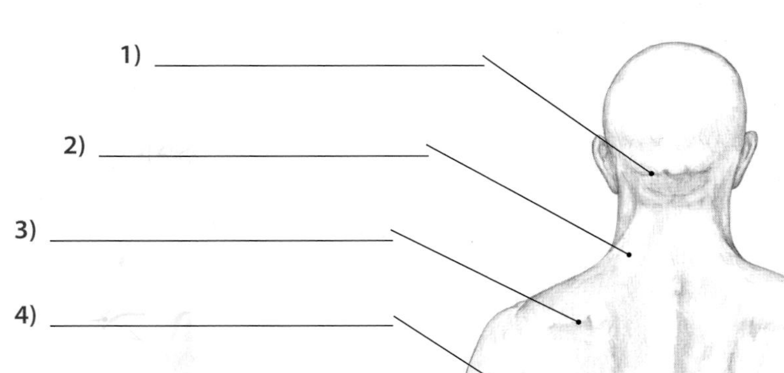

Posterior view

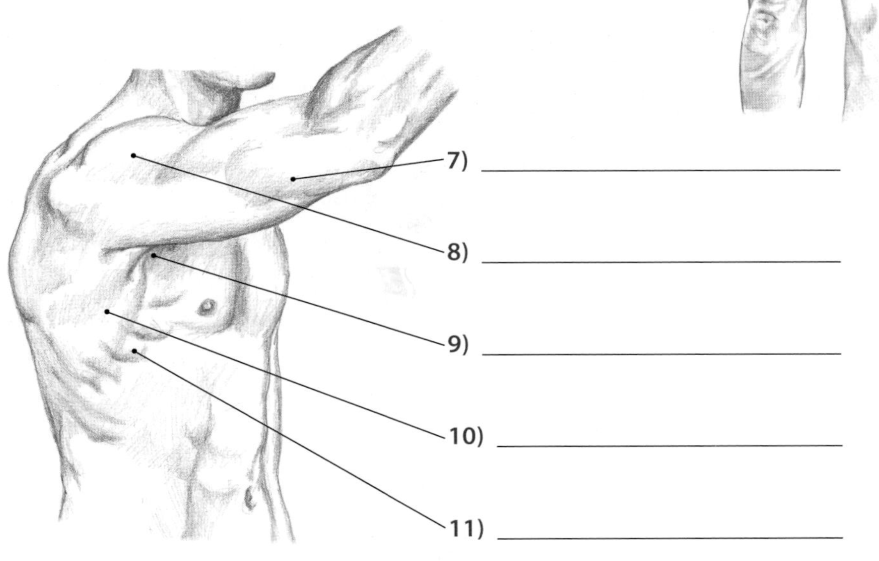

Anterior/lateral view

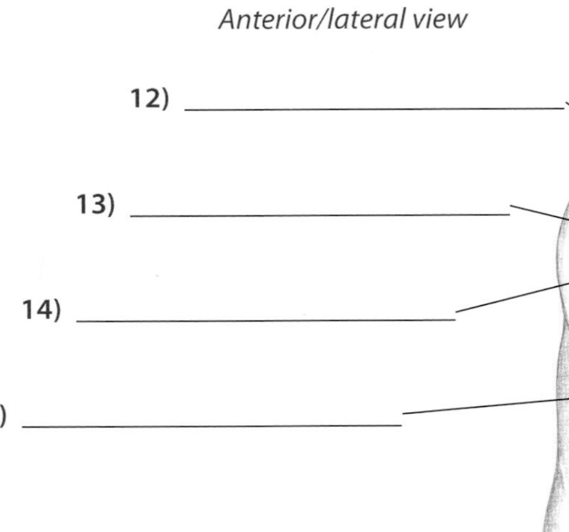

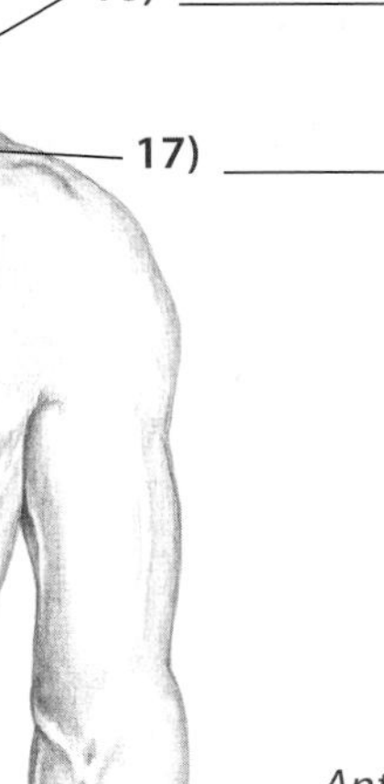

Anterior view

Bones and Bony Landmarks #1

p. 48-54

Please answer the following questions.

1) Three bones that make up the shoulder complex are the ______________________, ______________________ and ______________________.

2) The acromioclavicular and sternoclavicular are what types of joints? ______________________

3) The single attachment between the axial and upper appendicular skeletons is the ______________________ joint.

4) The humerus and scapula form the ______________________ joint.

5) A great base camp for locating other bony landmarks of the shoulder is the ______________________.

6) With your partner prone, how can you best position your partner's hand to locate the inferior angle of the scapula?

____

7) A "winged scapula" often indicates weakness in which muscle? ______________________

8) The superior angle of the scapula serves as an attachment site for the ______________________ muscle and is deep to the ______________________ muscle.

9) When accessing the lateral border of the scapula, through which two muscle bellies will you have to palpate?

______________________ ______________________

10) Accessing the infraglenoid tubercle can elicit tenderness. How can you palpate this landmark without causing pain?

____

Let's Palpate!

Remember—there are no right or wrong answers here

Locate and explore the **coracoid process of the scapula** on three individuals. Then write three words that describe what you feel. (See p. 59 in *Trail Guide*)

Person #1 ______________

____

____

____

Person #2 ______________

____

____

____

Person #3 ______________

____

____

____

Shoulder and Arm
Bones and Bony Landmarks #2

Please answer the following questions.

1) The three scapular fossae contain which three muscles?

_______________ _______________ _______________

2) The infraspinous fossa can be isolated by setting your fingers on which three bony landmarks?

_______________ _______________ _______________

3) Palpating laterally along the supraspinous fossa, your fingers will bump into which two bony structures?

_______________ _______________

4) To locate the subscapular fossa in a side lying position, you slowly sink your thumb onto the fossa's surface. What can your other hand do to help access the fossa? _______________

5) To access the medial portion of the subscapular fossa, how would you position your partner?

____

6) The acromion serves as an attachment site for which two muscles?

_______________ _______________

7) When palpating the clavicle, the _______________ end rises superiorly while the _______________ end curves inferiorly.

8) To feel the acromioclavicular joint space widen slightly and then diminish, you can ask your partner to do which two movements of the scapula? _______________ _______________

9) The coracoid process is often located in the _______________ groove.

10) Sculpting a circle around the edges of the coracoid process can help you get a better understanding of its _______________ and _______________.

11) What are the three muscles that attach to the greater tubercle of the humerus?

_______________ _______________ _______________

12) Within the intertubercular groove lies the tendon of which muscle? _______________

Extra Credit

How many muscles attach to the scapula? __________ Try listing them below.

Bones of the Shoulder and Arm #1

p. 48-49

*Please identify the following structures. Numbers in **black** indicate bones, numbers in **red** are bony landmarks.*

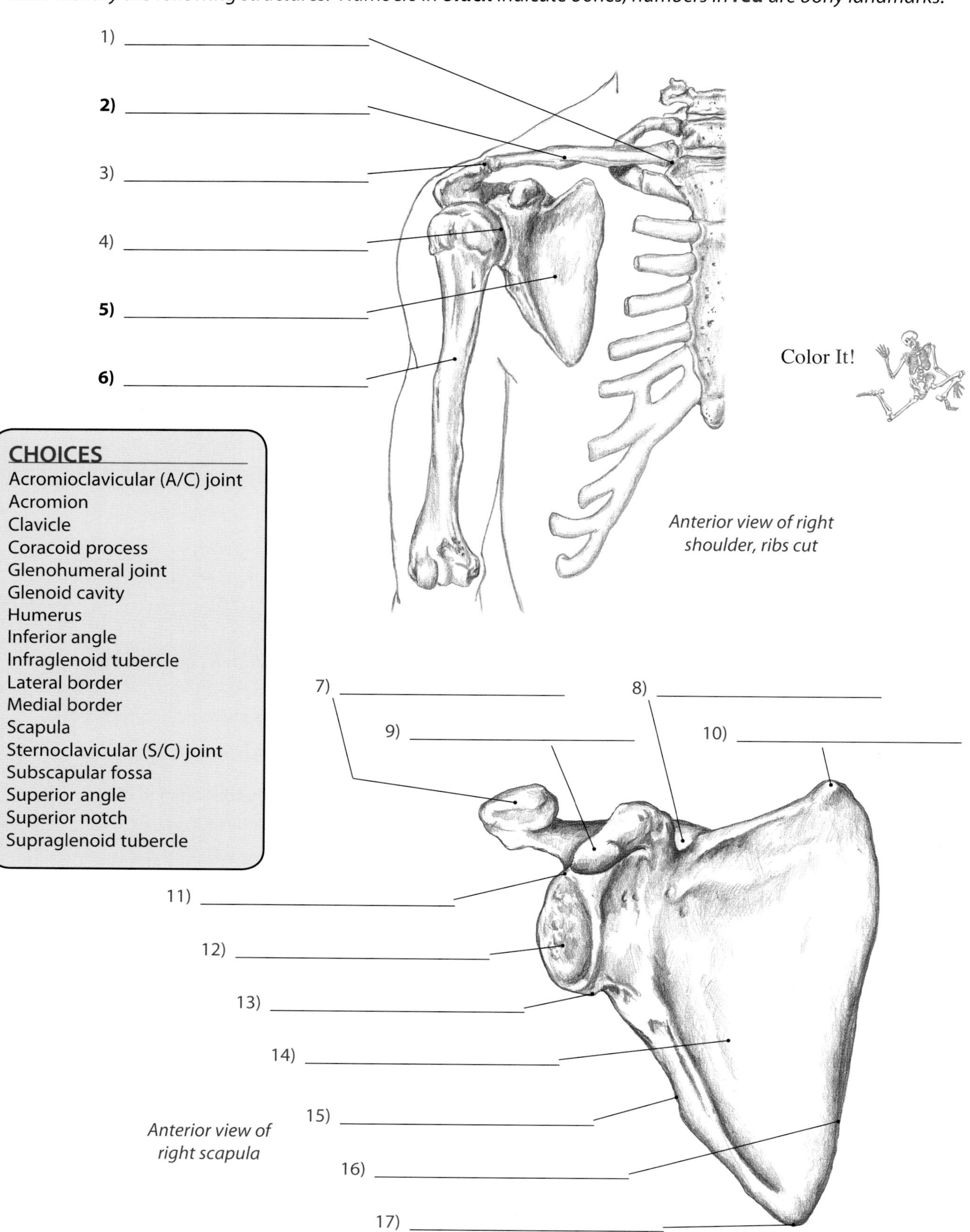

Anterior view of right shoulder, ribs cut

Anterior view of right scapula

CHOICES

- Acromioclavicular (A/C) joint
- Acromion
- Clavicle
- Coracoid process
- Glenohumeral joint
- Glenoid cavity
- Humerus
- Inferior angle
- Infraglenoid tubercle
- Lateral border
- Medial border
- Scapula
- Sternoclavicular (S/C) joint
- Subscapular fossa
- Superior angle
- Superior notch
- Supraglenoid tubercle

Shoulder and Arm

Bones of the Shoulder and Arm #2

Please identify the following structures.

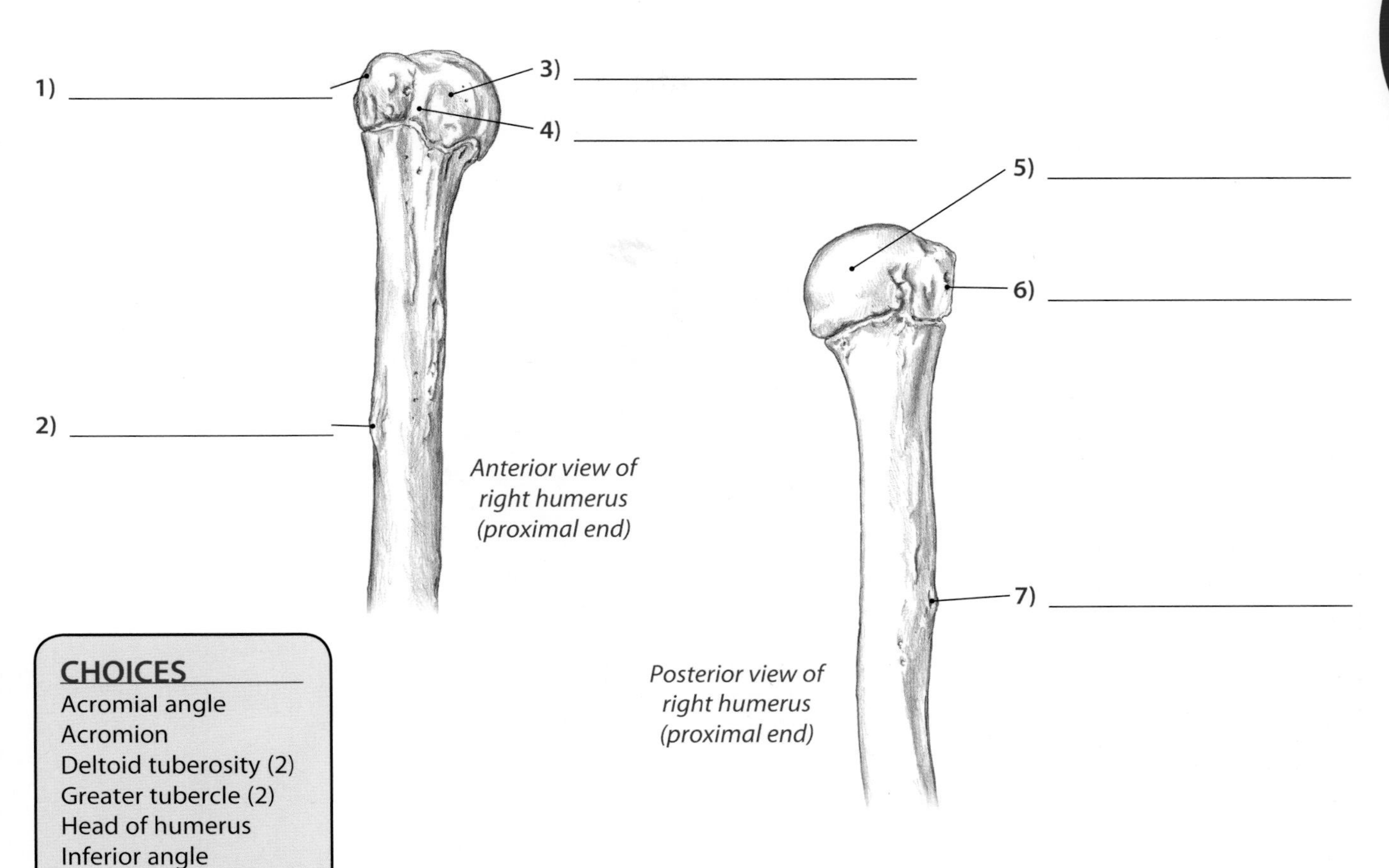

Anterior view of right humerus (proximal end)

Posterior view of right humerus (proximal end)

CHOICES

- Acromial angle
- Acromion
- Deltoid tuberosity (2)
- Greater tubercle (2)
- Head of humerus
- Inferior angle
- Infraspinous fossa
- Intertubercular groove
- Lateral border
- Lesser tubercle
- Medial border
- Spine of the scapula
- Superior angle
- Supraspinous fossa

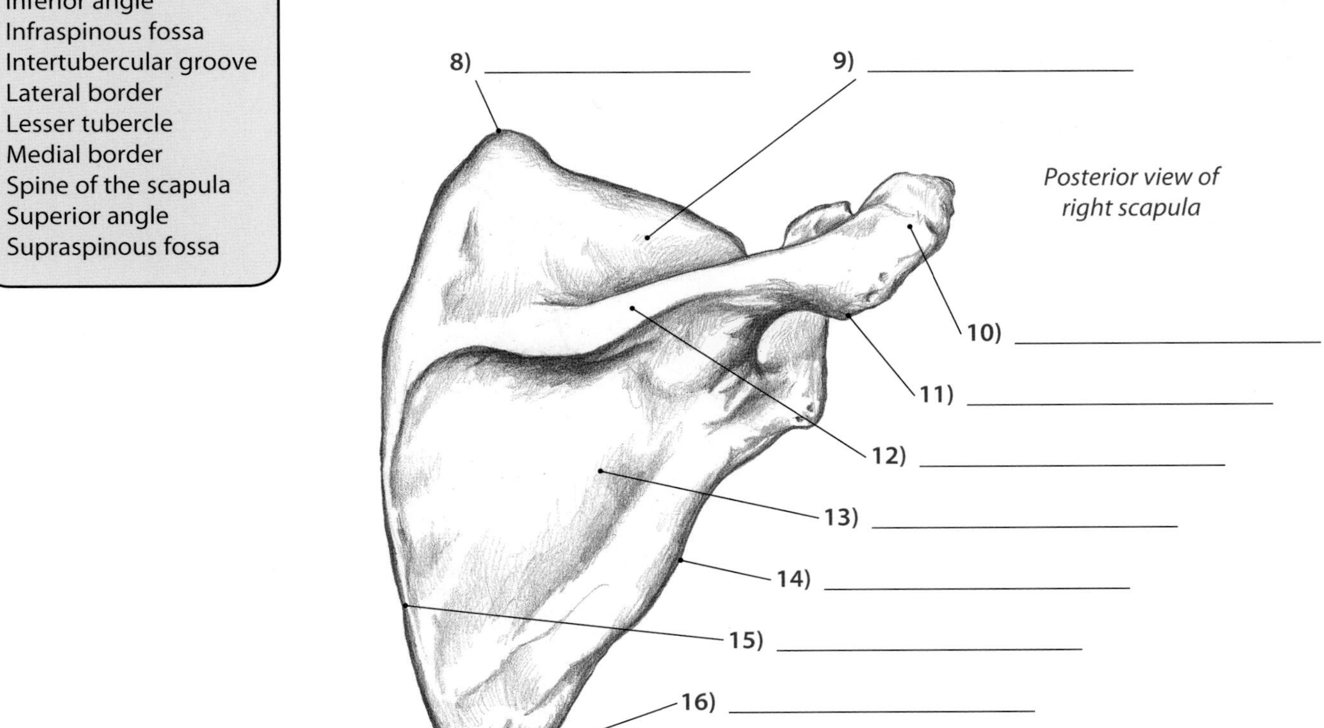

Posterior view of right scapula

Muscles of the Shoulder and Arm #1

Please identify the following structures.

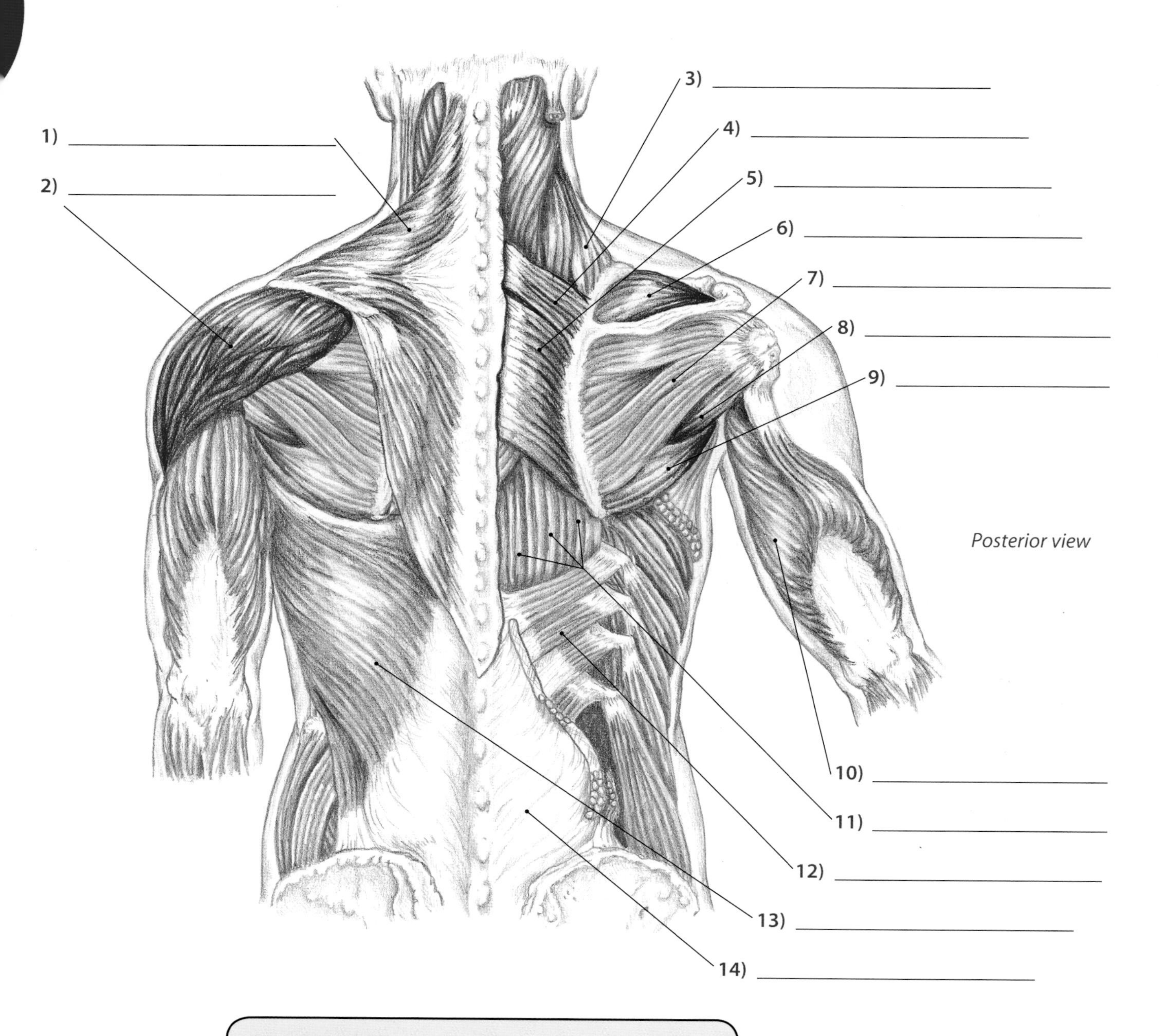

CHOICES

Deltoid	Serratus posterior inferior
Erector spinae group	Supraspinatus
Infraspinatus	Teres major
Latissimus dorsi	Teres minor
Levator scapula	Thoracolumbar aponeurosis
Rhomboid major	Trapezius
Rhomboid minor	Triceps brachii

Shoulder and Arm

Muscles of the Shoulder and Arm #2

Please identify the following structures.

CHOICES

- Biceps brachii (2)
- Brachialis
- Coracobrachialis
- Deltoid (2)
- External oblique
- Infraspinatus
- Latissimus dorsi
- Levator scapula (2)
- Pectoralis major
- Pectoralis minor
- Serratus anterior (2)
- Teres major
- Teres minor
- Trapezius (2)
- Triceps brachii

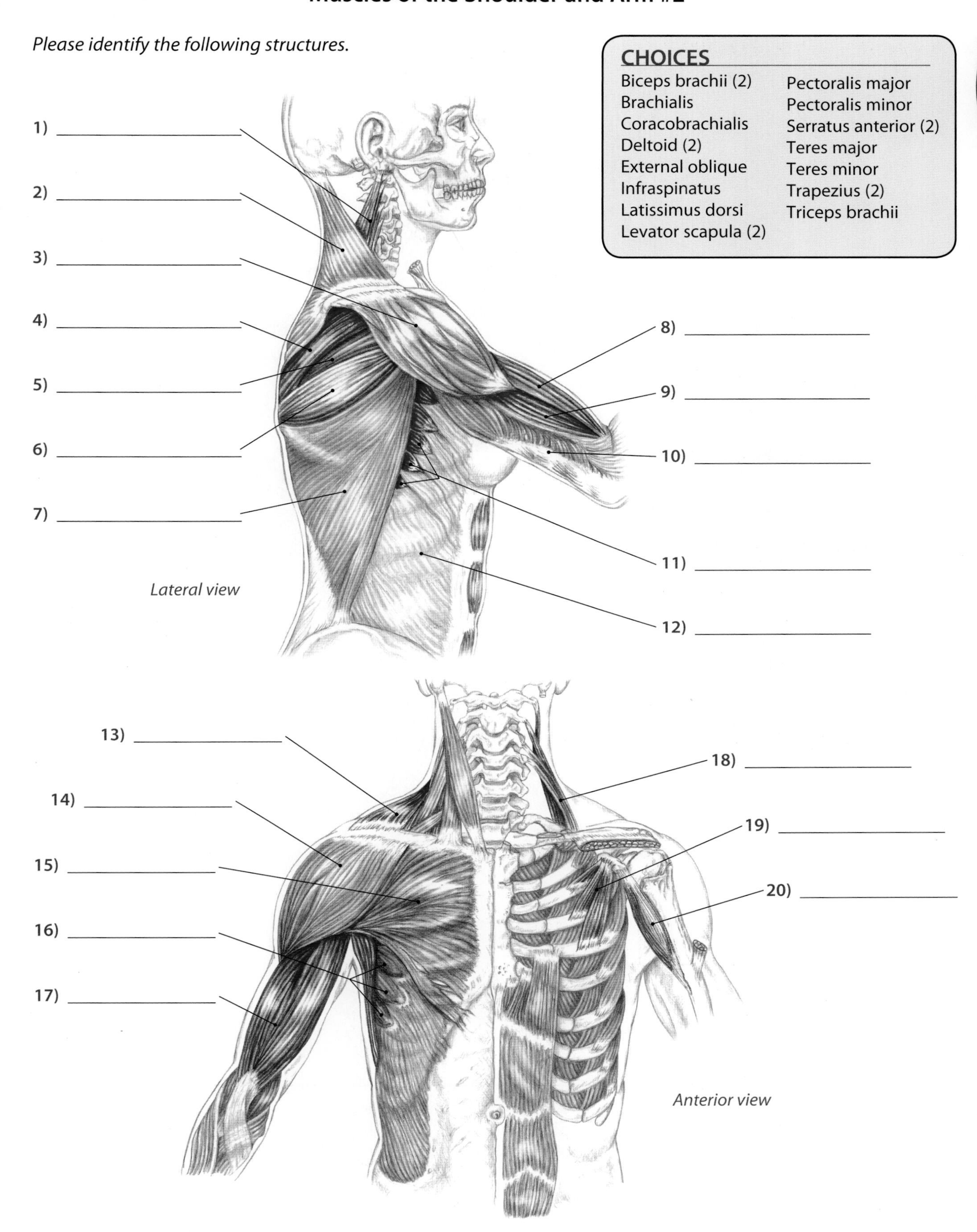

Color the Muscles #1

Using different colors, please fill in and label the muscles and other structures listed below.

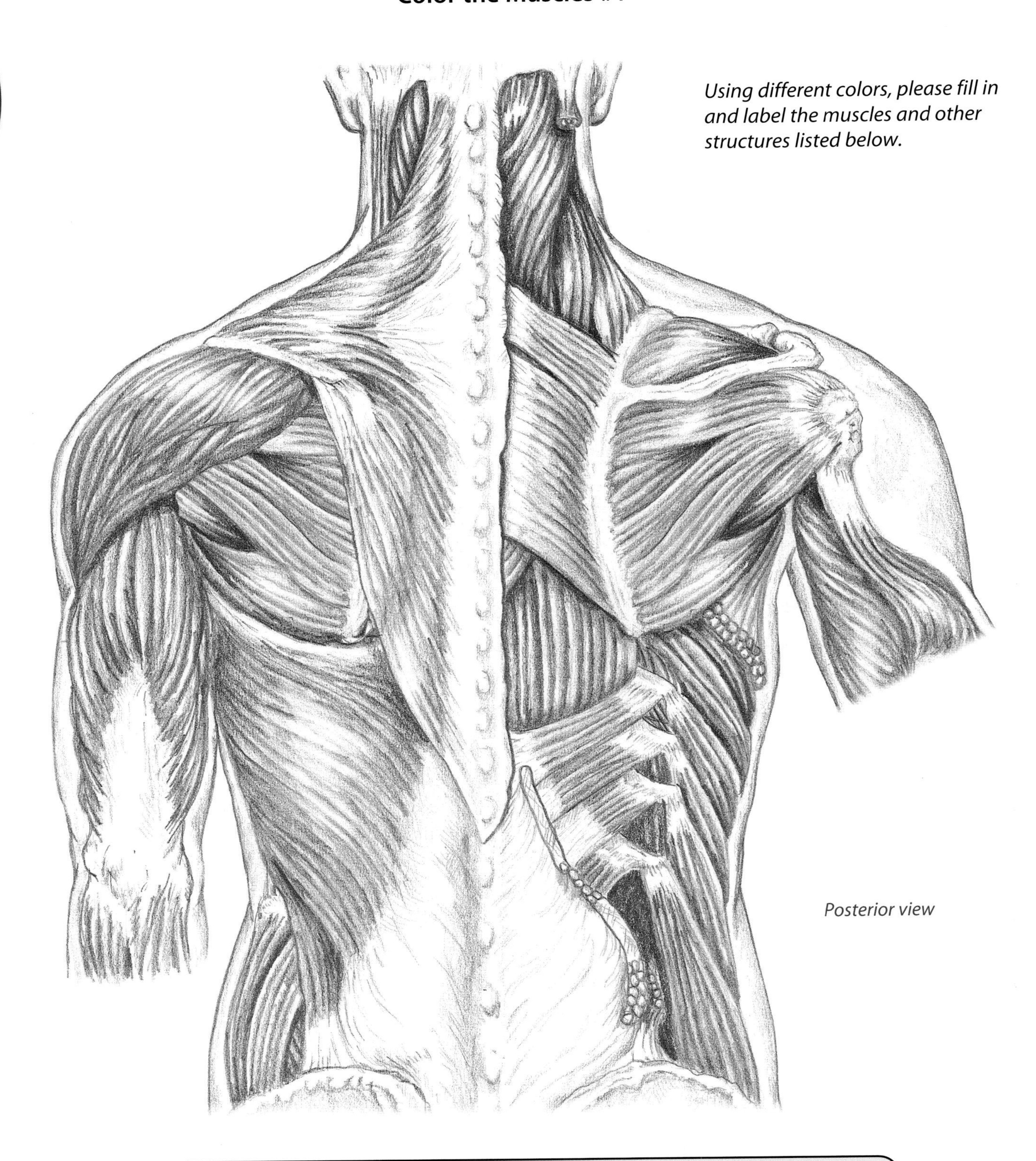

Posterior view

- Deltoid
- Erector spinae group
- Infraspinatus
- Latissimus dorsi
- Levator scapula
- Rhomboid major
- Rhomboid minor
- Serratus posterior inferior
- Supraspinatus
- Teres major
- Teres minor
- Thoracolumbar aponeurosis
- Trapezius
- Triceps brachii

Shoulder and Arm
Color the Muscles #2

Using different colors, please fill in and label the muscles listed below.

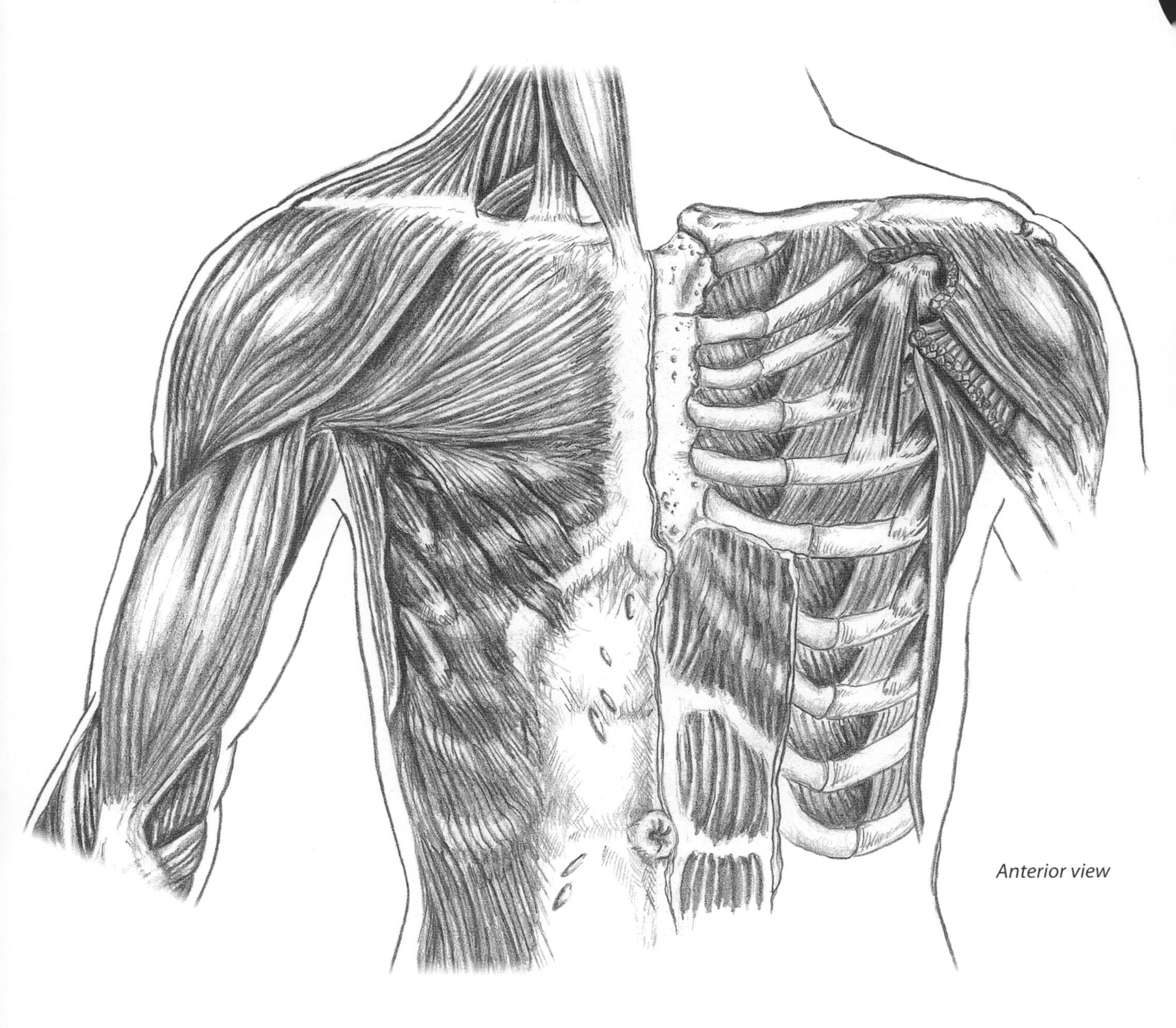

Biceps brachii	Pectoralis major	Serratus anterior
Coracobrachialis	Pectoralis minor	Sternocleidomastoid
Deltoid	Rectus abdominis	Trapezius
Latissimus dorsi		

Shoulder and Arm

p. 63-66

Muscles and Movements #1

Please list the action demonstrated, two synergists and one antagonist.
The first letter of each muscle has been provided.

1) This action happens at which joint?

____

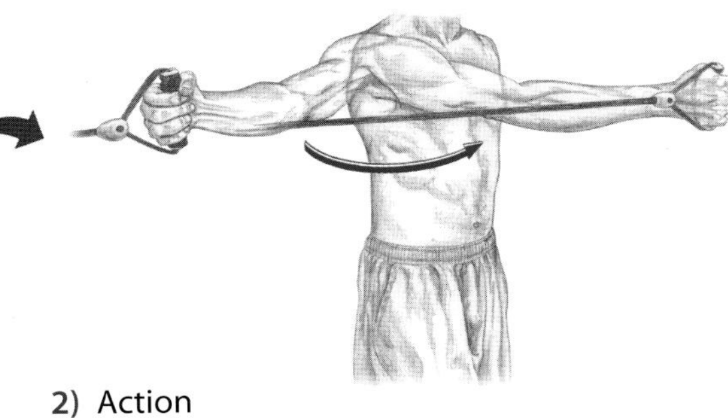

2) Action

____

3) Synergists

D ______________________________

P ______________________________

4) Antagonist

D ______________________________

5) Action

____

6) Synergists

T ______________________________

R ______________________________

7) Antagonist

P ______________________________

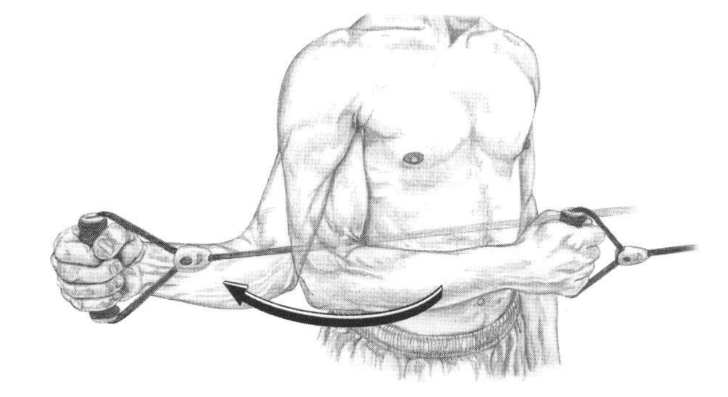

8) Action

____

9) Synergists

D ______________________________

T ______________________________

10) Antagonist

S ______________________________

Shoulder and Arm

Muscles and Movements #2

Please list the action demonstrated, synergist(s) and antagonist(s). The first letter of each muscle has been provided.

1) This action happens at which joint?

____

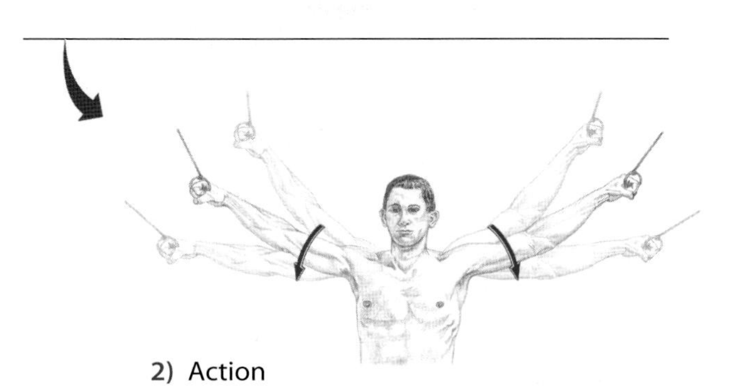

2) Action

____

3) Synergists

I ______________________________

T ______________________________

4) Antagonist

S ______________________________

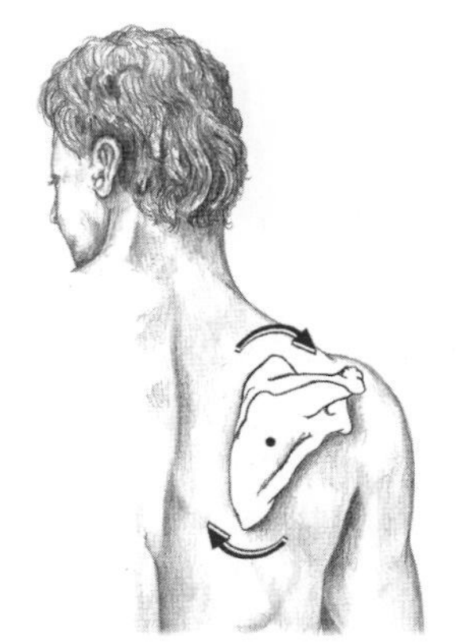

5) Action

____

6) Synergists

R ______________________________

L ______________________________

7) Antagonist

T ______________________________

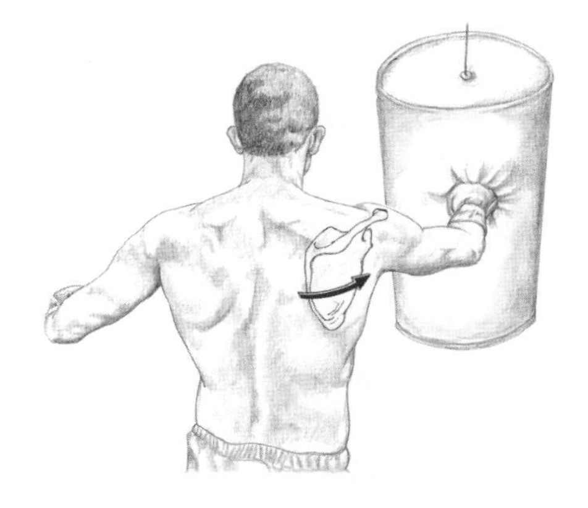

8) Action

____

9) Synergists

S ______________________________

P ______________________________

10) Antagonist

R ______________________________

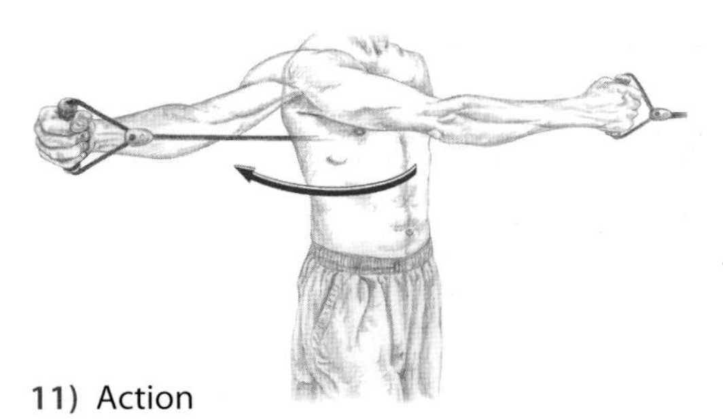

11) Action

____

12) Synergists

D ______________________________

13) Antagonist

P ______________________________

D ______________________________

Muscles and Movements #3

Please list the action demonstrated, two synergists and one antagonist. The first letter of each muscle has been provided.

1) This action happens at which joint?

____

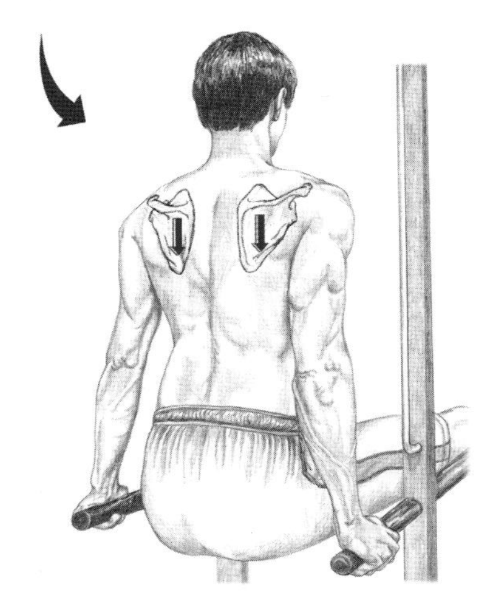

2) Action

____

3) Synergists

S ___________________________

P ___________________________

4) Antagonist

R ___________________________

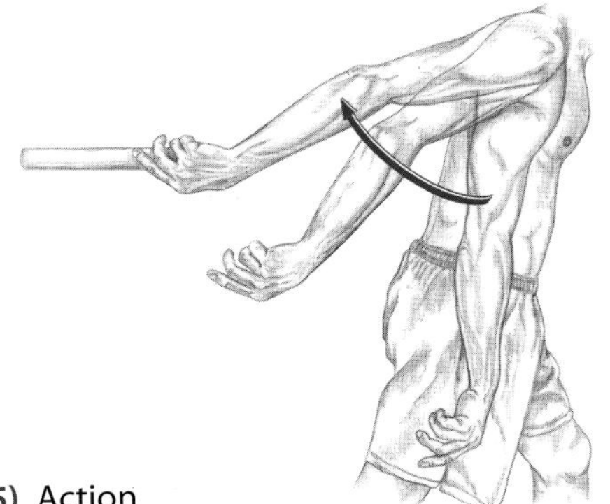

5) Action

____

6) Synergists

D ___________________________

L ___________________________

7) Antagonist

B ___________________________

8) Action

____

9) Synergists

D ___________________________

S ___________________________

10) Antagonist

P ___________________________

Shoulder and Arm

Muscles and Movements #4

Please list the action demonstrated, two synergists and one antagonist. The first letter of each muscle has been provided.

1) This action happens at which joint?

____

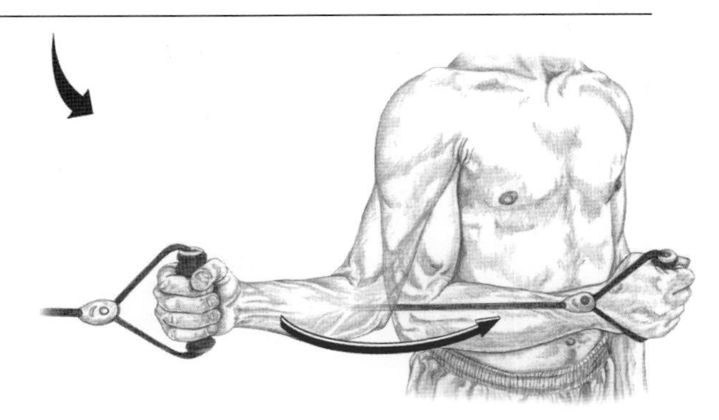

2) Action

____

3) Synergists

S ______________________

P ______________________

4) Antagonist

I ______________________

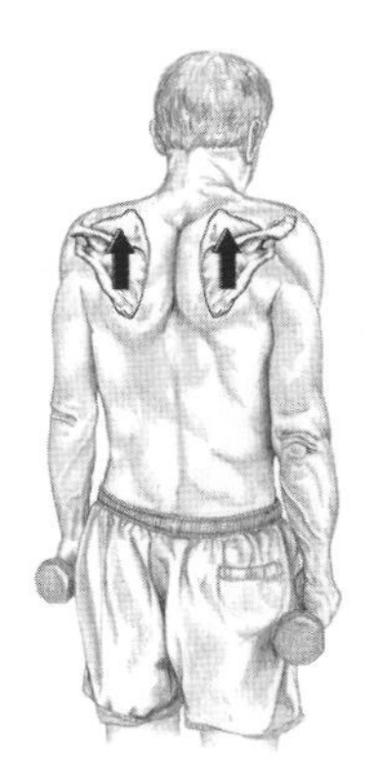

5) Action

____

6) Synergists

R ______________________

L ______________________

7) Antagonist

S ______________________

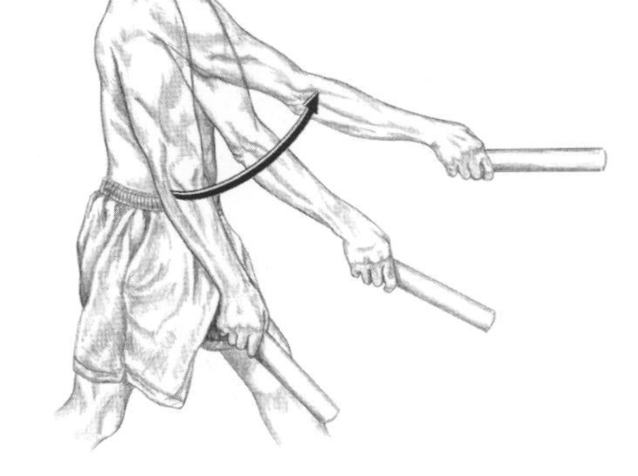

8) Action

____

9) Synergists

B ______________________

C ______________________

10) Antagonist

L ______________________

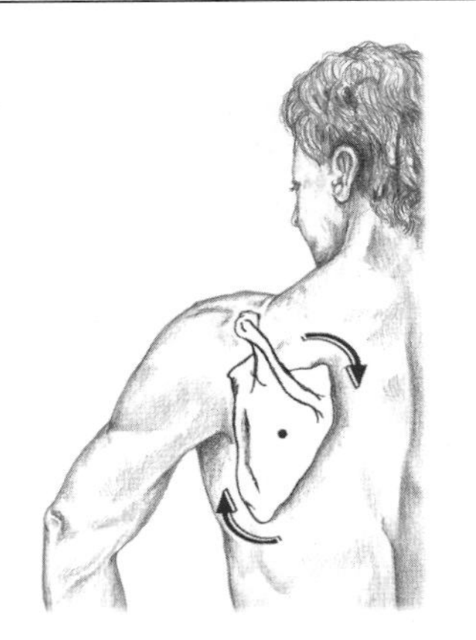

11) Action

____

12) Synergists

T ______________________

S ______________________

13) Antagonist

R ______________________

What's the Muscle?

Please identify the following muscles.

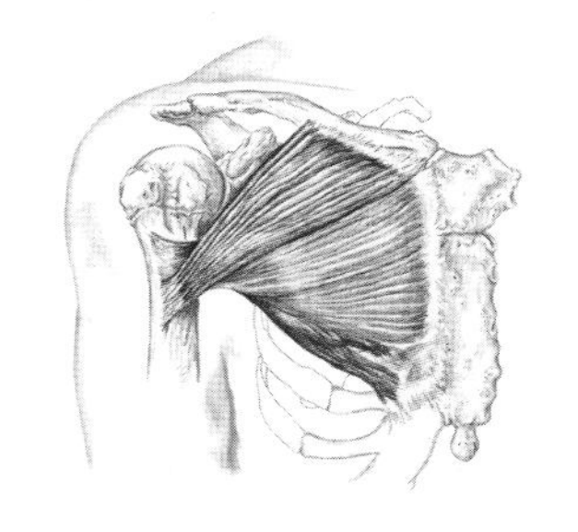

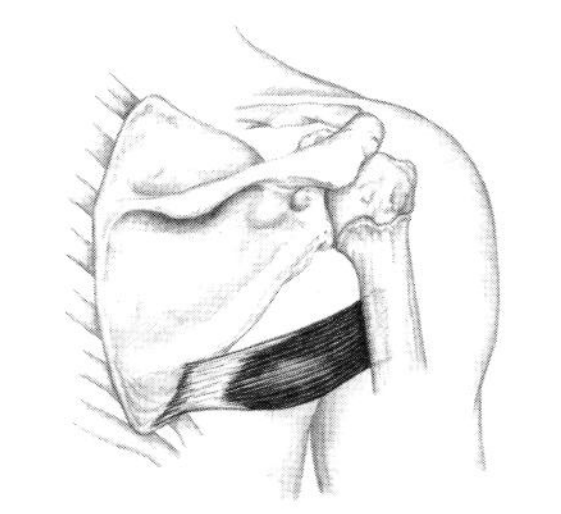

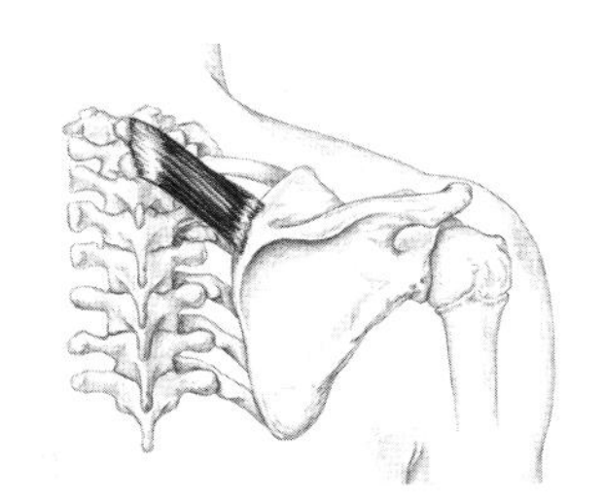

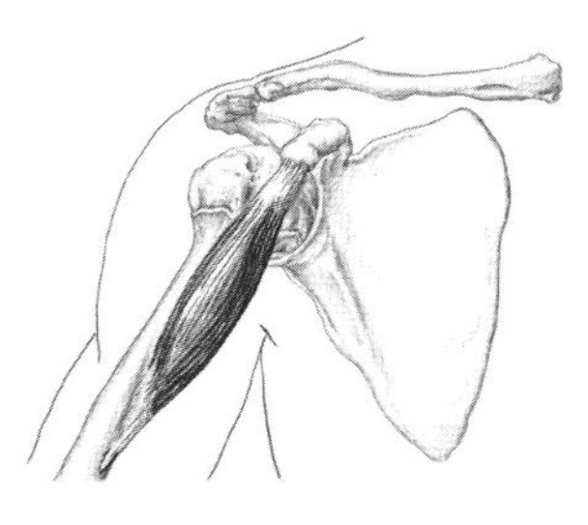

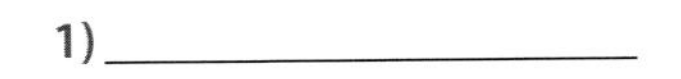

1) ____________________

2) ____________________

3) ____________________

4) ____________________

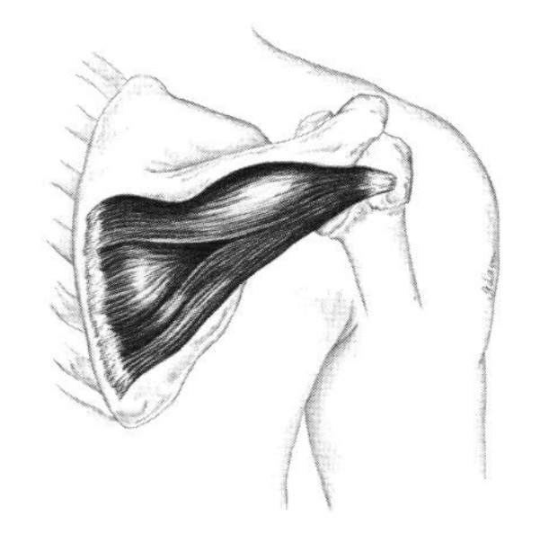

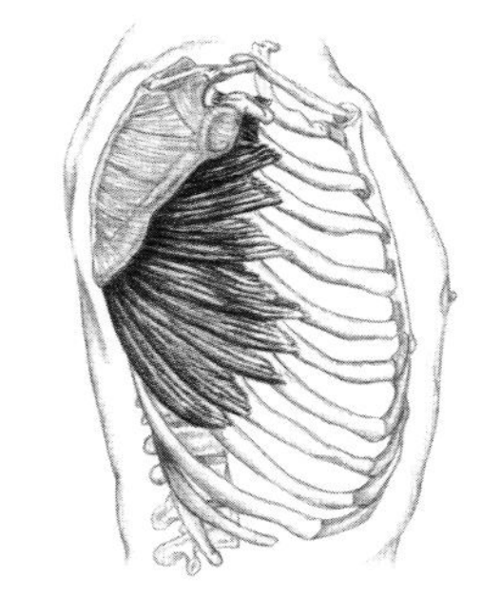

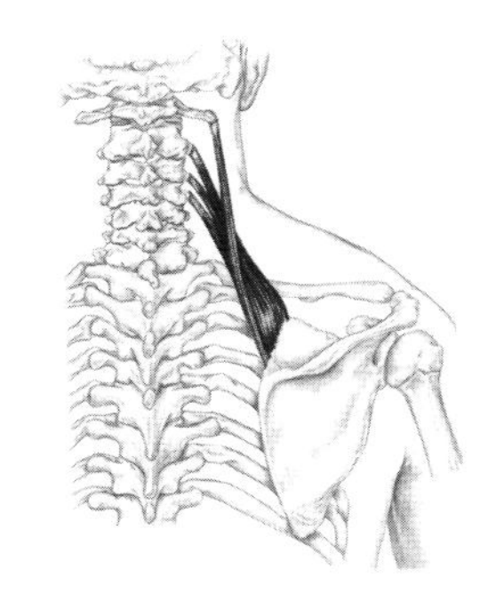

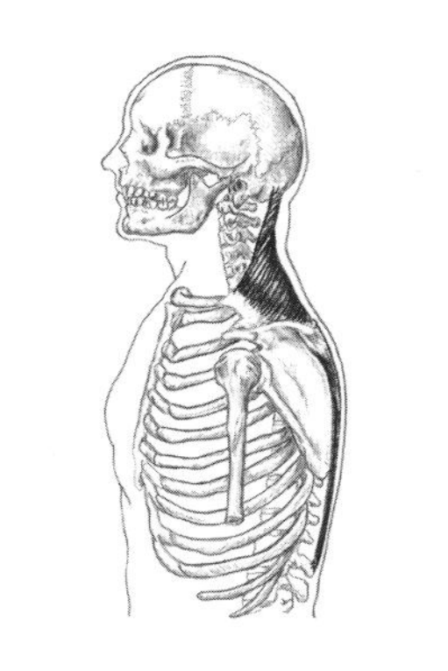

5) ____________________

6) ____________________

7) ____________________

8) ____________________

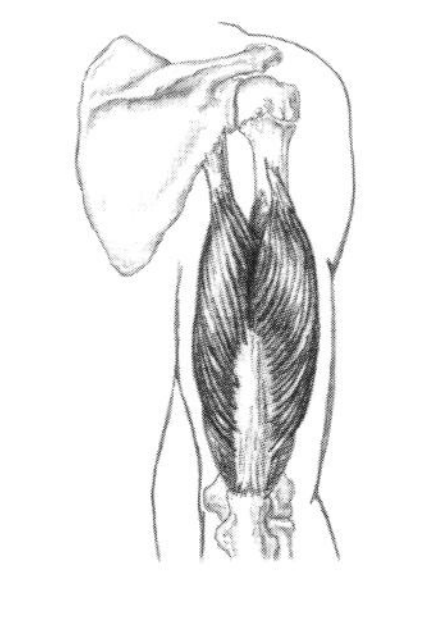

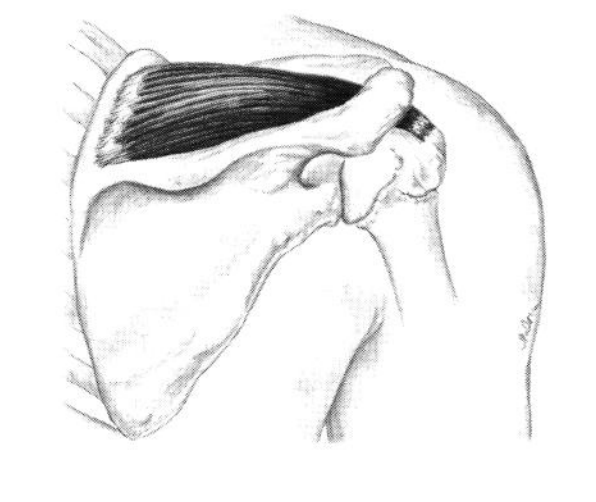

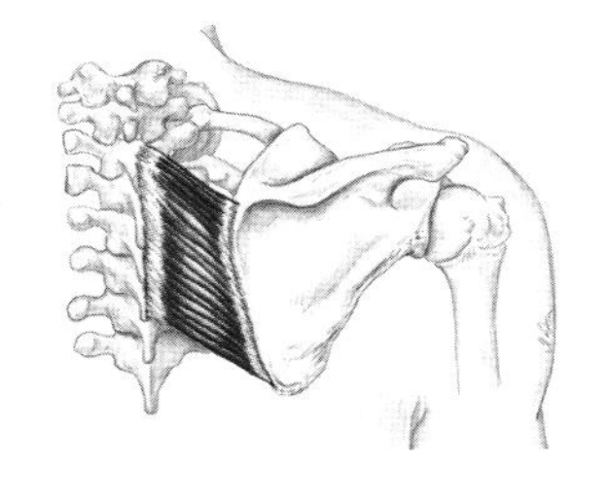

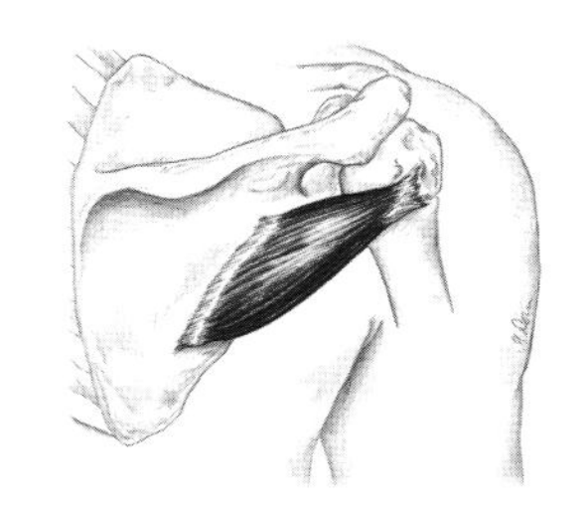

9) ____________________

10) ____________________

11) ____________________

12) ____________________

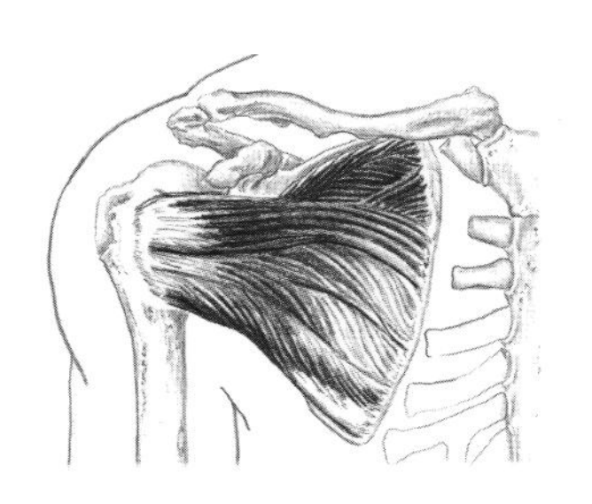

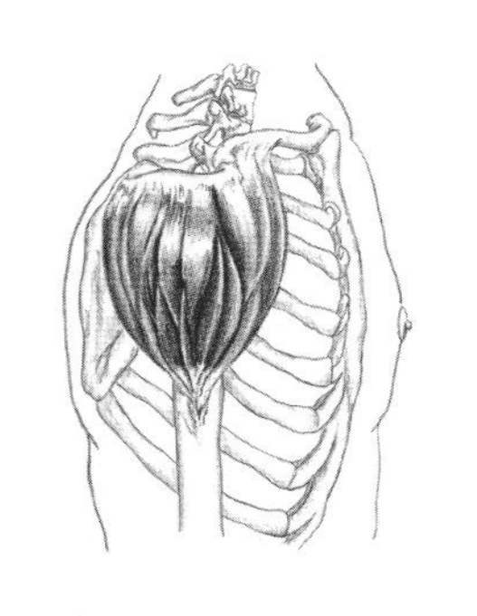

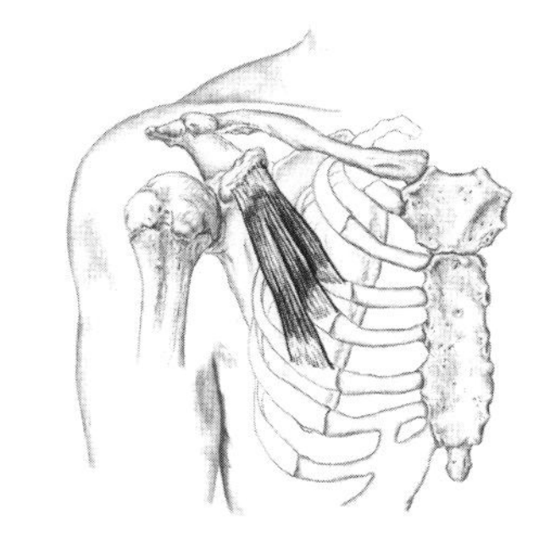

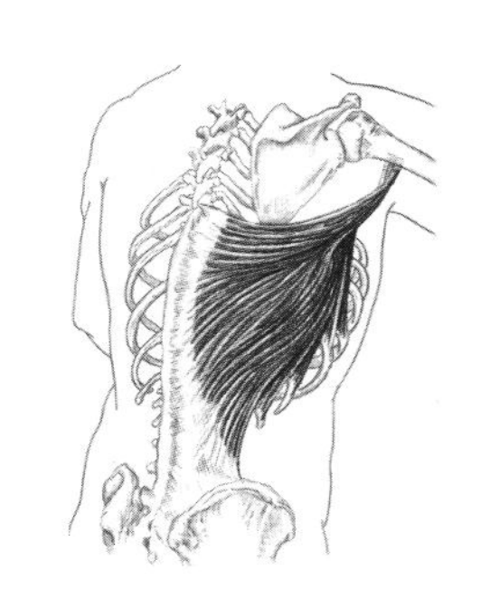

13) ____________________

14) ____________________

15) ____________________

16) ____________________

Shoulder and Arm, Muscle Group #1

Deltoid, Trapezius, Latissimus Dorsi and Teres Major

Please answer the following questions.

1) The *origin* of the deltoid is identical to the *insertion* of which muscle? ______________________

2) What action can you ask your partner to perform in order to contract all fibers of the deltoid?

____

3) The actions of the deltoid's anterior and posterior fibers make it an ______________________ to itself.

4) The upper fibers of the trapezius elevate the scapula, so the lower fibers must

______________________ the scapula.

5) Bilateral contraction of the upper fibers of the trapezius will create what movement of the head and neck?

____

6) To feel the middle fibers of the trapezius contract, you could ask your partner to perform which action?

____

7) Which portion of the latissimus dorsi is easy to grasp? ______________________

8) When palpating the latissimus dorsi, how can you discern the muscle tissue from the superficial skin?

____

9) One palpatory distinction between the teres major and latissimus dorsi is that the teres attaches to the

______________________ of the scapula.

Shorten or Lengthen?

10) Passive abduction of the scapula would ______________ the middle fibers of the trapezius.

11) Passive elevation of the scapula would ______________ the trapezius' upper fibers and

______________ its lower fibers.

12) Passive rotation of the head and neck to the left would ______________ the left trapezius' upper fibers.

13) Passive flexion of the shoulder would ______________ the anterior fibers of the deltoid.

14) Passive lateral rotation of the shoulder would ______________ the deltoid's posterior fibers.

15) Passive flexion of the shoulder would ______________ the latissimus dorsi.

16) Passive adduction of the shoulder would ______________ the teres major.

17) Passive medial rotation of the shoulder would ______________ the latissimus dorsi and teres major.

Shoulder and Arm, Muscle Group #1

p. 67-73

Deltoid, Trapezius, Latissimus Dorsi and Teres Major

Matching

Match the origin and insertion to the correct muscle.

Origins

1) External occipital protuberance, medial portion of superior nuchal line of the occiput, ligamentum nuchae and spinous processes of C-7 through T-12

2) Lateral one-third of clavicle, acromion and spine of scapula

3) Inferior angle and lower one-third of lateral border of the scapula

4) Inferior angle of scapula, spinous processes of last six thoracic vertebrae, last three or four ribs, thoracolumbar aponeurosis and posterior iliac crest

Insertions

5) Crest of the lesser tubercle of the humerus

6) Deltoid tuberosity

7) Lateral one-third of clavicle, acromion and spine of the scapula

8) Intertubercular groove of the humerus

Muscle	O	I
Deltoid	______	______
Latissimus dorsi	______	______
Teres major	______	______
Trapezius	______	______

Let's Palpate!

Remember—there are no right or wrong answers here

Locate and explore the **upper fibers of the trapezius** on three individuals. Then write three words that describe what you feel. (See p. 68-70 in *Trail Guide.*)

Person #1 ____________

____

____

____

Person #2 ____________

____

____

____

Person #3 ____________

____

____

____

p. 74-81

Shoulder and Arm, Muscle Group #2

Rotator Cuff Muscles

Please answer the following questions.

1) The four rotator cuff muscles encompass and stabilize the ______________ joint.

2) To locate the supraspinatus belly, you must palpate through which muscle? ______________

3) The only rotator cuff muscle not involved with rotation of the shoulder is the ______________.

4) The dense quality of the infraspinatus is due to its ______________.

5) The subscapularis is sandwiched between which fossa and which muscle?

______________ ______________

6) What is the best action to ask your partner to perform to feel the supraspinatus contract?

____

7) What three bony landmarks can you lay your fingers along to isolate the belly of the infraspinatus?

______________ ______________ ______________

8) If you follow the fibers of the infraspinatus laterally, they converge underneath which muscle?

____

9) You can distinguish the teres minor from the teres major by their sizes and actions. Explain.

____

10) When accessing the subscapularis from a side lying position, under what two muscles should you slide your thumb?

______________ ______________

11) What action could you ask your partner to do to gently contract the subscapularis?

____

12) To access the supraspinatus tendon, you need to sink your thumb tip through which muscle? ______________

13) To locate the infraspinatus and teres minor tendons in a supine position, how would you position your partner's shoulder?

____

14) The subscapularis tendon can be located between the two tendons of which muscle? ______________

15) In which two directions do you want to move from the coracoid process to locate the subscapularis tendon?

______________ ______________

Rotator Cuff Muscles

p. 74-81

Matching

Match the origin and insertion to the correct muscle.

Origins

1) Infraspinous fossa of the scapula

2) Subscapular fossa of the scapula

3) Upper two-thirds of lateral border of the scapula

4) Supraspinous fossa of the scapula

Muscle	O	I
Infraspinatus	______	______
Subscapularis	______	______
Supraspinatus	______	______
Teres minor	______	______

Insertions

5) Greater tubercle of the humerus (3)

6) Lesser tubercle of the humerus

Shorten or Lengthen?

7) Passive medial rotation of the shoulder would ____________________ the infraspinatus.

8) Passive abduction of the shoulder would ____________________ the supraspinatus.

9) Passive abduction of the shoulder would ____________________ the teres minor.

10) Passive lateral rotation of the shoulder would ____________________ the subscapularis.

Let's Palpate!

Remember—there are no right or wrong answers here

Locate and explore the **infraspinatus** on three individuals. Then write three words that describe what you feel. (See p. 77 in *Trail Guide*.)

Person #1 ______________

____

____

____

Person #2 ______________

____

____

____

Person #3 ______________

____

____

____

Shoulder and Arm, Muscle Group #3

Rhomboids, Levator, Serratus, Pectorals and Subclavius

Please answer the following questions.

1) The rhomboids are deep to the ________________ muscle and superficial to the ________________ muscles.

2) Can you name two actions in which the rhomboids and trapezius are synergists and one action in which they are antagonists? ________________ ________________ ________________

3) The levator scapula is situated between which two muscles on the lateral side of the neck?

________________ ________________

4) An action to ask your partner to perform to feel the levator scapula contract is ________________.

5) When accessing the levator scapula in a supine position, the benefits of rotating the head 45°away from the. side you are palpating include:

________________ ________________ ________________

6) The serratus anterior abducts the scapula, making it a direct antagonist to the ________________.

7) Most of the serratus anterior is deep to the scapula and which two muscles?

________________ ________________

8) Accessing the medial portion of the serratus anterior by curling your fingers around the medial border of the scapula, your fingers will inherently have to work through the bellies of which two muscles?

________________ ________________

9) The pectoralis major is divided into three segments:

________________ ________________ ________________

10) Can you name an everyday action in which you use your pectoralis major? ________________

11) The most important aspect when palpating near breast tissue is:

____

12) If you follow the fibers of the pectoralis major laterally, they blend with the fibers of which muscle?

____

13) Flexing the shoulder and pulling it anteriorly while you palpate the pectoralis major in a side lying position has which two benefits?

________________ ________________

14) The pectoralis minor has the potential to create neurovascular compression on which three vessels?

________________ ________________ ________________

Shoulder and Arm, Muscle Group #3

Rhomboids, Levator, Serratus, Pectorals and Subclavius

p. 82-94

Matching

Match the origin and insertion to the correct muscle.

Origins

1) First rib and cartilage
2) Medial half of clavicle, sternum and cartilage of first through sixth ribs
3) Spinous processes of C-7 and T-1
4) Spinous processes of T-2 to T-5
5) External surfaces of upper eight or nine ribs
6) Third, fourth and fifth ribs
7) Transverse processes of first through fourth cervical vertebrae

Insertions

8) Anterior surface of medial border of the scapula
9) Medial surface of coracoid process of the scapula
10) Crest of greater tubercle of humerus
11) Inferior surface of middle one-third of clavicle
12) Medial border of scapula, between superior angle and superior portion of spine of scapula
13) Medial border of the scapula between the spine of the scapula and inferior angle
14) Upper portion of medial border of the scapula, across from spine of the scapula

Muscle	O	I
Levator scapula	______	______
Pectoralis major	______	______
Pectoralis minor	______	______
Rhomboid major	______	______
Rhomboid minor	______	______
Serratus anterior	______	______
Subclavius	______	______

Let's Palpate!

Remember—there are no right or wrong answers here

Locate and explore the **pectoralis minor** on three individuals. Then write three words that describe what you feel. (See p. 92-93 in *Trail Guide.*)

Person #1 ____________

____

____

____

Person #2 ____________

____

____

____

Person #3 ____________

____

____

____

Shoulder and Arm, Muscle Group #4

Biceps Brachii, Triceps Brachii and Coracobrachialis

Please answer the following questions.

1) Which head of the biceps brachii passes through the intertubercular groove? ____________________

2) Can you name an everyday action in which the biceps brachii's ability to supinate the forearm would come in handy?

____

3) As you follow the biceps brachii belly proximally, it becomes deep to which muscle? ____________________

4) The thin sheet of fascia extending from the distal biceps brachii tendon is called the

____________________.

5) The long head of the triceps brachii weaves between which two muscles before attaching at the infraglenoid tubercle?

____________________ ____________________

6) To outline the distal tendon of the triceps brachii, which bony landmark do you want to locate?

____

7) What action could you ask your partner to perform to feel the contraction of the long head of the triceps brachii?

____

8) In anatomical position, the coracobrachialis is deep to which two muscles?

____________________ ____________________

9) How can you position the shoulder to bring the belly of the coracobrachialis to a superficial position?

____

10) To locate the belly of the coracobrachialis, from which muscle would you slide off and into the axilla?

____

Shorten or Lengthen?

11) Passive abduction of the shoulder would ____________________ the coracobrachialis.

12) Passive extension of the shoulder would ____________________ the biceps brachii.

13) Passive flexion of the shoulder would ____________________ the triceps brachii.

14) Passive pronation of the forearm would ____________________ the biceps brachii.

Biceps Brachii, Triceps Brachii and Coracobrachialis

p. 95-99

Matching

Match the origin and insertion to the correct muscle.

Muscle	O	I
Biceps brachii	_______	_______
Coracobrachialis	_______	_______
Triceps brachii	_______	_______

Origins

1) Coracoid process of the scapula
2) Coracoid process of scapula, supraglenoid tubercle of scapula
3) Infraglenoid tubercle of the scapula, posterior surface of proximal half of the humerus and posterior surface of distal half of the humerus

Insertions

4) Medial surface of mid-humeral shaft
5) Olecranon process of the ulna
6) Tuberosity of the radius and aponeurosis of the biceps brachii

Let's Palpate!

Remember—there are no right or wrong answers here

Locate and explore the **triceps brachii** on three individuals. Then write three words that describe what you feel. (See p. 97-98 in *Trail Guide*.)

Person #1 ____________	Person #2 ____________	Person #3 ____________
____________________	____________________	____________________
____________________	____________________	____________________
____________________	____________________	____________________

Shoulder and Arm

Other Structures

Please identify the following structures.

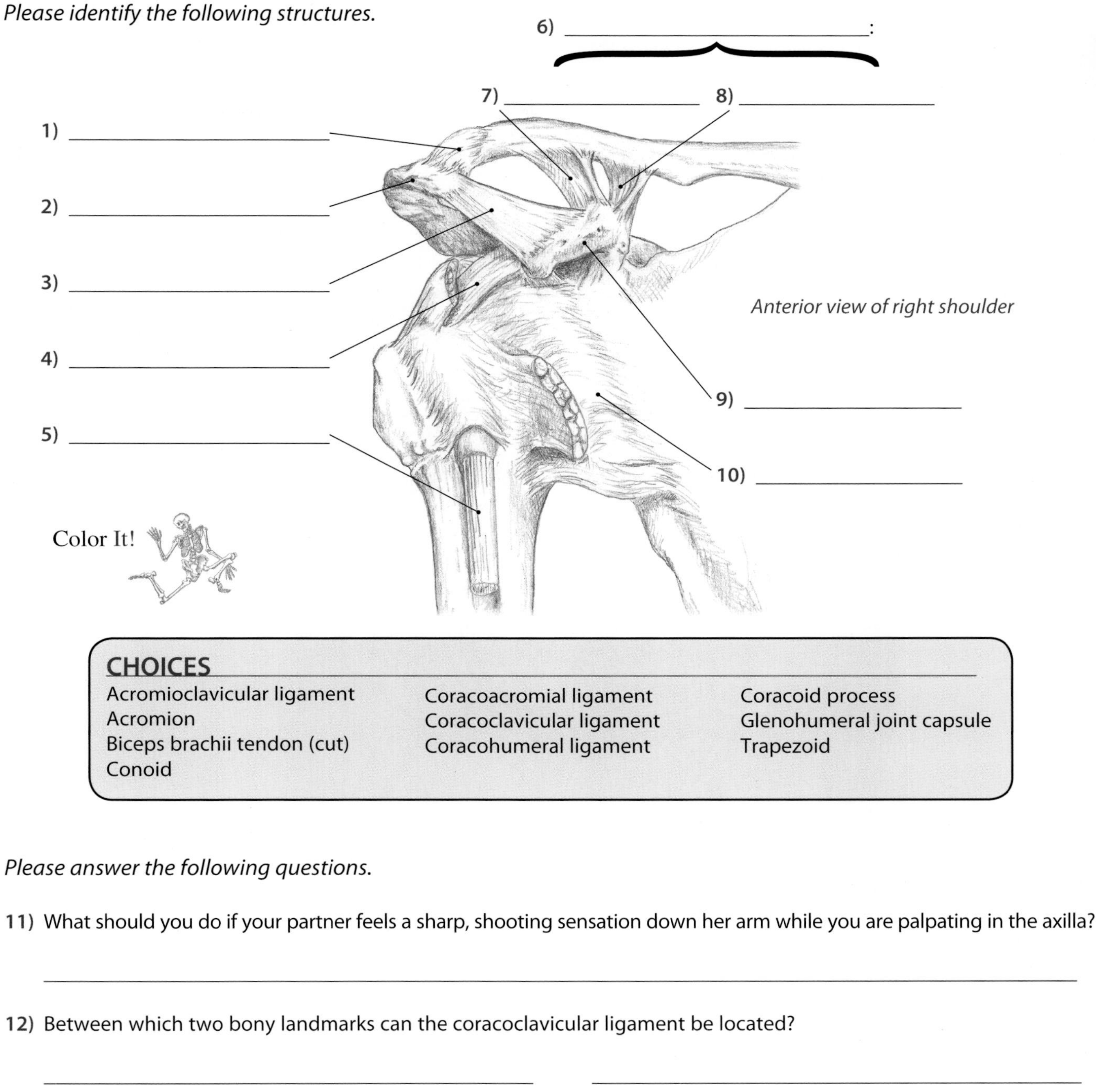

CHOICES

Acromioclavicular ligament	Coracoacromial ligament	Coracoid process
Acromion	Coracoclavicular ligament	Glenohumeral joint capsule
Biceps brachii tendon (cut)	Coracohumeral ligament	Trapezoid
Conoid		

Please answer the following questions.

11) What should you do if your partner feels a sharp, shooting sensation down her arm while you are palpating in the axilla?

____

12) Between which two bony landmarks can the coracoclavicular ligament be located?

______________________ ______________________

13) The ligamentous arch that protects the rotator cuff tendons and subacromial bursa from direct trauma is formed by the ______________________ ligament.

14) To bring the coracoacromial ligament closer to the surface, ______________________ the arm.

15) How can you position the arm to bring the subacromial bursa forward? ______________________

16) The brachial artery can be located on the medial side of the arm between which two muscles?

______________________ ______________________

Glenohumeral Joint #1

p. 102

Please identify the following structures.

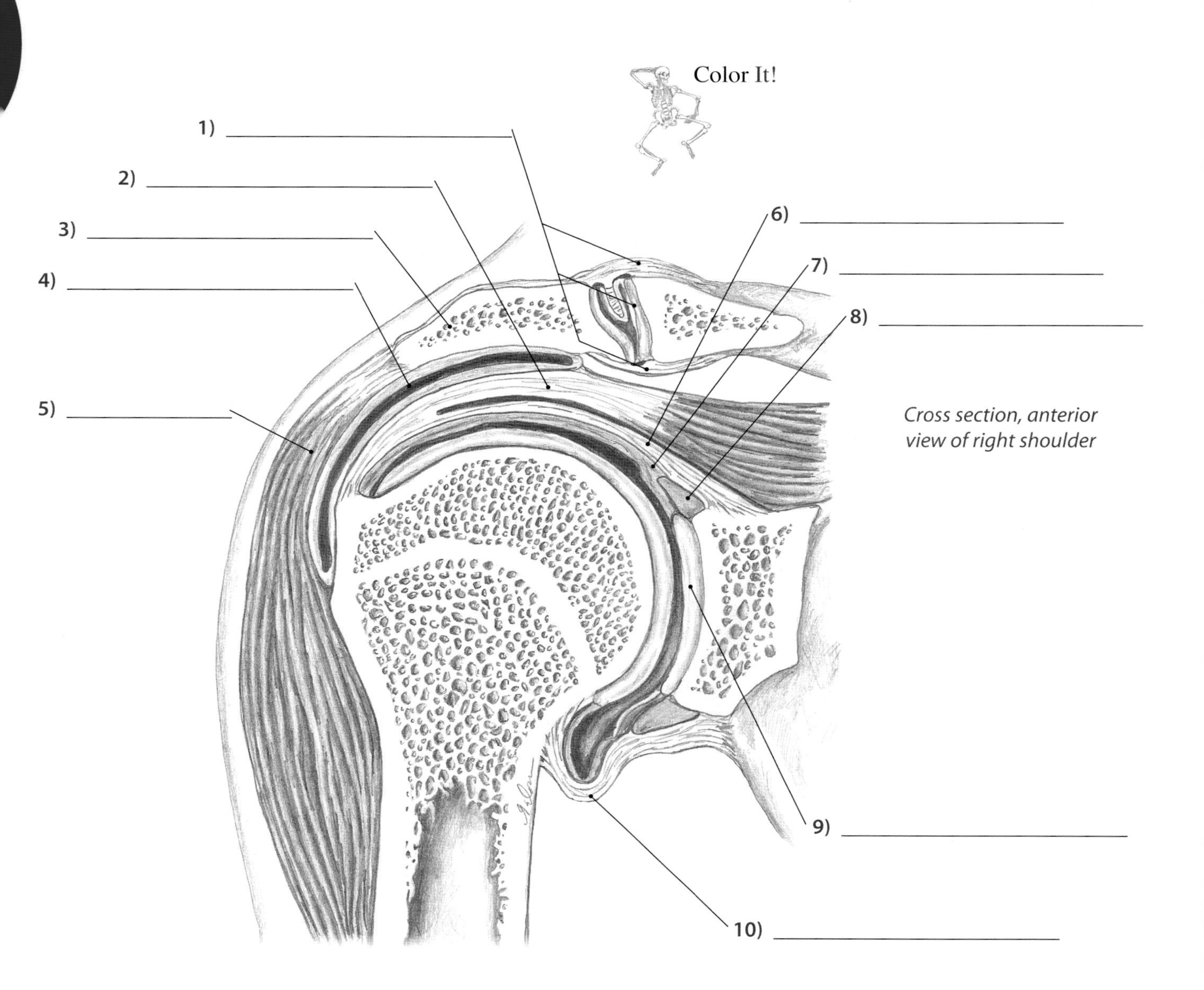

CHOICES

Acromioclavicular joint and ligament
Acromion
Articular capsule
Capsular ligament
Cartilage of glenoid cavity
Deltoid
Glenoid labrum
Subacromial bursa
Supraspinatus tendon
Synovial membrane

Shoulder and Arm

Glenohumeral Joint #2

Please identify the following structures.

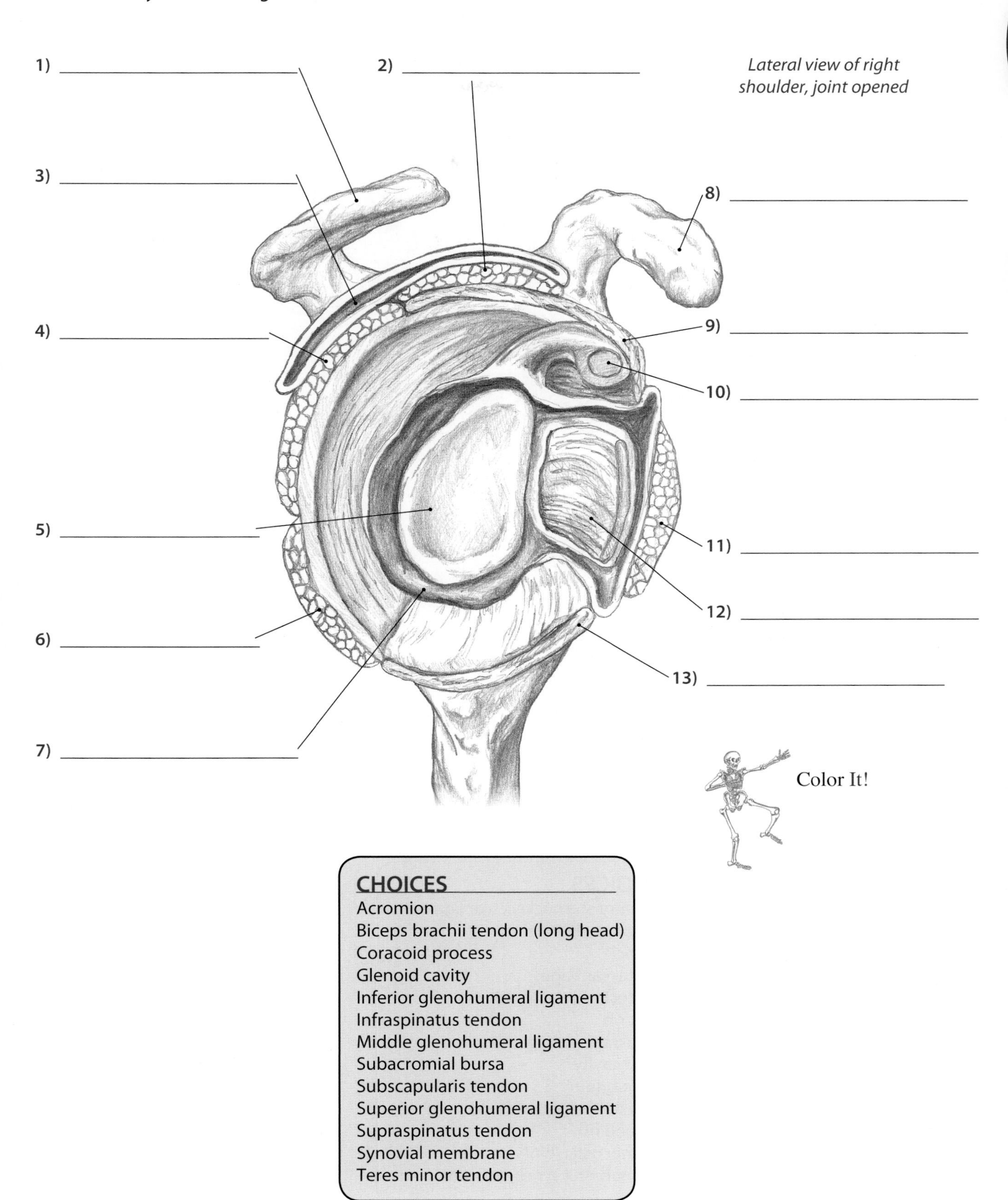

Sternoclavicular Joint

Please identify the following structures.

Color It!

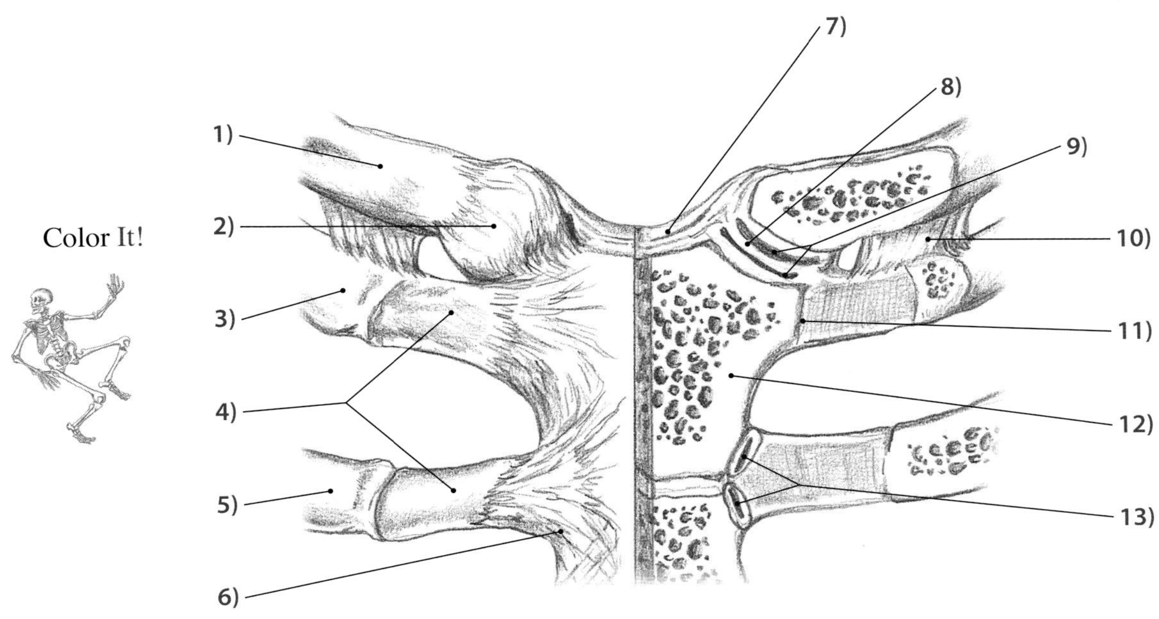

Anterior view, right side of illustration shown in coronal section

1) ______________________________
2) ______________________________
3) ______________________________
4) ______________________________
5) ______________________________
6) ______________________________
7) ______________________________
8) ______________________________
9) ______________________________
10) ______________________________
11) ______________________________
12) ______________________________
13) ______________________________

CHOICES

Anterior sternoclavicular ligament
Articular disc
Clavicle
Costal cartilages
Costoclavicular ligament
First rib
Interclavicular ligament
Joint cavity
Manubrium
Radiate sternocostal ligament
Second rib
Sternocostal joints
Sternocostal synchondrosis

Notes

Forearm and Hand
Topographical Views

Please identify the following structures.

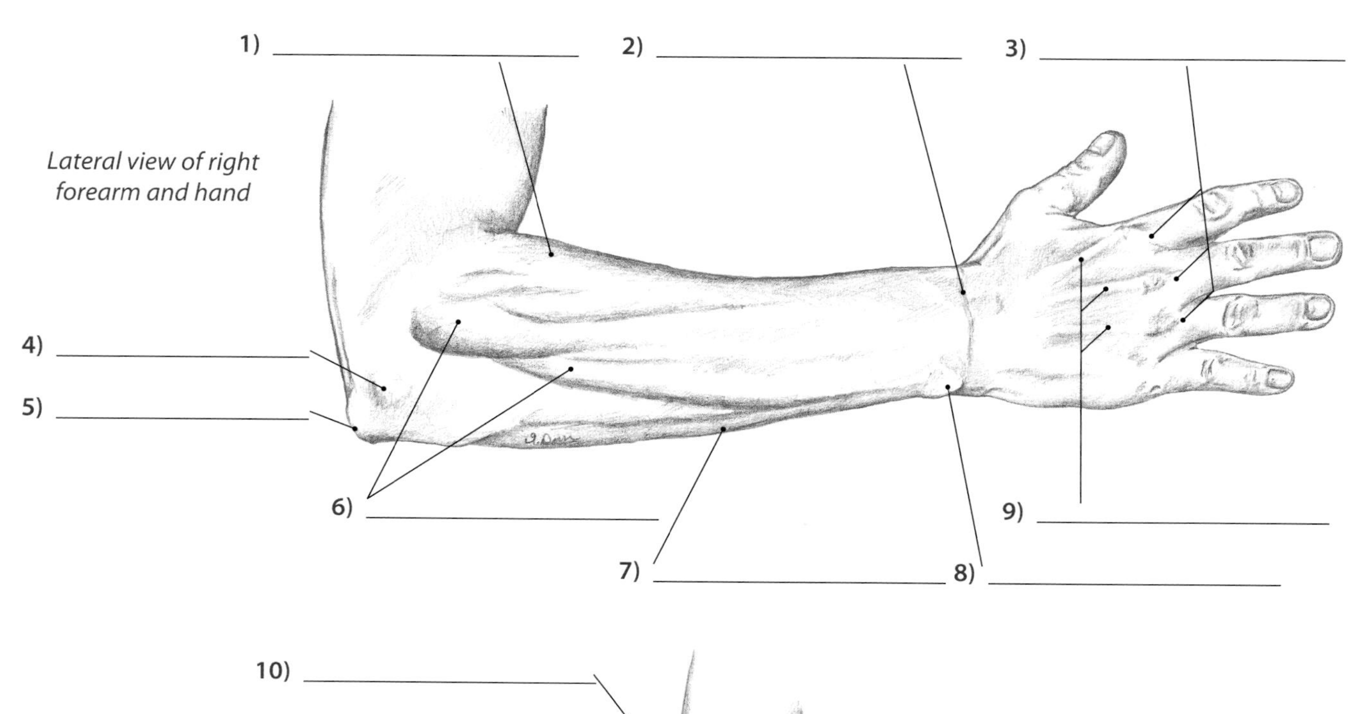

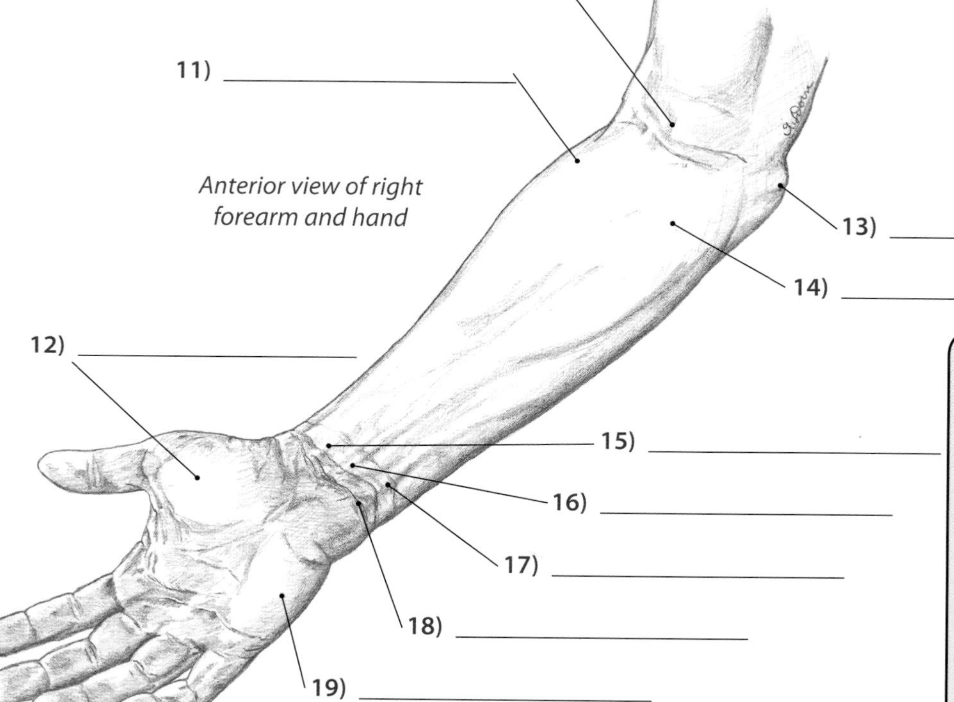

CHOICES
- Biceps brachii tendon
- Brachioradialis (2)
- Extensor bellies
- Extensor crease of the wrist
- Extensor digitorum tendons
- Flexor bellies
- Flexor carpi radialis tendon
- Flexor carpi ulnaris tendon
- Flexor crease of the wrist
- Head of the ulna
- Hypothenar eminence
- Lateral epicondyle
- Medial epicondyle
- Metacarpophalangeal joints
- Olecranon process
- Palmaris longus tendon
- Shaft of the ulna
- Thenar eminence

Forearm and Hand

Bones and Bony Landmarks

Please answer the following questions.

1) The palpable edge of which bone runs the length of the forearm? ______________________

2) Which two movements occur when the radius pivots back and forth around the ulna?

______________________ ______________________

3) The elbow is comprised of two joints, the ______________________ and ______________________.

4) The eight carpals are located just distal to which topographical landmark? ______________________

5) The olecranon process serves as an attachment site for which muscle? ______________________

6) Which bony landmark serves as an attachment site for the tendons of the wrist and finger extensors?

____

7) Which superficial, bony knob is visible along the posterior, medial side of the wrist?

____

8) The head of the radius is stabilized by which ligament? ______________________

9) Which bony landmark of the radius serves as the attachment site for brachioradialis?

____

10) Lister's tubercle is directly across—perhaps an inch away—from which bony landmark? ______________________

11) The styloid processes of the radius and ulna serve as important jumping off points for locating which group of bones?

____

Let's Palpate!

Remember—there are no right or wrong answers here

Locate and explore the **olecranon process and both epicondyles of the humerus** on three individuals. Then write three words that describe what you feel. (See p. 114-115 in *Trail Guide*)

Person #1 ______	Person #2 ______	Person #3 ______
______	______	______
______	______	______
______	______	______

Humerus

Please identify the following structures.

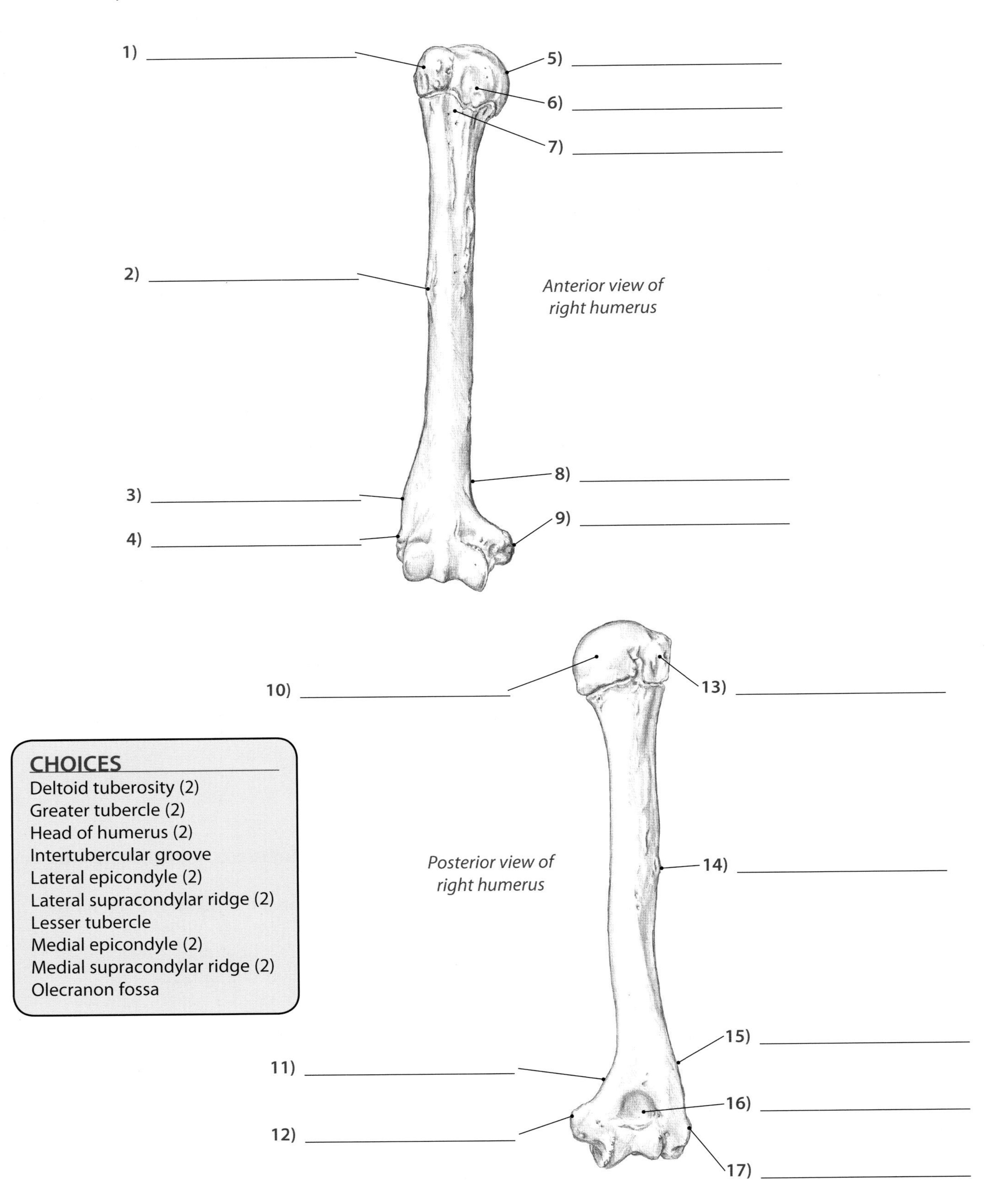

CHOICES

Deltoid tuberosity (2)
Greater tubercle (2)
Head of humerus (2)
Intertubercular groove
Lateral epicondyle (2)
Lateral supracondylar ridge (2)
Lesser tubercle
Medial epicondyle (2)
Medial supracondylar ridge (2)
Olecranon fossa

Forearm and Hand

Ulna and Radius

*Please identify the following **bones**. (Questions 1-5)*

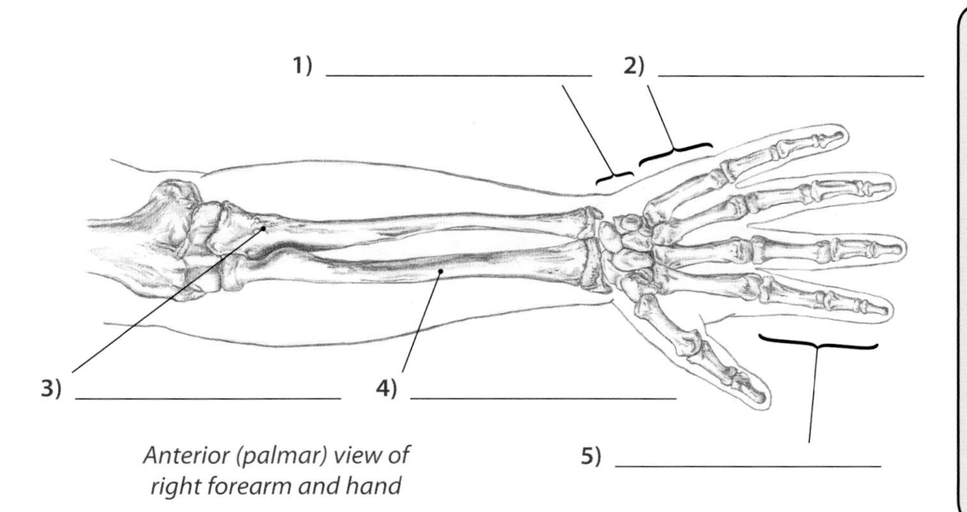

Anterior (palmar) view of right forearm and hand

CHOICES

Carpals
Coronoid process
Head of the radius (2)
Lister's tubercle
Metacarpals
Olecranon process
Phalanges
Radial tuberosity
Radius
Shaft of the radius
Shaft of the ulna
Styloid process of the radius (2)
Styloid process of the ulna
Trochlear notch
Ulna

*Please identify the following **bony landmarks**. (Questions 6-17)*

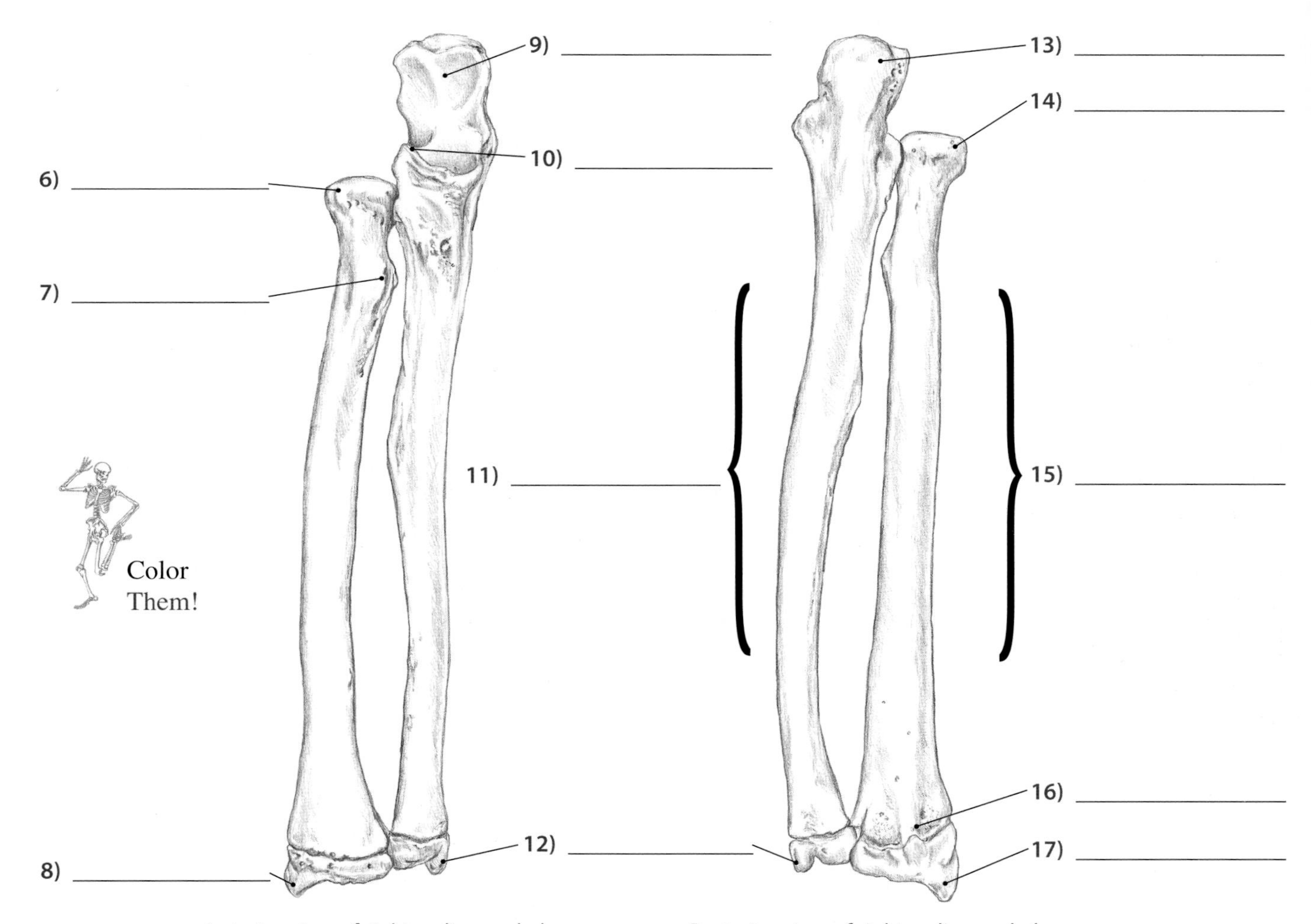

Anterior view of right radius and ulna

Posterior view of right radius and ulna

Forearm and Hand

Carpals

Please identify the following structures.

1) ____________
2) ____________
3) ____________
4) ____________
5) ____________
6) ____________
7) ____________
8) ____________
9) ____________
10) ____________
11) ____________

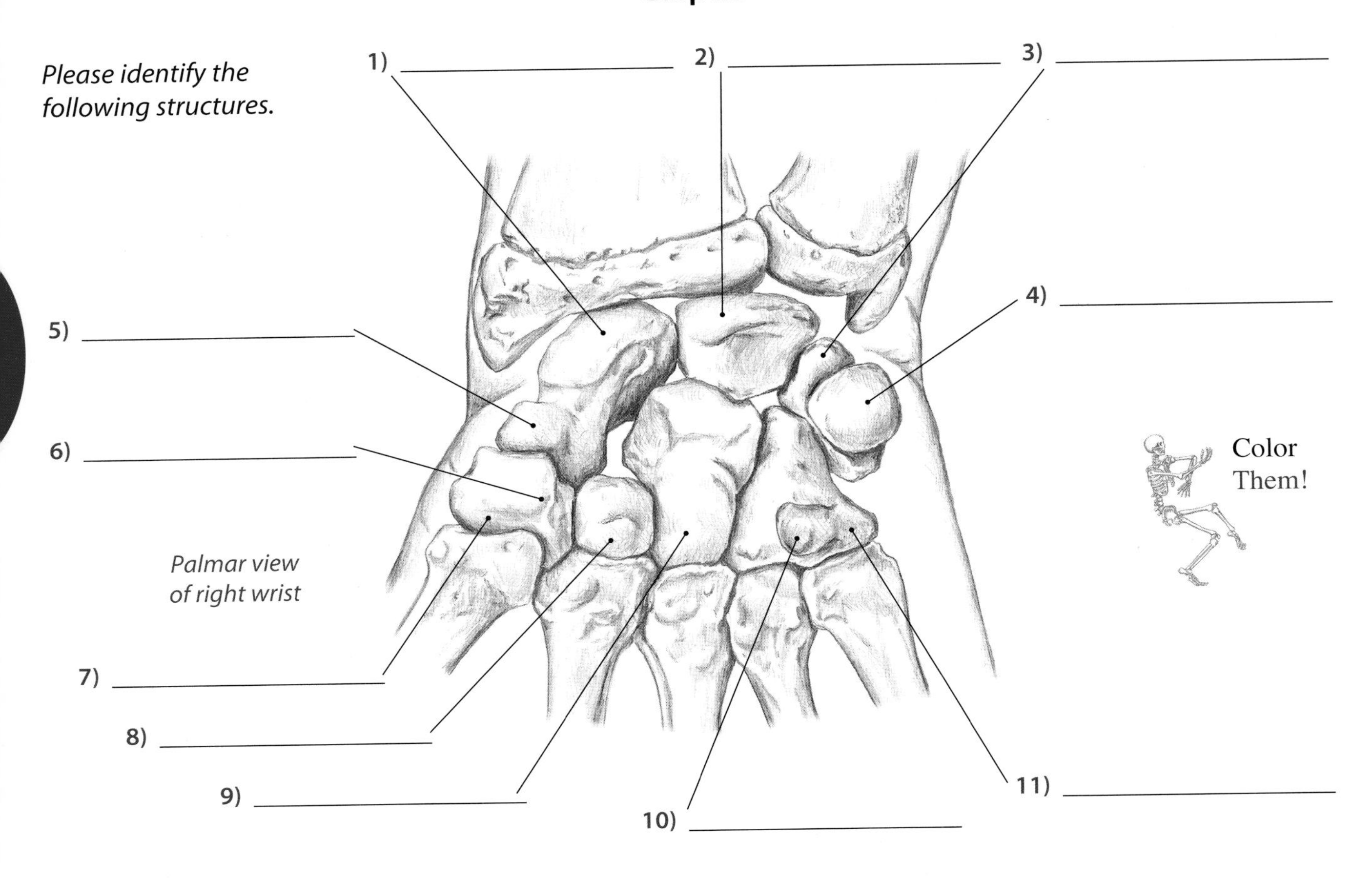

Palmar view of right wrist

Color Them!

12) ____________
13) ____________
14) ____________
15) ____________
16) ____________
17) ____________
18) ____________
19) ____________

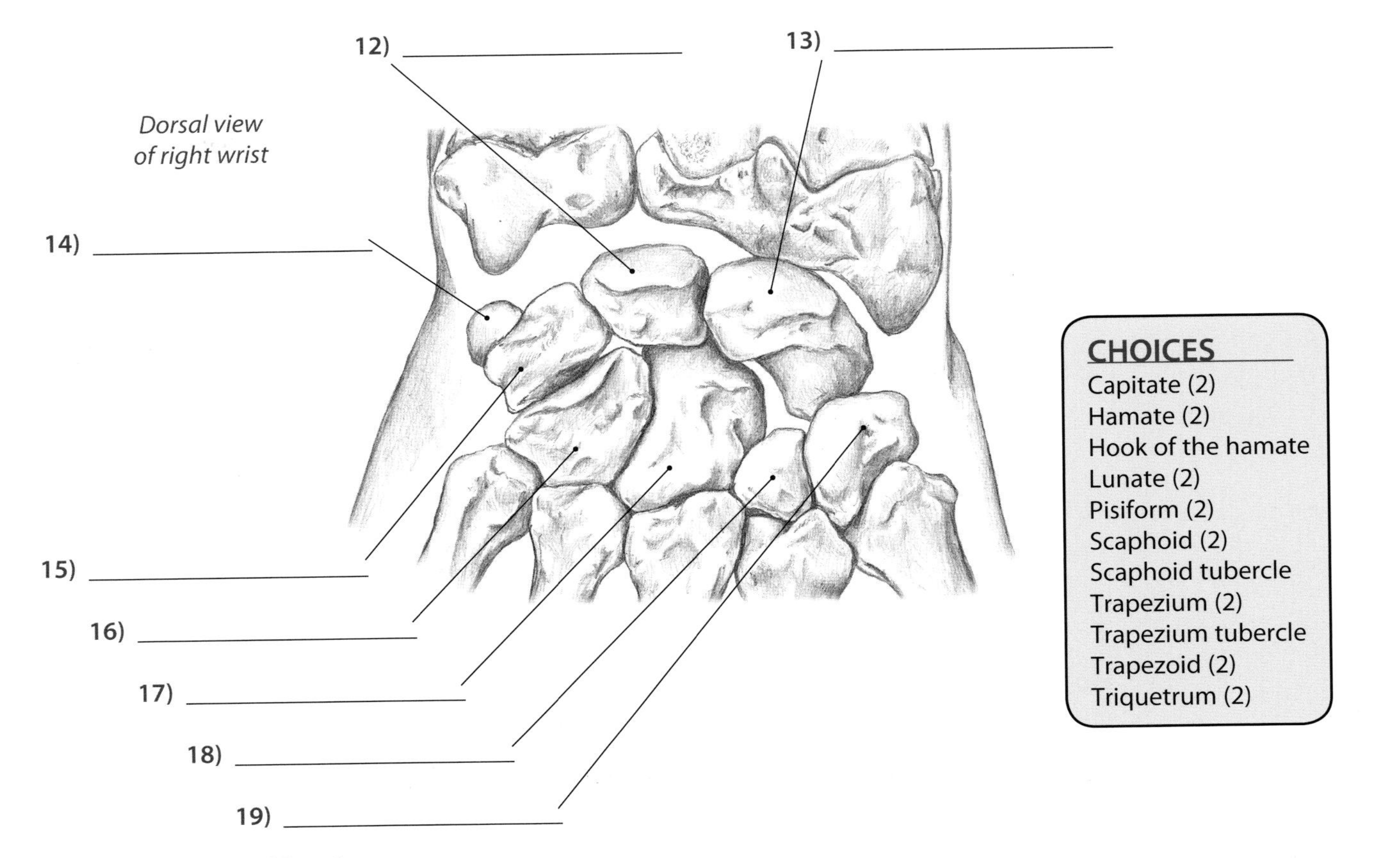

Dorsal view of right wrist

CHOICES
- Capitate (2)
- Hamate (2)
- Hook of the hamate
- Lunate (2)
- Pisiform (2)
- Scaphoid (2)
- Scaphoid tubercle
- Trapezium (2)
- Trapezium tubercle
- Trapezoid (2)
- Triquetrum (2)

Forearm and Hand

Bones and Bony Landmarks of the Wrist and Hand #1

Please answer the following questions.

1) What are the four surface sides of the carpals that can be palpated?

_______________ _______________ _______________ _______________

2) The carpals are located distal to which topographical feature of the palmar side? _______________

3) Which carpal can be felt on the ulnar/palmar side of the hand, just distal to the flexor crease?

____

4) The pisiform acts as an attachment site for which muscle? _______________

5) Which carpal can best be palpated by asking your partner to abduct and adduct her wrist as you palpate just distal to the styloid process of the ulna? _______________

6) A hook-shaped protuberance is the distinct landmark used to isolate which carpal? _______________

7) Which band of connective tissue forms the "roof" of the carpal tunnel? _______________

8) Which two structures pass through the Tunnel of Guyon?

_______________ _______________

9) Which four carpals serve as attachment sites for the flexor retinaculum?

_______________ _______________

_______________ _______________

10) Which carpal forms the floor of the "anatomical snuffbox"? _______________

11) Which bone articulates with the first metacarpal and is the source of the thumb's unique movements?

____

12) Which carpal can be located just distal to the styloid process of the radius and felt upon adduction of the wrist?

____

13) Which two carpals are located between Lister's tubercle and the base of the third metacarpal and are best palpated from the dorsal surface?

_______________ _______________

14) Anatomically speaking, the proper name for a "knuckle" joint is the _______________ joint.

Extra Credit

List all eight carpal bones below. Also create your own mnemonic device to remember their order.

Bones and Bony Landmarks of the Wrist and Hand #2

Please identify the following structures.

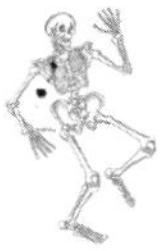

1) ______________________

Color It!

2) ______________________

3) ______________________

4) ______________________

5) ______________________

6) ______________________

7) ______________________

8) ______________________

9) ______________________

10) ______________________

11) ______________________

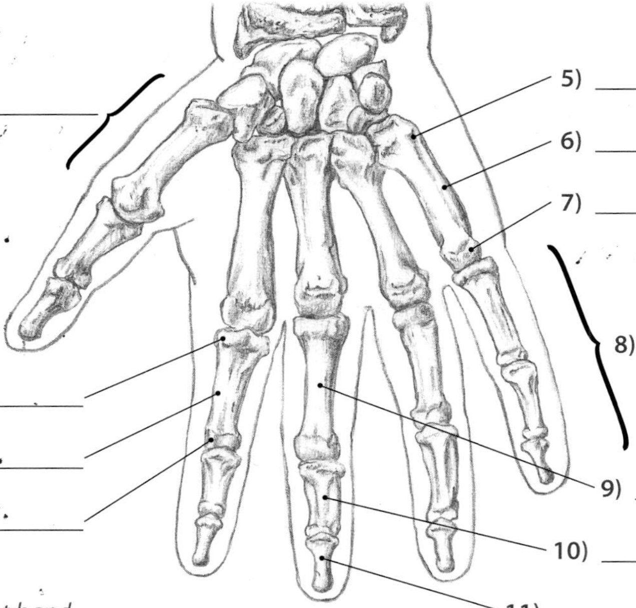

Palmar view of right hand

CHOICES

- Base (2)
- Distal phalanx
- Head (2)
- Metacarpals
- Middle phalanx
- Phalanges
- Proximal phalanx
- Shaft (2)

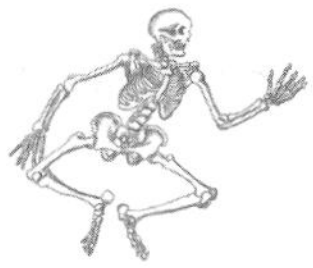

Let's Palpate!

Remember—there are no right or wrong answers here

Locate and explore the **carpals** on three individuals. Then write three words that describe what you feel. (See p. 119-120 in *Trail Guide*.)

Person #1 ______________

____

____

____

Person #2 ______________

____

____

____

Person #3 ______________

____

____

____

Muscles of the Forearm #1

Please identify the following structures.

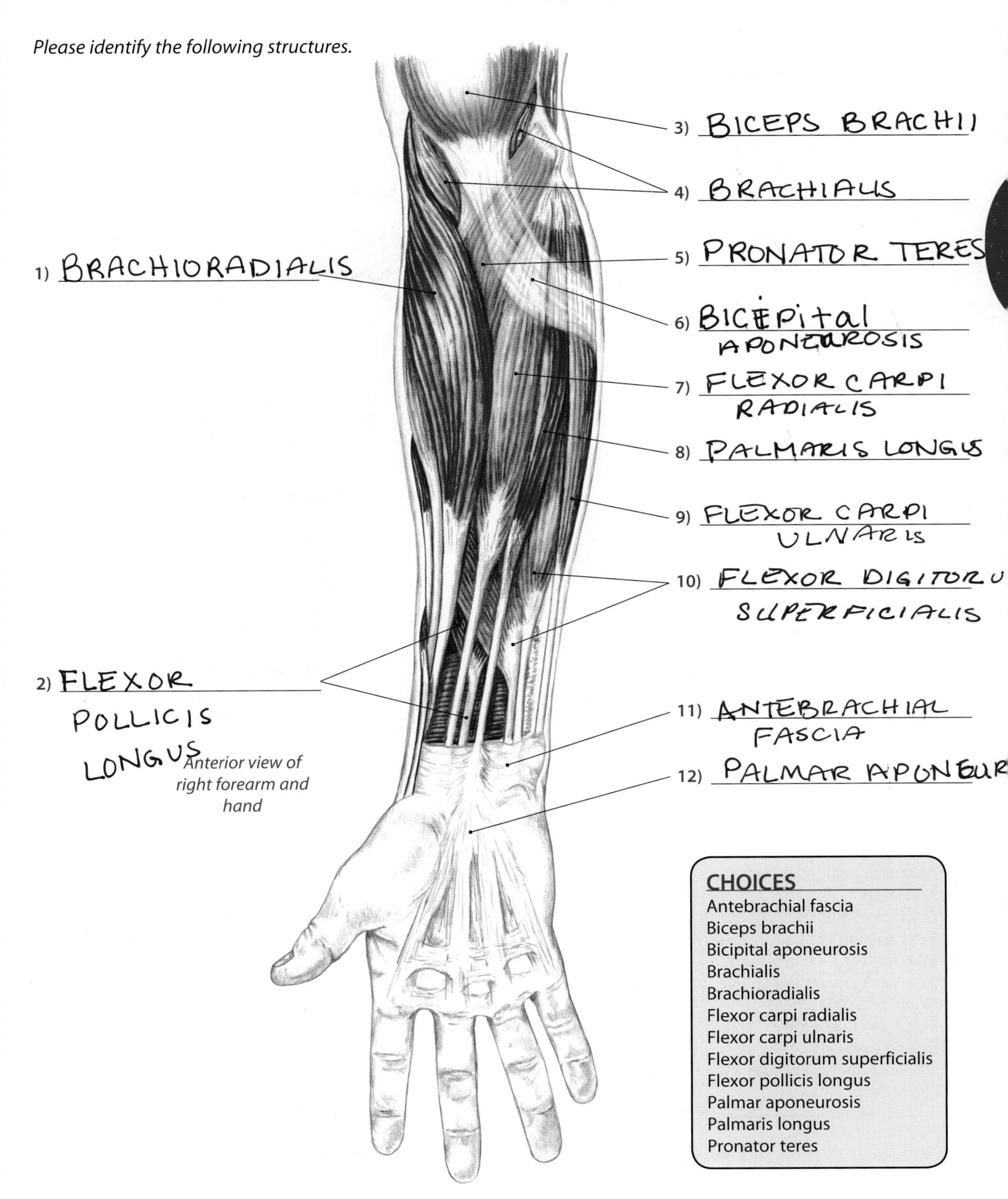

Anterior view of right forearm and hand

CHOICES

- Antebrachial fascia
- Biceps brachii
- Bicipital aponeurosis
- Brachialis
- Brachioradialis
- Flexor carpi radialis
- Flexor carpi ulnaris
- Flexor digitorum superficialis
- Flexor pollicis longus
- Palmar aponeurosis
- Palmaris longus
- Pronator teres

Muscles of the Forearm #2

Please identify the following structures.

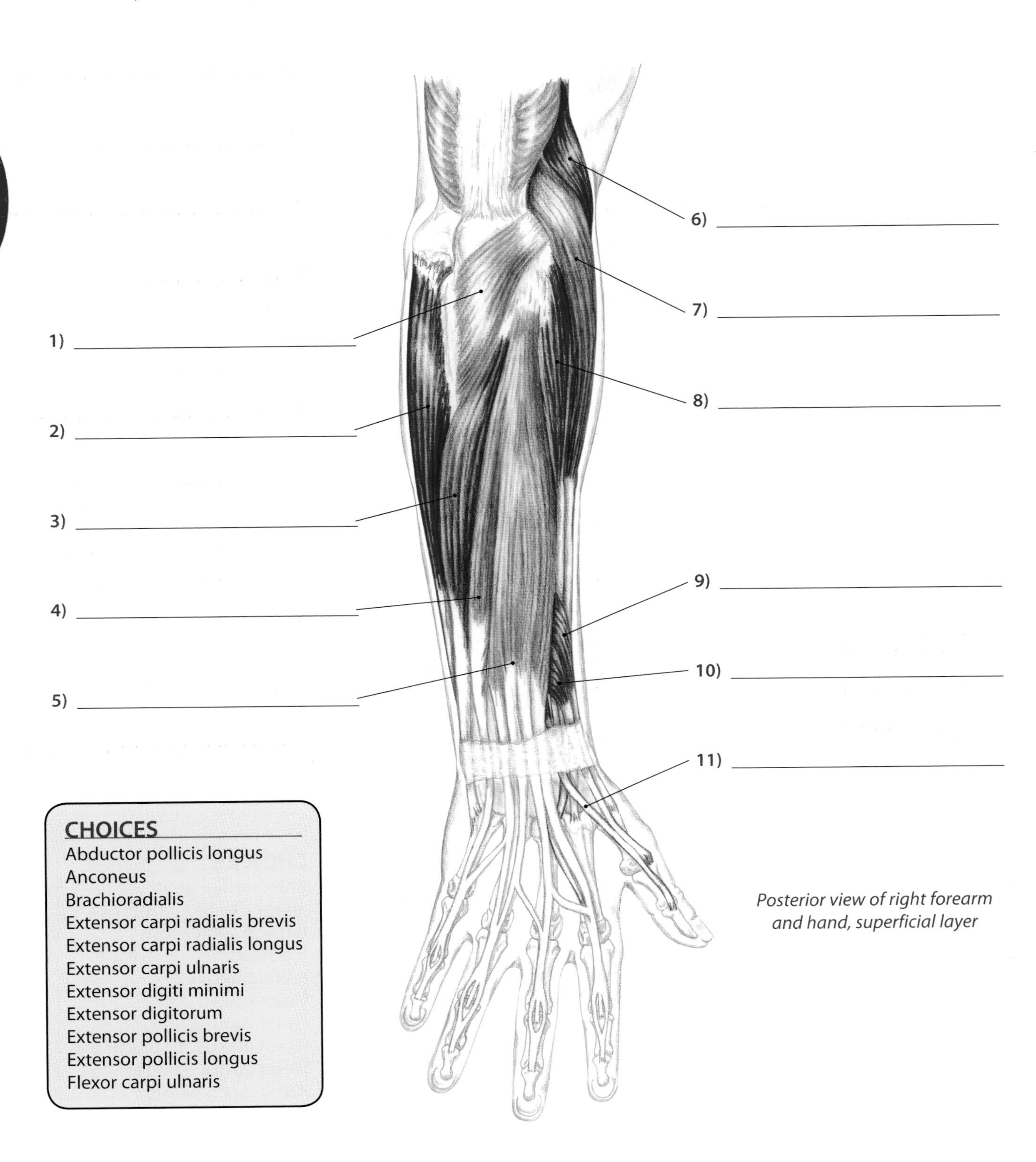

Posterior view of right forearm and hand, superficial layer

Forearm and Hand

Muscles of the Forearm #3

Please identify the following structures.

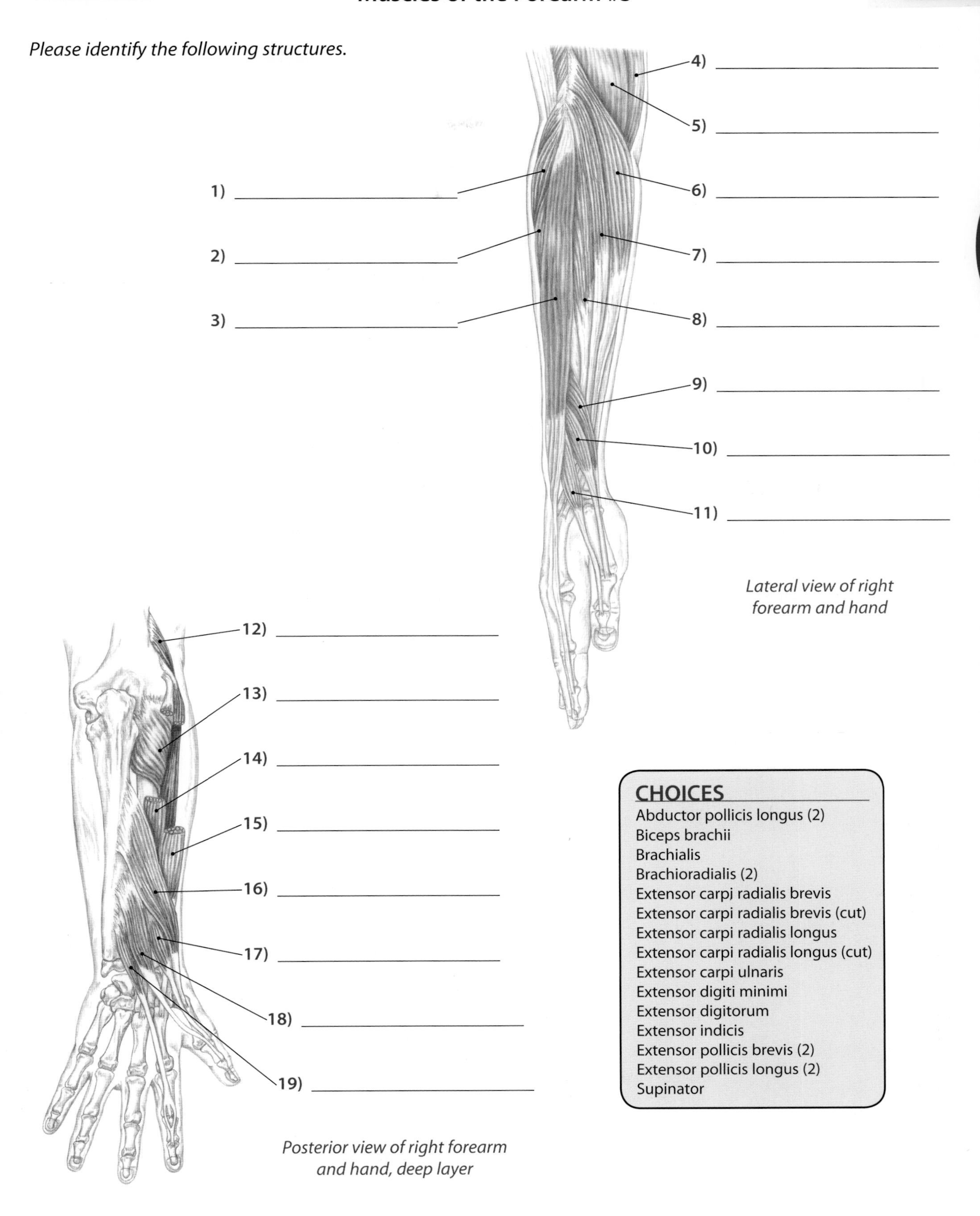

Lateral view of right forearm and hand

Posterior view of right forearm and hand, deep layer

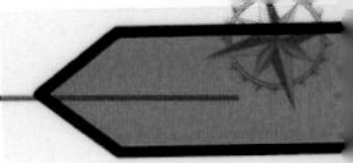

CHOICES

Abductor pollicis longus (2)
Biceps brachii
Brachialis
Brachioradialis (2)
Extensor carpi radialis brevis
Extensor carpi radialis brevis (cut)
Extensor carpi radialis longus
Extensor carpi radialis longus (cut)
Extensor carpi ulnaris
Extensor digiti minimi
Extensor digitorum
Extensor indicis
Extensor pollicis brevis (2)
Extensor pollicis longus (2)
Supinator

Color the Muscles #1

Using different colors, please fill in and label the muscles and other structures listed below.

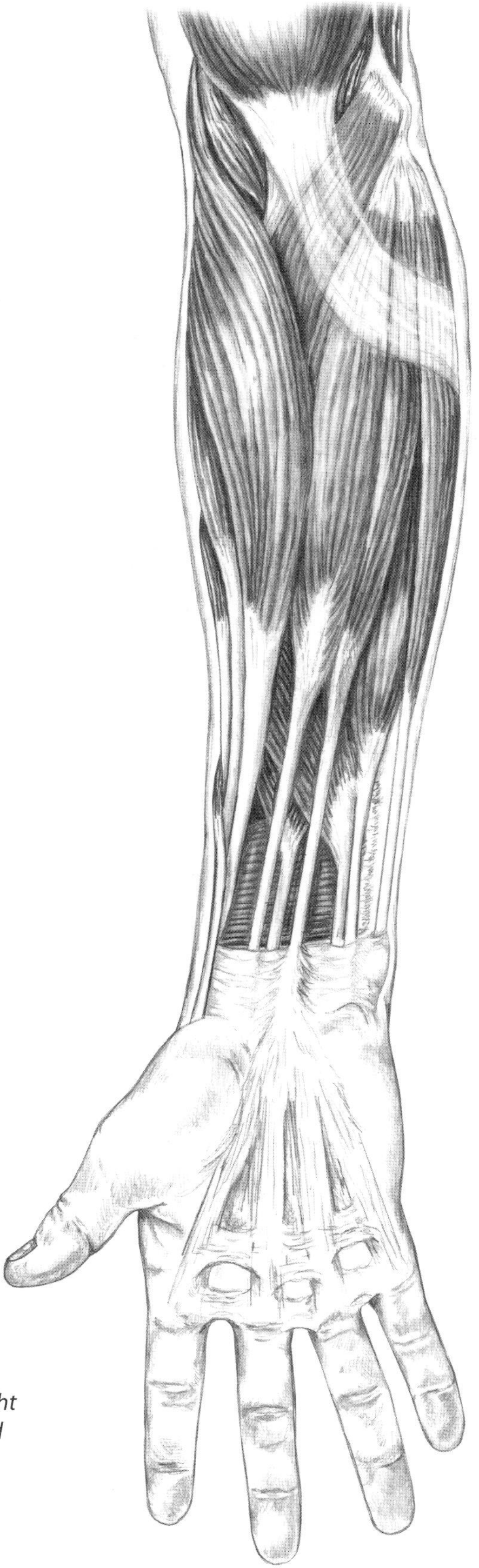

CHOICES

Antebrachial fascia
Biceps brachii
Bicipital aponeurosis
Brachialis
Brachioradialis
Flexor carpi radialis
Flexor carpi ulnaris
Flexor digitorum superficialis
Flexor pollicis longus
Palmar aponeurosis
Palmaris longus
Pronator teres

Anterior view of right forearm and hand

Forearm and Hand
Color the Muscles #2

Using different colors, please fill in and label the muscles listed below.

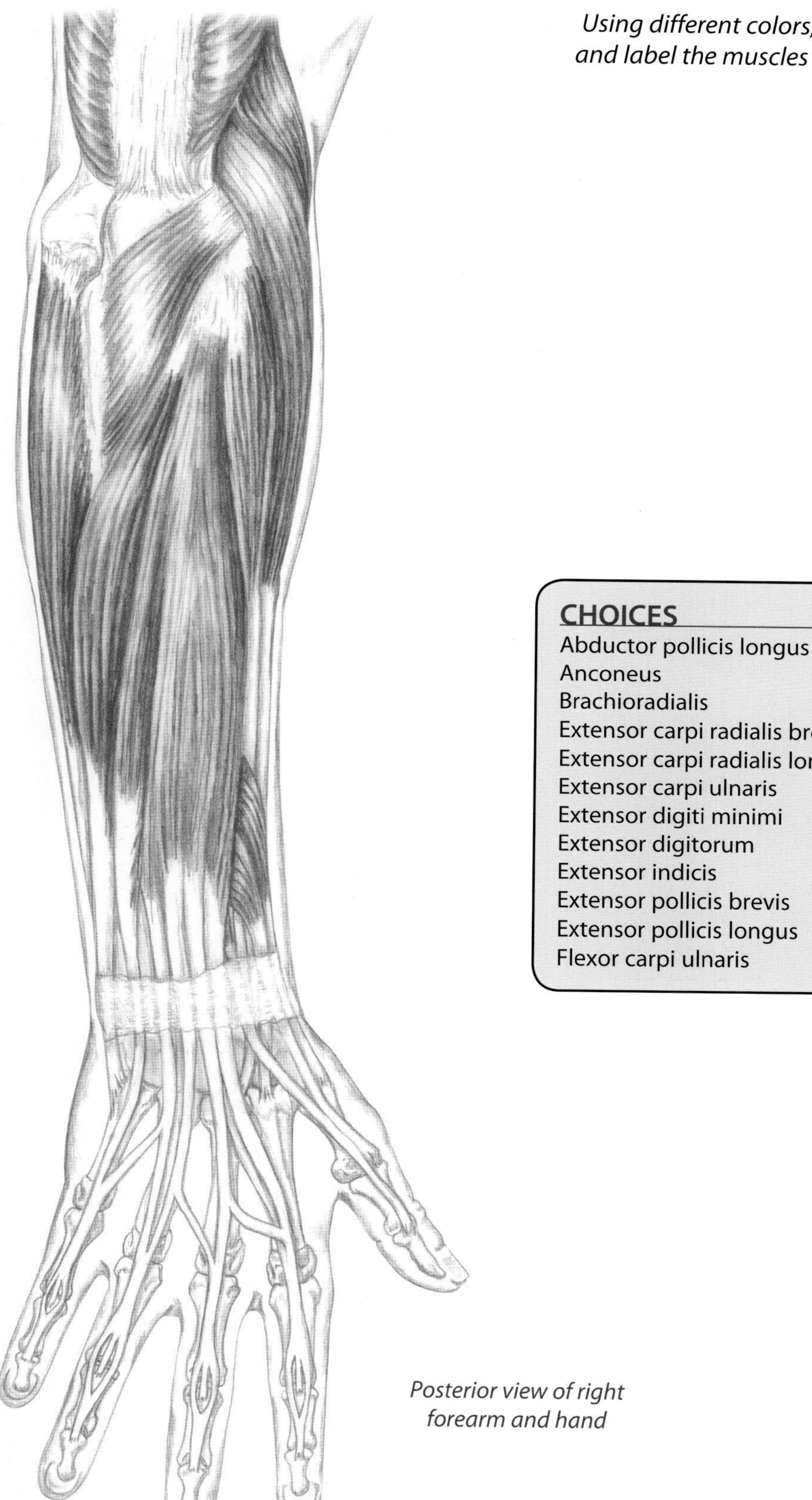

CHOICES

Abductor pollicis longus
Anconeus
Brachioradialis
Extensor carpi radialis brevis
Extensor carpi radialis longus
Extensor carpi ulnaris
Extensor digiti minimi
Extensor digitorum
Extensor indicis
Extensor pollicis brevis
Extensor pollicis longus
Flexor carpi ulnaris

Posterior view of right forearm and hand

Color the Muscles #3

p. 128-129

Using different colors, please fill in and label the muscles and other structures listed below.

CHOICES

- Abductor pollicis longus
- Biceps brachii
- Brachialis
- Brachioradialis
- Extensor carpi radialis brevis
- Extensor carpi radialis longus
- Extensor carpi ulnaris
- Extensor digiti minimi
- Extensor digitorum
- Extensor indicis
- Extensor pollicis brevis
- Extensor pollicis longus
- Supinator

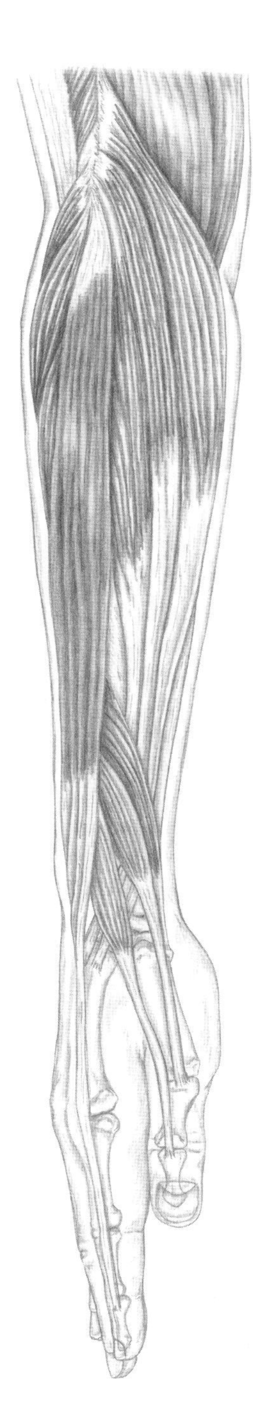

Lateral view of right forearm and hand

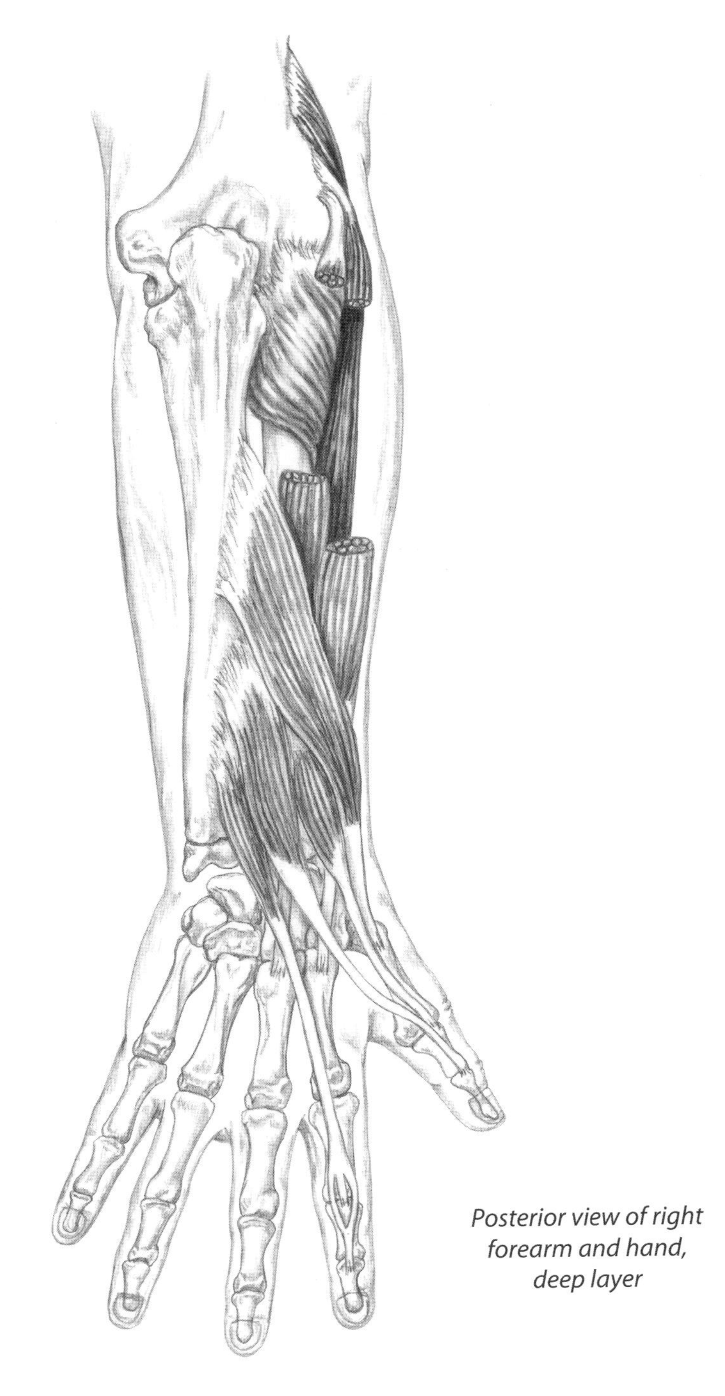

Posterior view of right forearm and hand, deep layer

Forearm and Hand

Muscles and Movements #1

Please list the action demonstrated, synergist(s) and antagonist. The first letter of the muscles has been provided.

1) This action happens at which two joints?

____

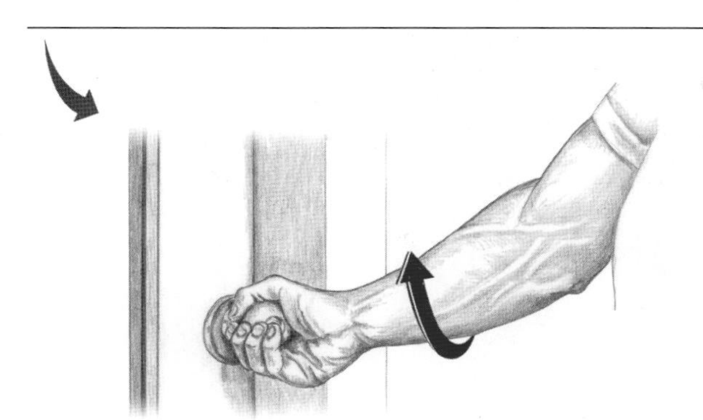

2) Action

____

3) Synergists

B ____________________

S ____________________

4) Antagonist

P ____________________

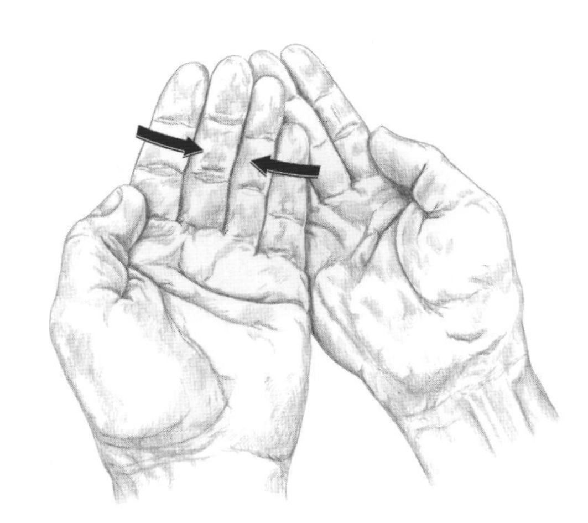

5) Action

____

6) Muscle group that performs this action

____

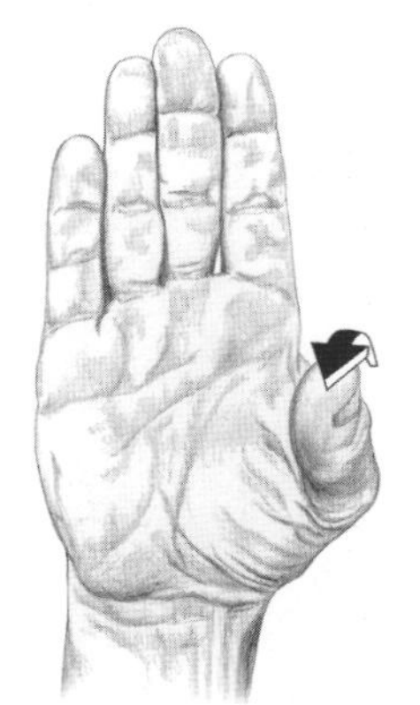

7) Action

____

8) Synergists

A ____________________

A ____________________

9) Antagonist

A ____________________

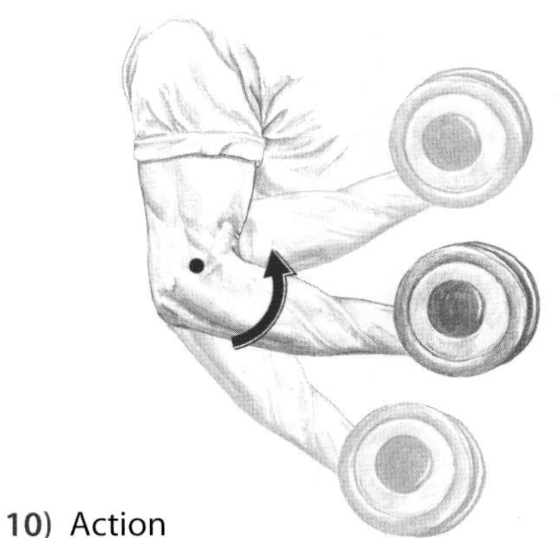

10) Action

____

11) Synergists

F ____________________

P ____________________

12) Antagonist

T ____________________

Forearm and Hand

Muscles and Movements #2

Please list the action demonstrated, synergists and antagonist. The first letter of each muscle has been provided.

1) This action happens at which joint?

____

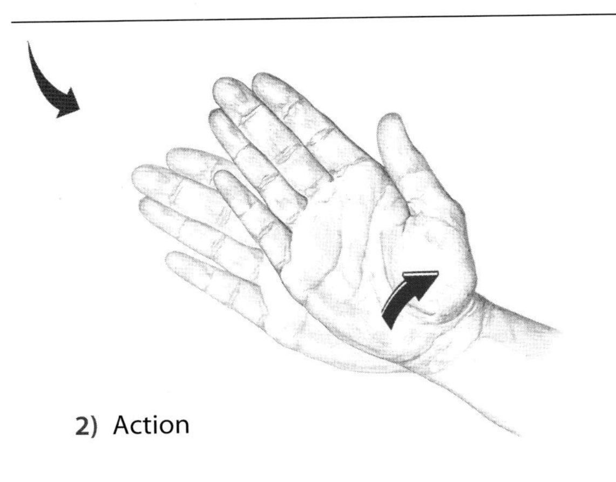

2) Action

____

3) Synergists

E ______________________________

F ______________________________

4) Antagonist

E ______________________________

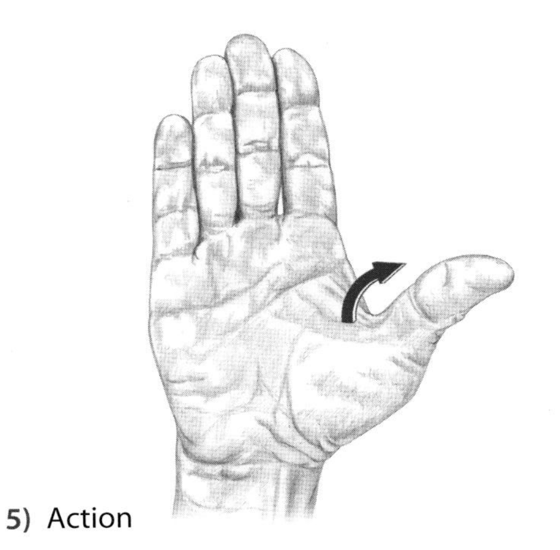

5) Action

____

6) Synergists

E ______________________________

A ______________________________

7) Antagonist

F ______________________________

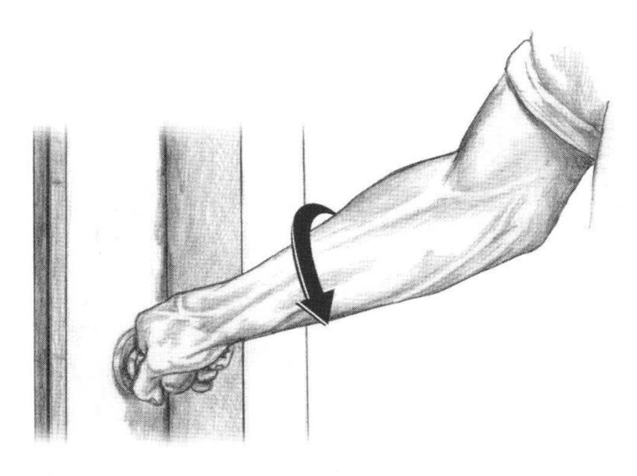

8) Action

____

9) Synergists

P ______________________________

B ______________________________

10) Antagonist

B ______________________________

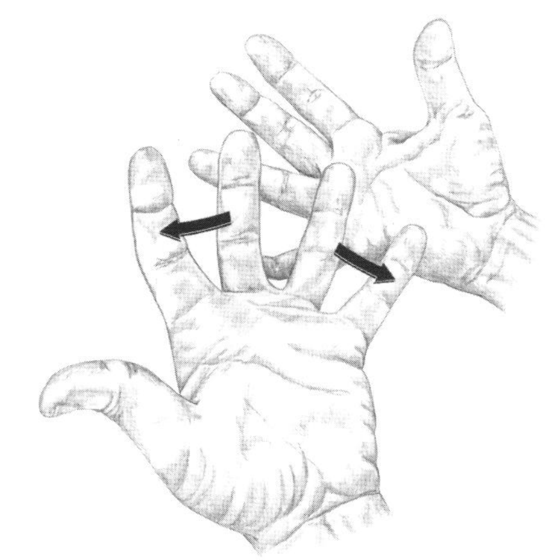

11) Action

____

12) Muscle group that performs this action

____

Forearm and Hand

Muscles and Movements #3

Please list the action demonstrated, synergists and antagonist. The first letter of each muscle has been provided.

1) This action happens at which joint?

____

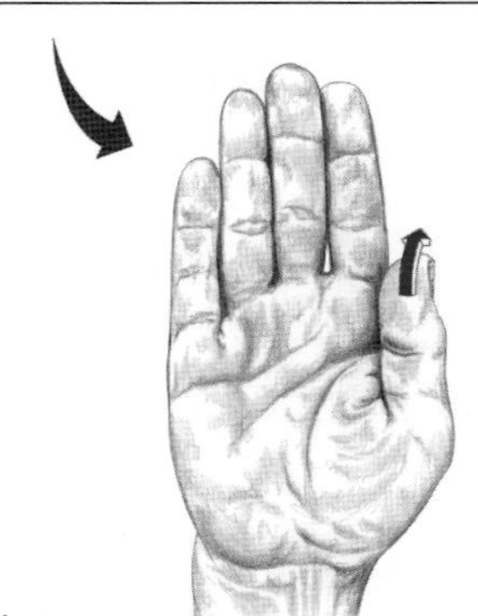

2) Action

____

3) Synergists

A ______________________________

P ______________________________

4) Antagonist

A ______________________________

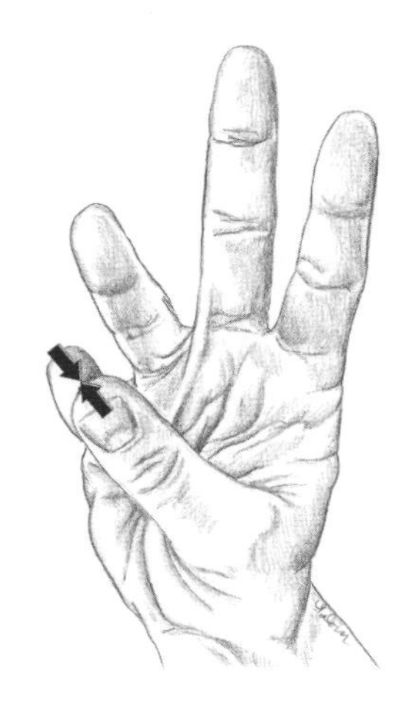

5) Action at the thumb

____

6) Synergists

O ______________________________

F ______________________________

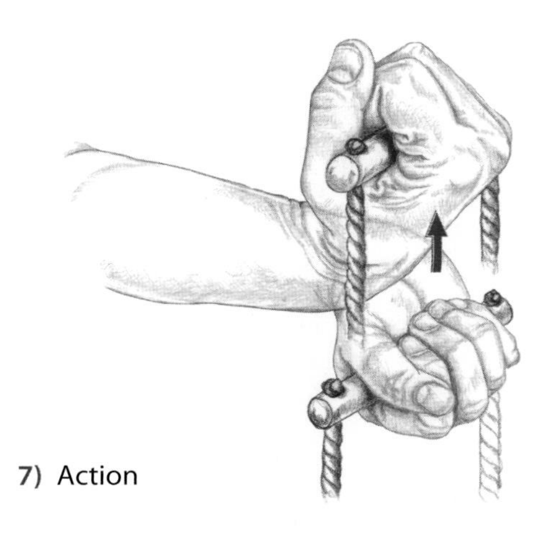

7) Action

____

8) Synergists

F ______________________________

F ______________________________

9) Antagonist

E ______________________________

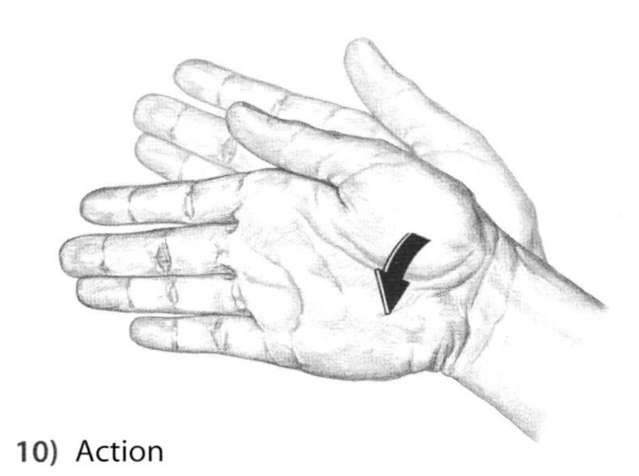

10) Action

____

11) Synergists

E ______________________________

F ______________________________

12) Antagonist

F ______________________________

Muscles and Movements #4

p. 130-131

Please list the action demonstrated, synergists and antagonist. The first letter of each muscle has been provided.

1) This action happens at which joint?

____

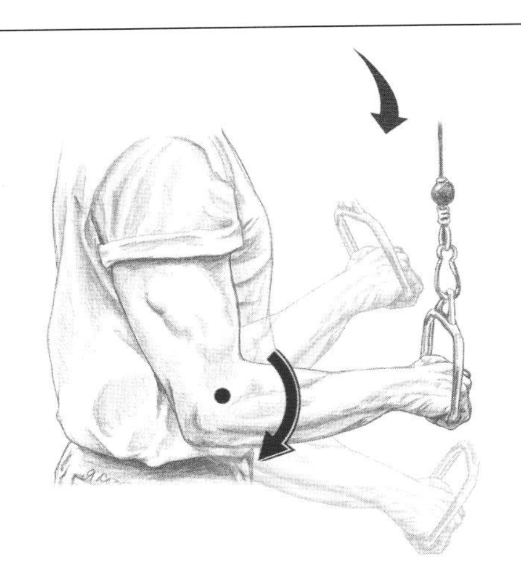

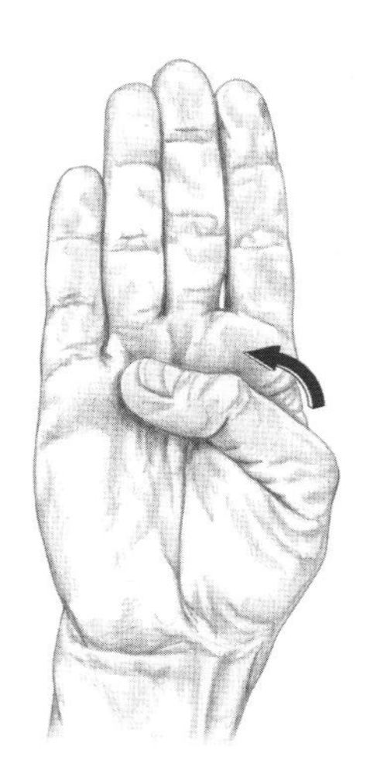

2) Action

____

3) Synergists

T ______________________________

A ______________________________

4) Antagonist

B ______________________________

5) Action

____

6) Synergists

F ______________________________

A ______________________________

7) Antagonist

A ______________________________

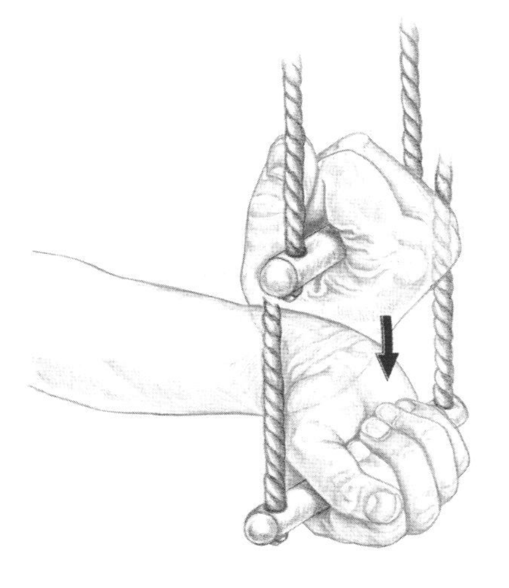

8) Action

____

9) Synergists

E ______________________________

E ______________________________

10) Antagonist

P ______________________________

Forearm and Hand
What's the Muscle? #1

Please identify the following muscles.

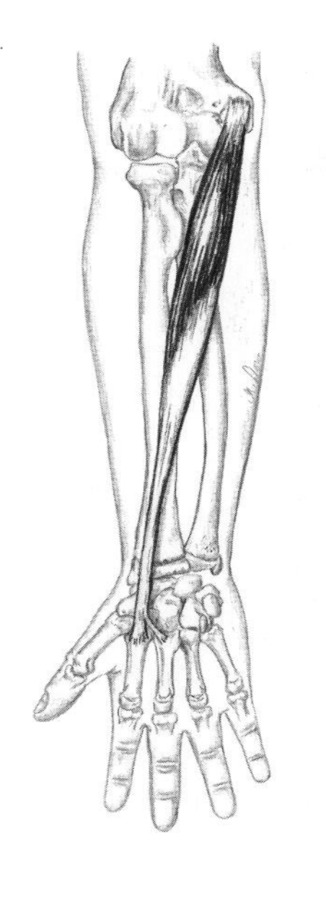

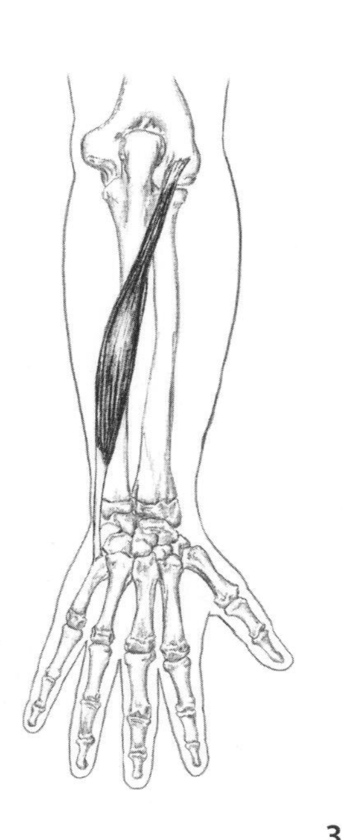

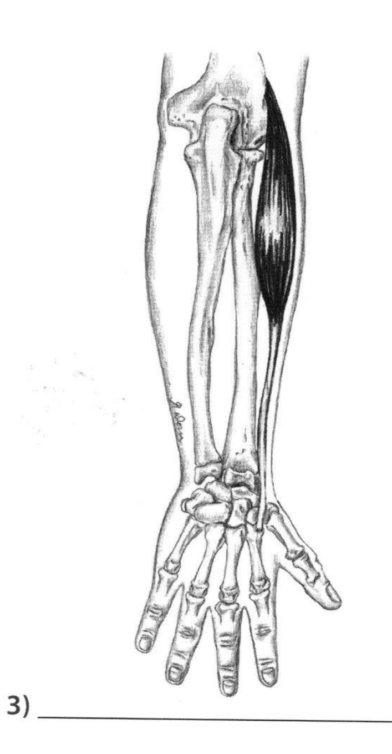

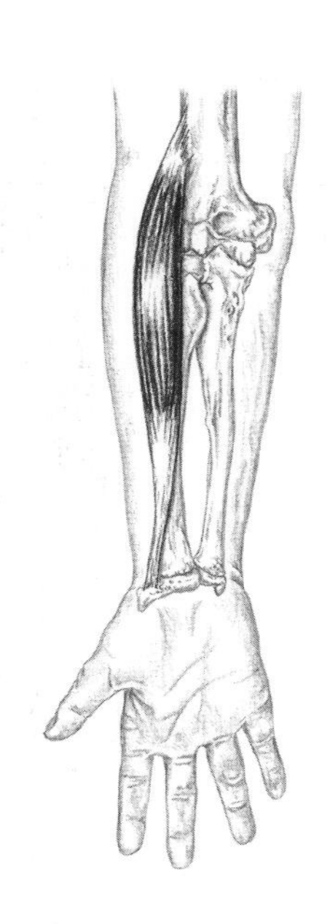

1) ______________________________

3) ______________________________

2) ______________________________

4) ______________________________

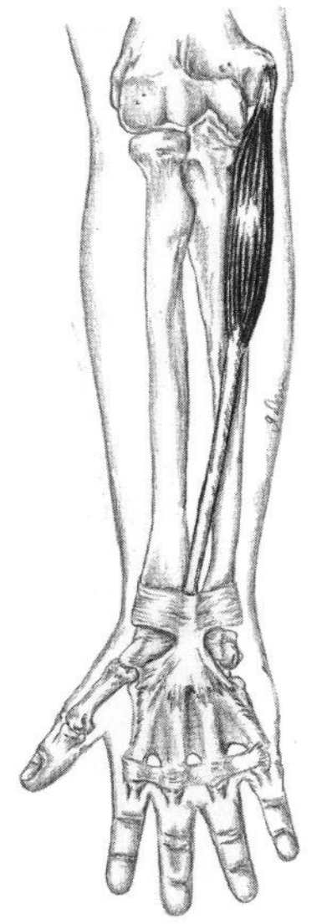

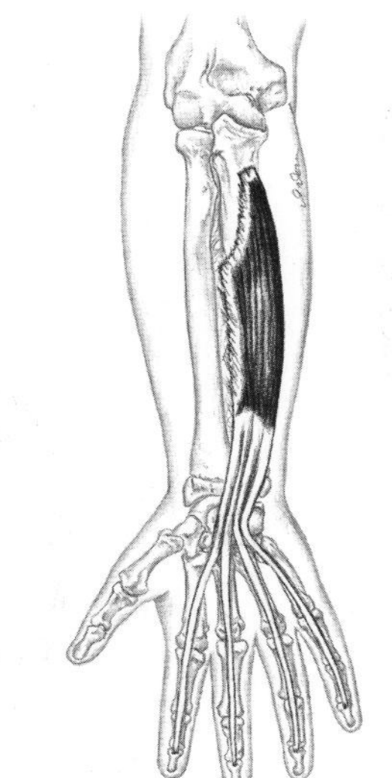

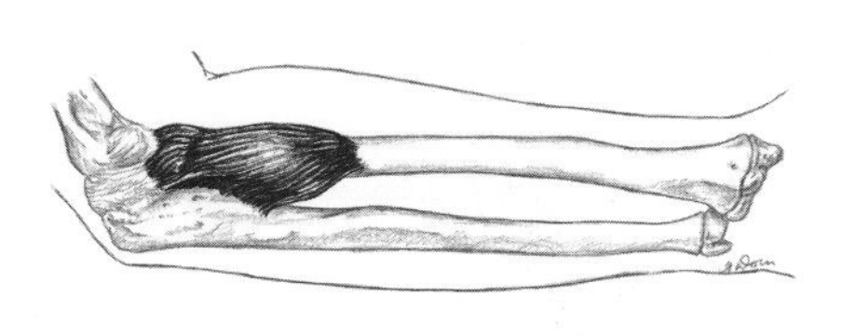

5) ______________________________

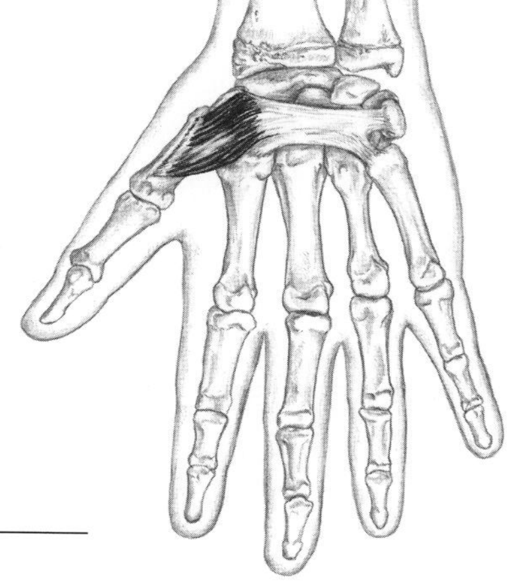

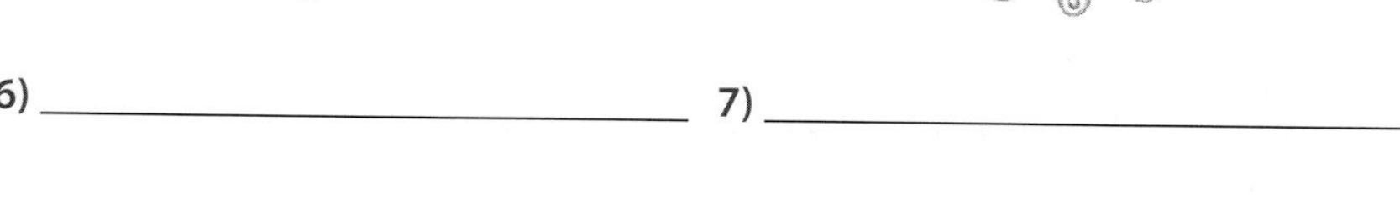

6) ______________________________

7) ______________________________

8) ______________________________

What's the Muscle? #2

Please identify the following muscles.

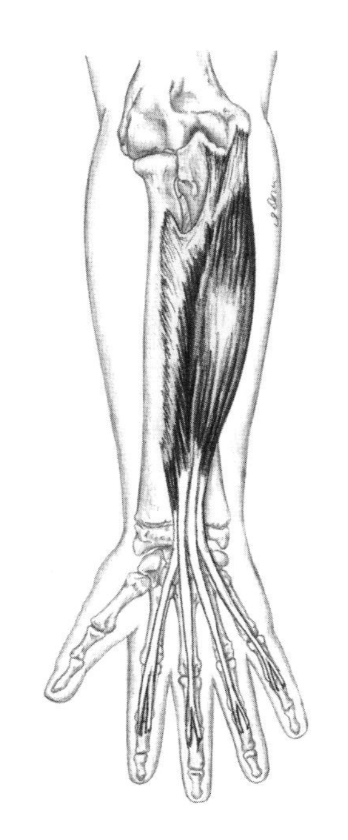

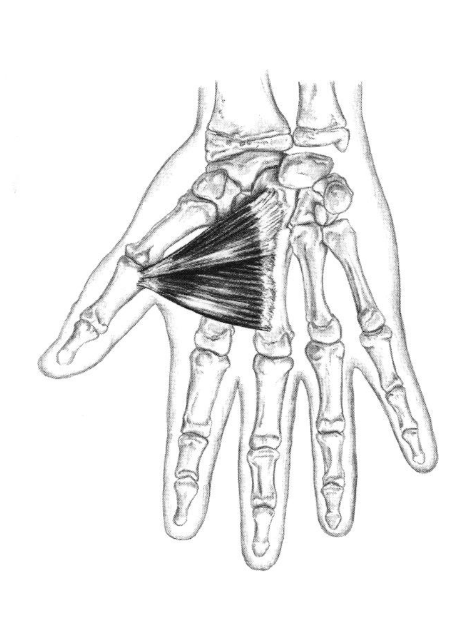

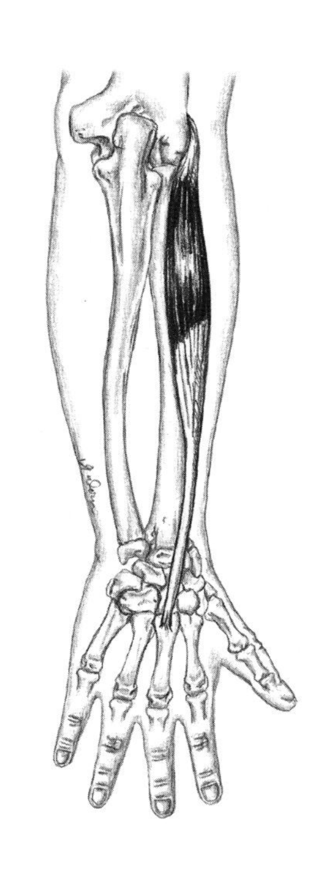

1) ______________________

2) ______________________

3) ______________________

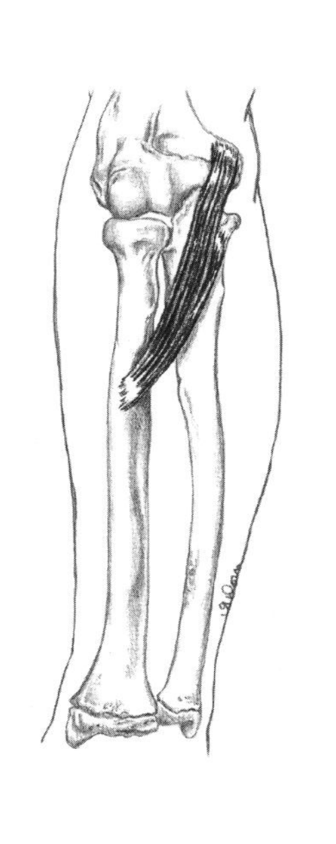

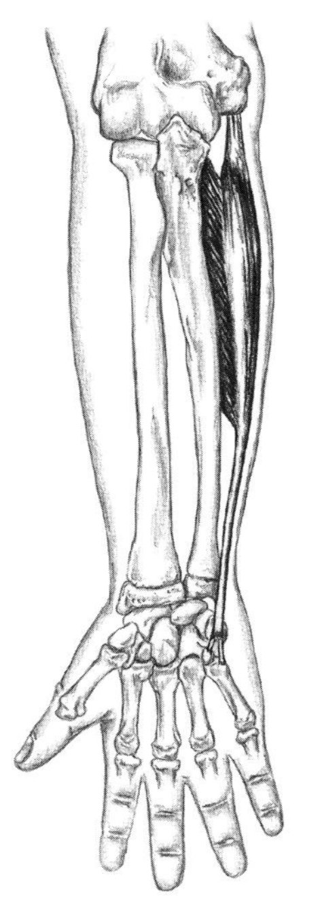

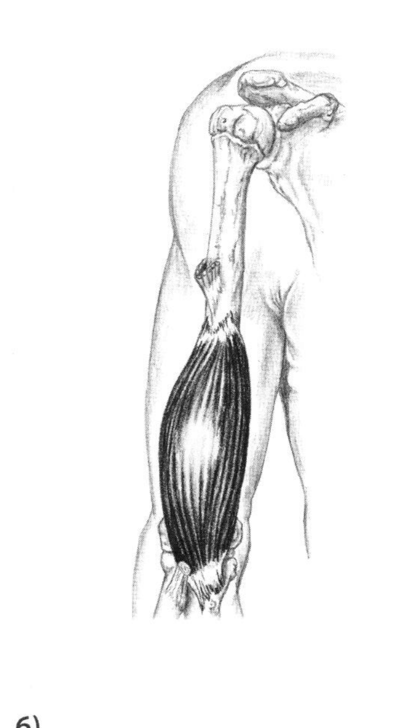

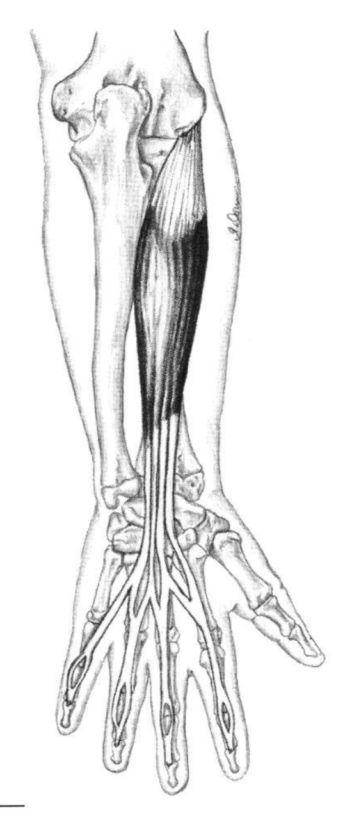

4) ______________________

6) ______________________

5) ______________________

7) ______________________

Forearm and Hand, Muscle Group #1

Brachialis, Brachioradialis, Pronators and Supinator

Please answer the following questions.

1) Which muscle is a strong elbow flexor located deep to the biceps brachii? ____________________

2) The brachioradialis creates a helpful dividing line between which two muscle groups?

____________________ ____________________

3) Which muscle runs the length of the forearm but does not cross the wrist joint? ____________________

4) The pronator quadratus is deep to the ____________________ tendons and is accessible only on the quadratus' ____________________ portion.

5) The ____________________ muscle is an antagonist to both the biceps brachii and supinator.

6) Palpating medial to the distal tendon of which muscle can help you locate the pronator teres?

____

7) To access the supinator, you must palpate deep to which muscle group? ____________________

Shorten or Lengthen?

8) Passive pronation of the forearm would ____________________ the supinator.

9) Passive flexion of the elbow would ____________________ the brachioradialis.

10) Passive extension of the elbow would ____________________ the brachialis.

11) Passive pronation of the forearm would ____________________ the pronator teres.

12) ____________________

13) ____________________

14) ____________________

15) ____________________

16) ____________________

Forearm and Hand, Muscle Group #1

Brachialis, Brachioradialis, Pronators and Supinator

p. 132-133, 146-148

Matching

Match the origin and insertion to the correct muscle.

Origins

1) Distal half of anterior surface of humerus
2) Proximal two-thirds of the lateral supracondylar ridge of humerus
3) Medial, anterior surface of distal ulna
4) Common flexor tendon from medial epicondyle of humerus and coronoid process of the ulna
5) Lateral epicondyle of humerus, radial collateral ligament, annular ligament and supinator crest of the ulna

Muscle	O	I
Brachialis	______	______
Brachioradialis	______	______
Pronator quadratus	______	______
Pronator teres	______	______
Supinator	______	______

Insertions

6) Lateral, anterior surface of distal radius
7) Anterior, lateral surface of proximal one-third of radial shaft
8) Middle of lateral surface of the radius
9) Styloid process of radius
10) Tuberosity and coronoid process of ulna

Let's Palpate!

Remember—there are no right or wrong answers here

Locate and explore the **pronator teres** on three individuals. Then write three words that describe what you feel. (See p. 146 in *Trail Guide.*)

Person #1 ____________

____

____

____

Person #2 ____________

____

____

____

Person #3 ____________

____

____

____

Forearm and Hand, Muscle Group #2

Extensors and Flexors

Please answer the following questions.

1) With the forearm in anatomical position, the ________________ group is located on the posterior/lateral side of the forearm, while the ________________ group is located on the anterior/medial side.

2) The brachioradialis and the ________________ clearly divide the forearm flexors from the extensors.

3) Looking at its name, what information can you gather about this muscle—*flexor carpi radialis*?

________________ ________________

________________ ________________

4) Which extensor muscle can be palpated alongside the shaft of the ulna? ________________

5) The extensor digitorum creates movement at which fingers? ________________

6) When palpating the forearm, the muscle bellies of the ________________ group will feel smaller and more sinewy than the ________________ group.

7) Which forearm muscles comprise the "wad of three"? ________________

________________ ________________

8) Which action can you ask your partner to perform at the wrist to distinguish the brachioradialis from the extensor carpi radialis? ________________

9) What are the three superficial muscles in the flexor group? ________________

________________ ________________

10) Flexor digitorum superficialis and flexor digitorum profundus each have ________________ thin tendons which pass through which anatomical structure? ________________

11) Pinching the fingers together highlights the tendon of which muscle at the wrist? ________________

12) What muscle runs between the pisiform and the medial epicondyle? ________________

13) Although the flexor digitorum superficialis and profundus are deep to the other forearm flexors, they can be accessed along the medial side of which bony landmark? ________________

Let's Palpate!

Remember—there are no right or wrong answers here

Locate and explore the **extensor digitorum** on three individuals. Then write three words that describe what you feel. (See p. 138 in *Trail Guide*.)

Person #1 ________	Person #2 ________	Person #3 ________
________________	________________	________________
________________	________________	________________
________________	________________	________________

Matching

Match the origin and insertion to the correct muscle.

Origins

1) Anterior and medial surfaces of proximal three-quarters of ulna
2) Common extensor tendon from the lateral epicondyle of humerus (3)
3) Common flexor tendon from medial epicondyle of humerus (2)
4) Common flexor tendon from medial epicondyle of humerus, ulnar collateral ligament, coronoid process of ulna, interosseous membrane and proximal shaft of radius
5) Distal one-third of the lateral supracondylar ridge of humerus
6) Common flexor tendon from medial epicondyle of humerus and posterior surface of proximal two-thirds of ulna

Insertions

7) Base of fifth metacarpal
8) Bases of second and third metacarpals
9) Base of second metacarpal
10) Base of third metacarpal
11) Bases of distal phalanges, palmar surface of second through fifth fingers
12) Sides of middle phalanges of second through fifth fingers
13) Flexor retinaculum and palmar aponeurosis
14) Bases of middle and distal phalanges of second through fifth fingers
15) Pisiform, hook of the hamate and base of fifth metacarpal

Muscle	O	I
Extensor carpi radialis brevis	_______	_______
Extensor carpi radialis longus	_______	_______
Extensor carpi ulnaris	_______	_______
Extensor digitorum	_______	_______
Flexor carpi radialis	_______	_______
Flexor carpi ulnaris	_______	_______
Flexor digitorum profundus	_______	_______
Flexor digitorum superficialis	_______	_______
Palmaris longus	_______	_______

Shorten or Lengthen?

16) Passive abduction of the wrist would ____________________ the extensor carpi radialis longus.
17) Passive flexion of fingers 2-5 would ____________________ the extensor digitorum.
18) Passive flexion of the wrist would ____________________ the palmaris longus.
19) Passive adduction of the wrist would ____________________ the flexor carpi radialis.
20) Passive extension of fingers 2-5 would ____________________ the flexor digitorum profundus.

p. 149

Forearm and Hand

Muscles of the Hand #1

Please identify the following structures.

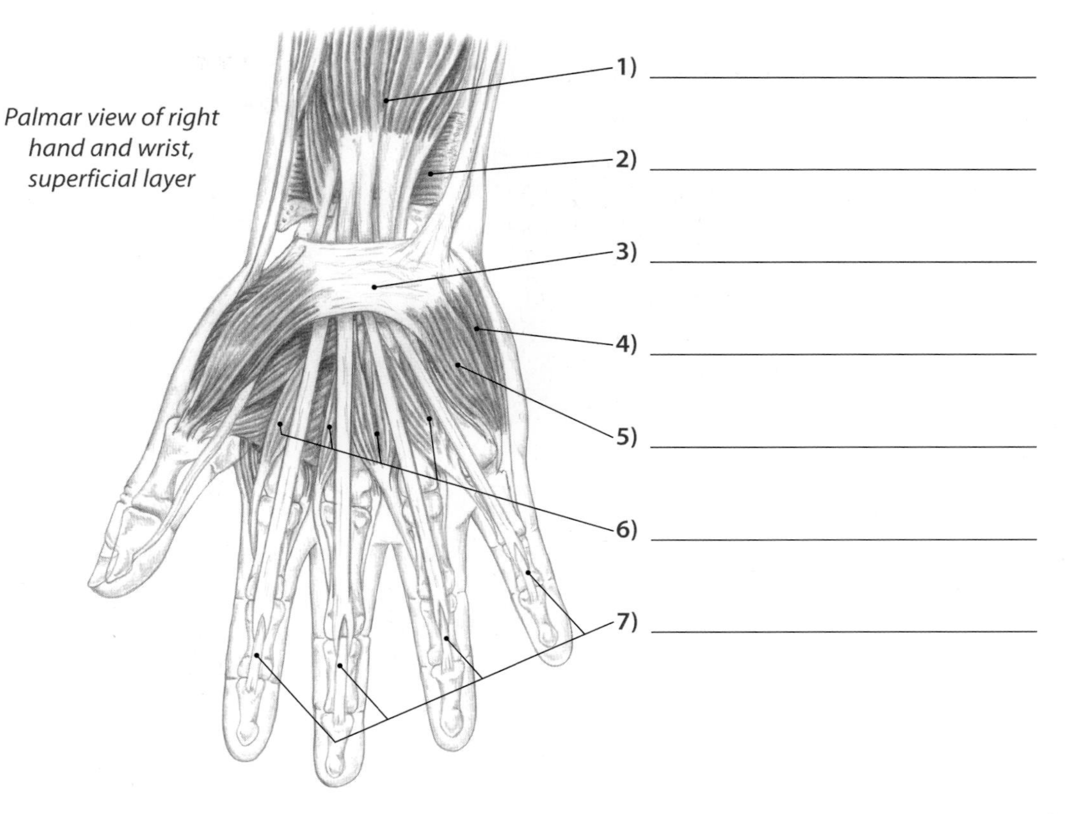

Palmar view of right hand and wrist, superficial layer

8) ______

9) ______

10) ______

11) ______

12) ______

Palmar view of right hand and wrist, intermediate layer

CHOICES

Abductor digiti minimi
Adductor pollicis
Flexor digiti minimi brevis
Flexor digitorum profundus (2)
Flexor digitorum superficialis
Flexor pollicis longus
Flexor retinaculum
Lumbricals
Opponens digiti minimi
Opponens pollicis
Pronator quadratus

Muscles of the Hand #2

Please identify the following structures.

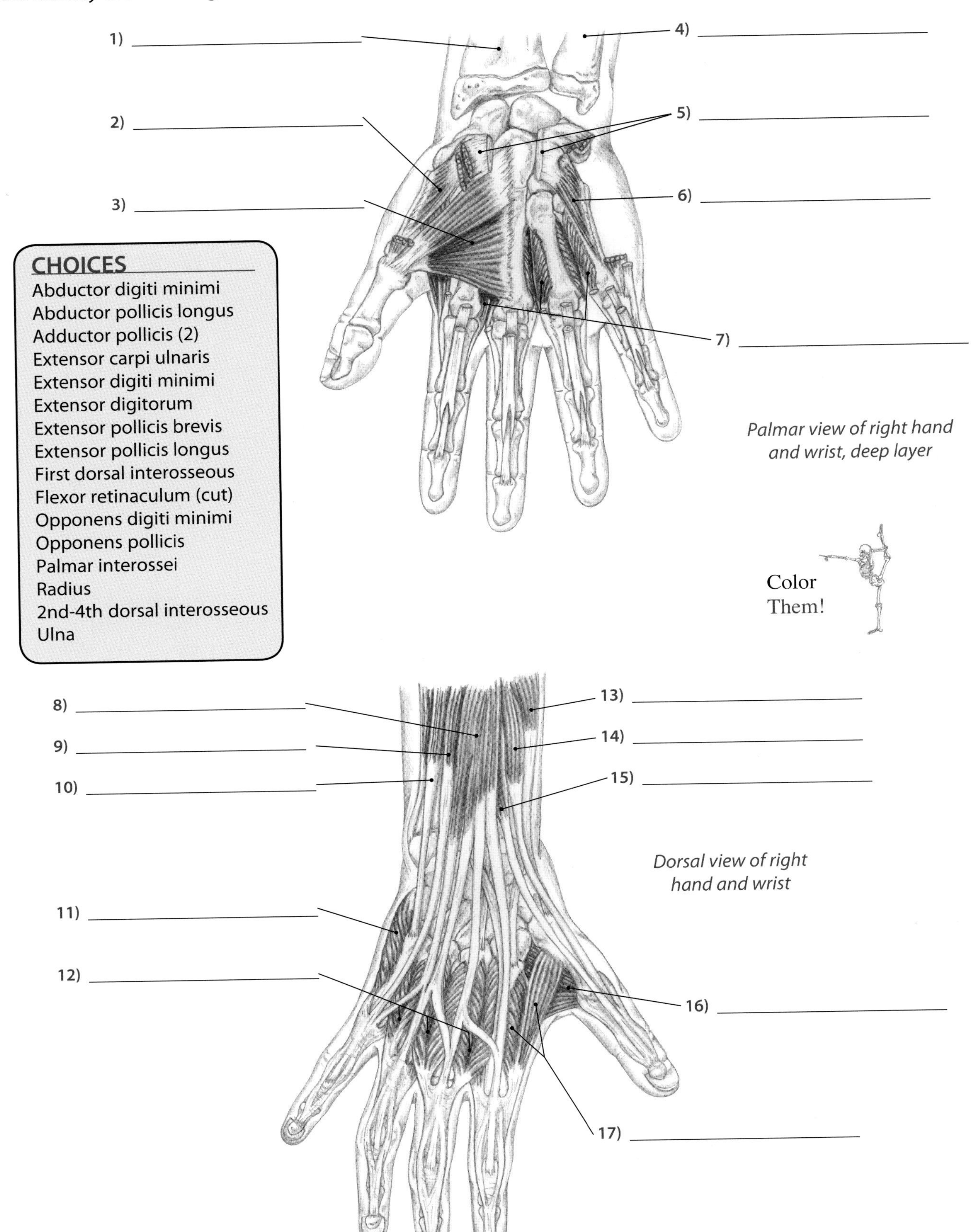

Palmar view of right hand and wrist, deep layer

Dorsal view of right hand and wrist

p. 151-159

Forearm and Hand, Muscle Group #3

Muscles of the Thumb and Hand

Please answer the following questions.

1) The ______________________________ eminence is located at the thumb's base, while the ______________________________ eminence is located along the ulnar side of the palm.

2) How many muscles act upon the thumb? ______________________________

   How many of these are located at the thenar eminence? ______________________________

3) Which muscle is responsible for creating opposition of the thumb? ______________________________

4) The distal tendons of which three muscles form the "anatomical snuffbox"?

   ______________________ ______________________ ______________________

5) The palmar interossei are difficult to access because they are deep to the ______________________________ muscles and situated between the ______________________________ bones.

6) The lumbricals sprout from the sides of the tendons of which muscle? ______________________________

7) Which muscle is located between the pisiform and the base of the fifth finger? ______________________________

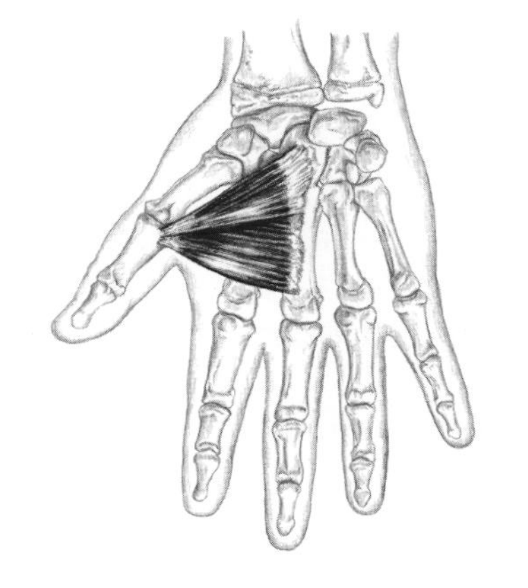

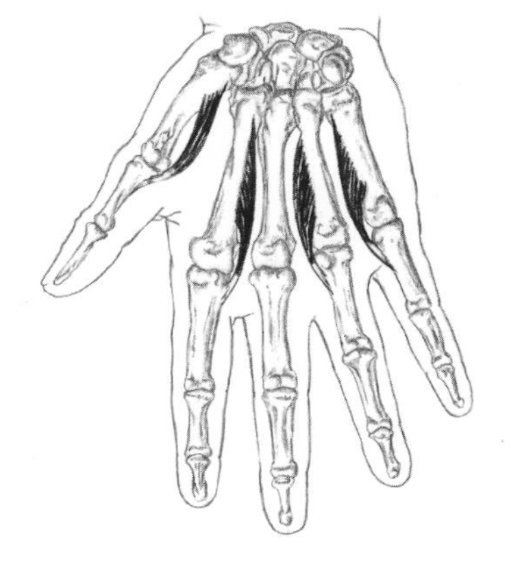

8) ______________________________

9) ______________________________

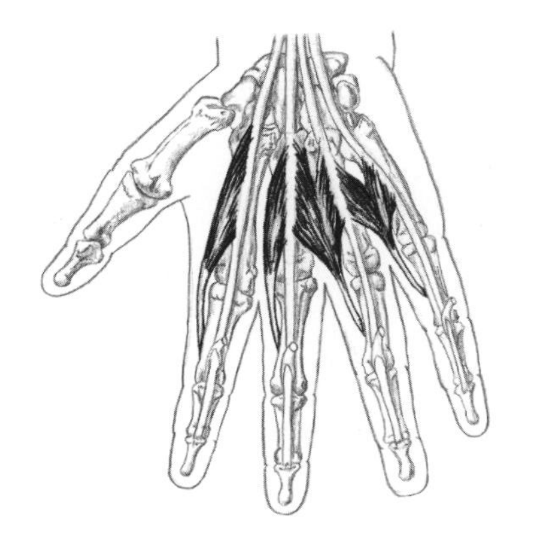

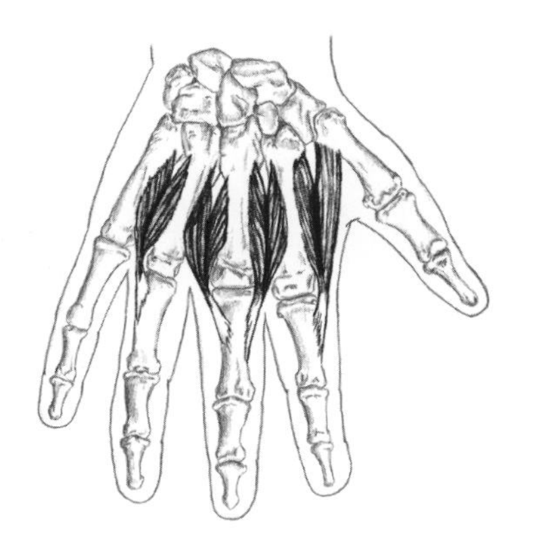

10) ______________________________

11) ______________________________

Muscles of the Thumb and Hand

p. 151-159

Matching

Match the origin and insertion to the correct muscle.

Origins

1) Anterior surface of radius and interosseous membrane
2) Capitate, second and third metacarpals
3) Flexor retinaculum and tubercle of the trapezium
4) Posterior surface of radius and ulna, and interosseous membrane
5) Posterior surface of ulna and interosseous membrane

Insertions

6) Base of first metacarpal
7) Base of proximal phalanx of thumb
8) Base of distal phalanx of thumb (2)
9) Entire length of first metacarpal bone, radial surface

Muscle	O	I
Abductor pollicis longus	_______	_______
Adductor pollicis	_______	_______
Extensor pollicis longus	_______	_______
Flexor pollicis longus	_______	_______
Opponens pollicis	_______	_______

Let's Palpate!

Remember—there are no right or wrong answers here

Locate and explore the **thenar eminence** on three individuals. Then write three words that describe what you feel. (See p. 156 in *Trail Guide*.)

Person #1 _______________

____

____

____

Person #2 _______________

____

____

____

Person #3 _______________

____

____

____

Forearm and Hand

Other Structures

Please answer the following questions.

1) Which two structures reinforce the elbow joint by spanning from their respective epicondyles to the bones of the forearm?

_______________________________ _______________________________

2) During pronation and supination, which ligament stabilizes the proximal end of the radius against the ulna?

____

3) Between which two bony landmarks is the ulnar nerve particularly accessible and superficial?

_______________________________ _______________________________

4) Which structure pads the space between the olecranon process and the skin of the elbow?

____

5) The carpal tunnel is a passageway for many _______________ and the _______________ nerve.

6) The transverse fibers of the _______________________ and carpal bones form the carpal tunnel.

7) Which span of connective tissue is a continuation of the antebrachial fascia into the palm of the hand?

____

8) Which artery is often used for taking a pulse at the wrist? _______________________

Please identify the following structures.

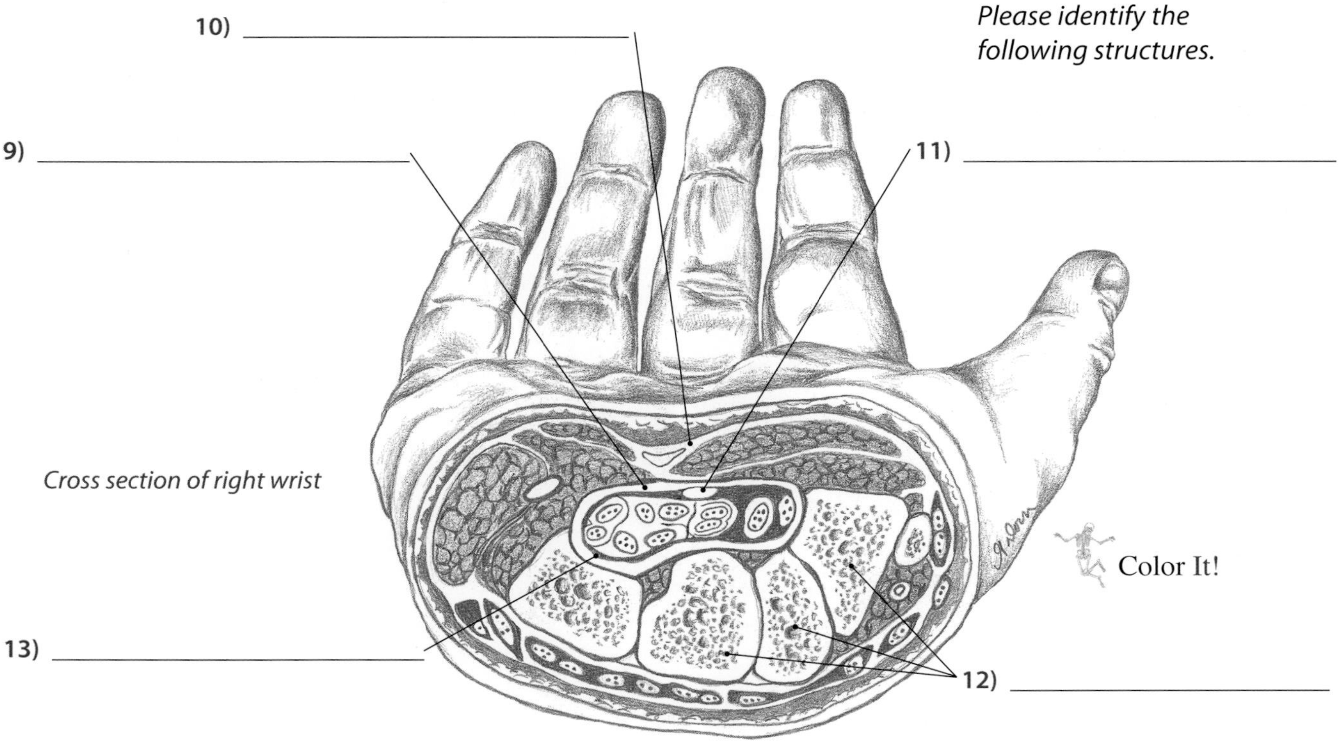

9) _______________________

10) _______________________

11) _______________________

12) _______________________

13) _______________________

Cross section of right wrist

Color It!

Humeroulnar and Proximal Radioulnar Joints

Please identify the following structures.

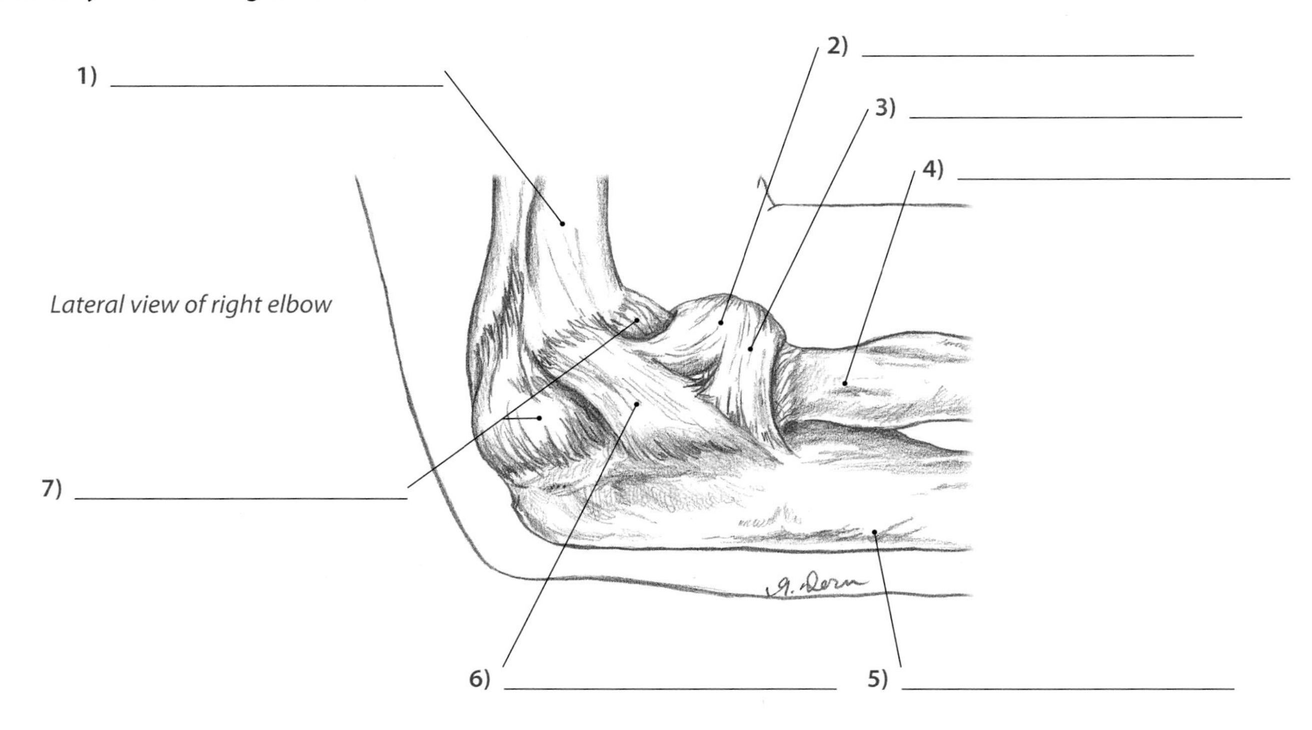

CHOICES

Annular ligament (2)	Olecranon process
Articular capsule (2)	Radial collateral ligament
Head of radius (deep)	Radius (2)
Humerus (2)	Ulna (2)
Medial epicondyle	Ulnar collateral ligament

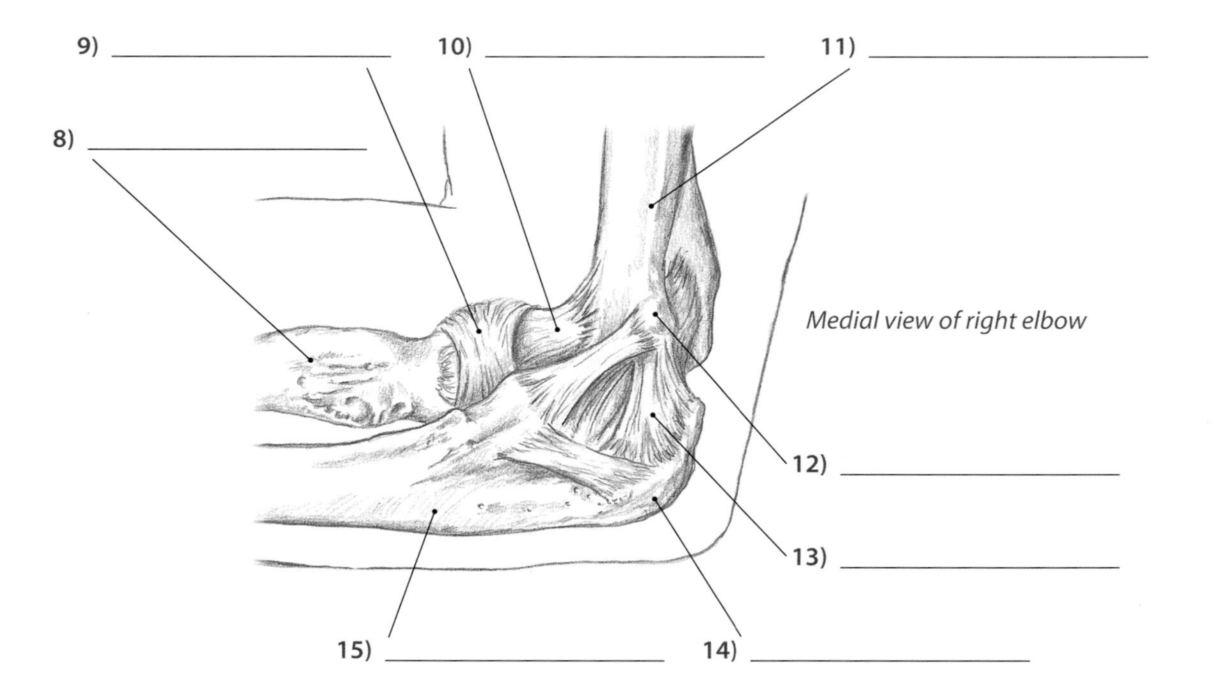

Radiocarpal Joint

Please identify the following structures.

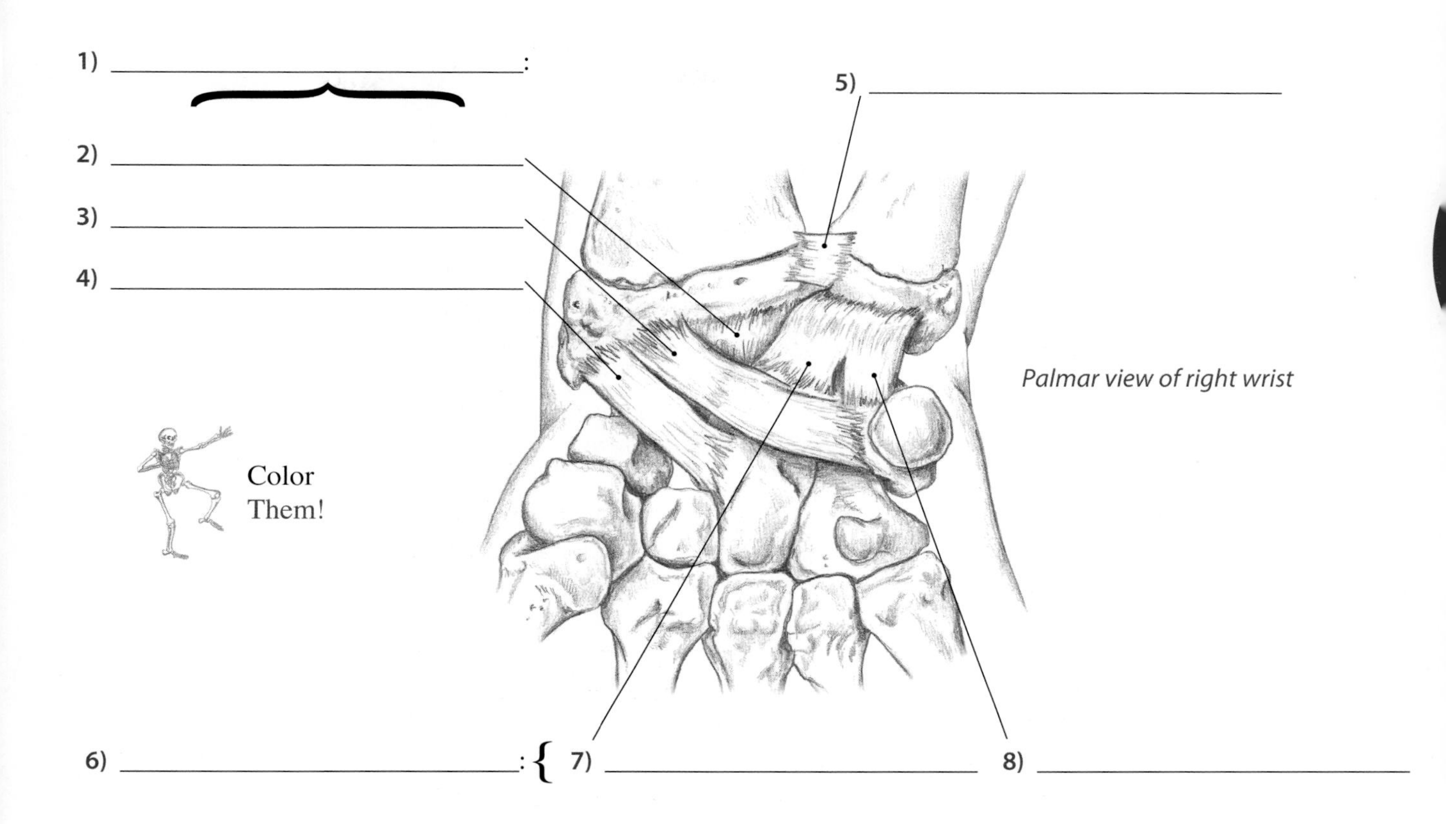

Palmar view of right wrist

CHOICES

Dorsal radiocarpal ligament
Dorsal radioulnar ligament
Palmar radiocarpal ligament
Palmar radioulnar ligament
Palmar ulnocarpal ligament
Radial collateral ligament
Radiocapitate part
Radioscapholunate part
Radiotriquetral part
Ulnar collateral ligament
Ulnolunate part
Ulnotriquetral part

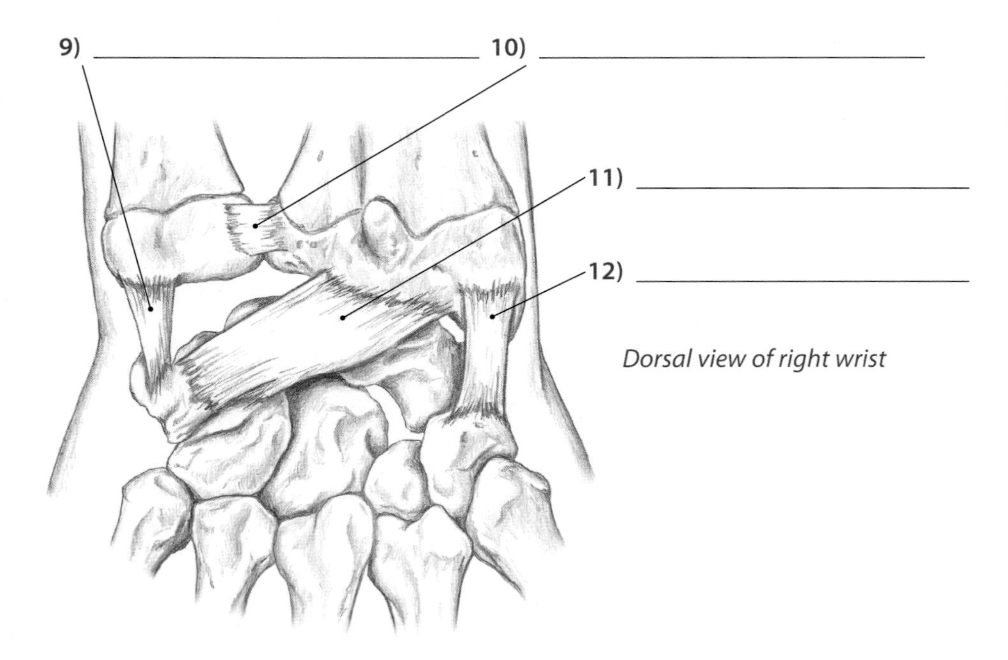

Dorsal view of right wrist

Intercarpal, Carpometacarpal and Metacarpal Joints

p. 165-166

Please identify the following structures.

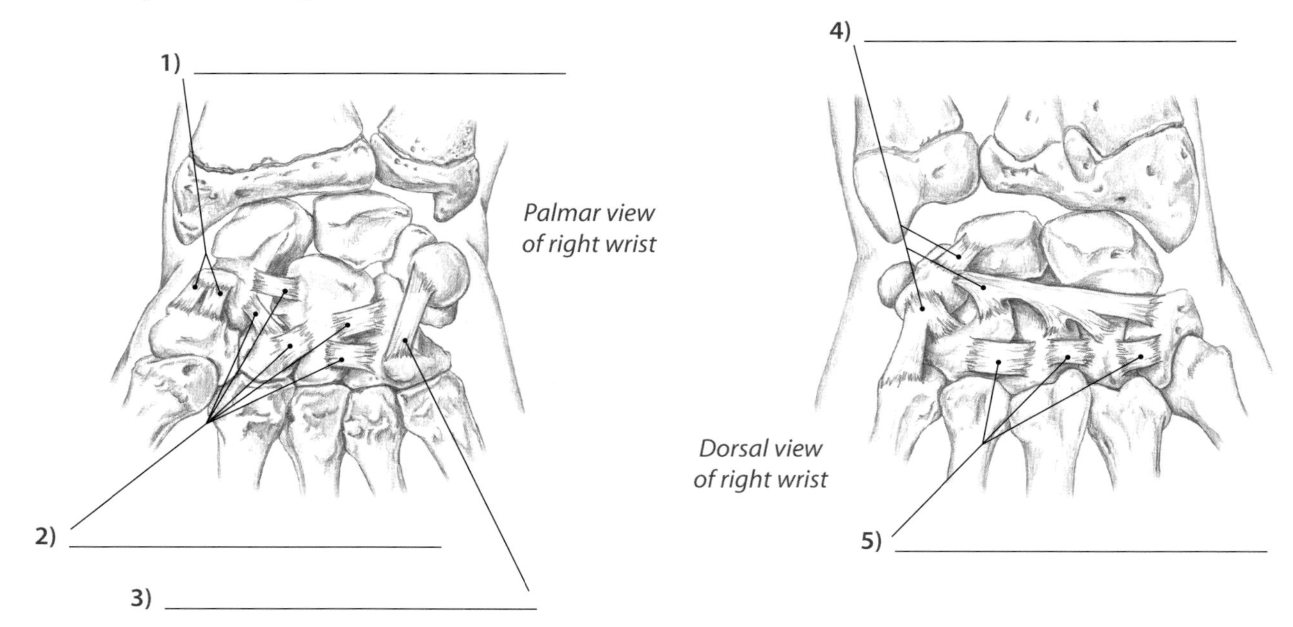

CHOICES

Distal intercarpal ligaments	Palmar intercarpal ligaments
Dorsal carpometacarpal ligaments	Palmar metacarpal ligaments
Dorsal intercarpal ligaments	Pisohamate ligament
Dorsal metacarpal ligaments	Pisometacarpal ligament
Palmar carpometacarpal ligaments	Radiate carpal ligaments

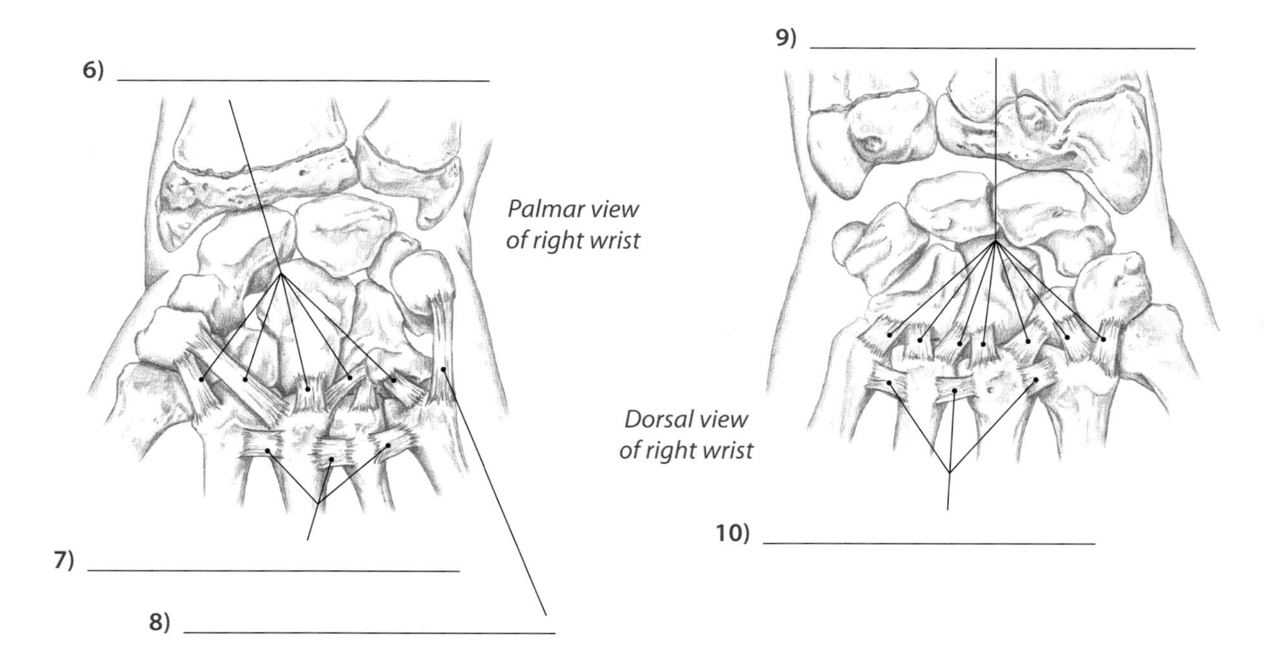

Topographical Views

Please identify the following structures.

CHOICES

- Edge of rib cage
- Erector spinae group
- External oblique
- Iliac crest (2)
- Jugular notch
- Medial border of the scapula
- Posterior superior iliac spine
- Rectus abdominis
- Ribs
- Sacrum
- Spinous process of C-7
- Spinous processes of thoracic and lumbar vertebrae
- Sternum
- Twelfth rib
- Umbilicus

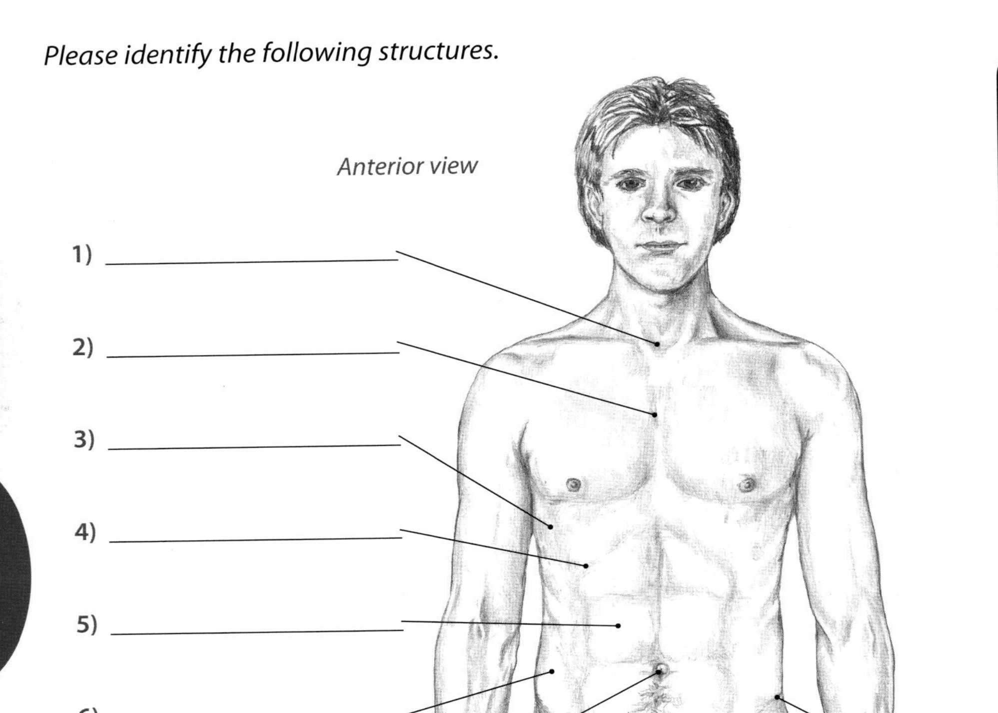

Posterior view

9) ____________

10) ____________

11) ____________

12) ____________

13) ____________

14) ____________

15) ____________

16) ____________

Spine and Thorax

Bones and Bony Landmarks #1

Please answer the following questions.

1) Which section of the vertebral column is capable of the most movement? ______________________

2) The thorax is comprised of which two structures?

______________________ ______________________

3) The visible row of bumps running down the center of the back are the ______________________.

4) Please match the bony landmark with the corresponding spinous process.

______ T-12	a) Top of the iliac crest
______ T-2	b) Base of the neck
______ L-4	c) Superior angle of the scapula
______ C-7	d) Twelfth rib
______ T-7	e) Inferior angle of the scapula

5) With your partner seated, what two movements at the spine could you ask your partner to perform to feel the movement of the spinous processes?

______________________ ______________________

6) The angles of the scapula and the corresponding spinous processes do not always line up. Name two factors that might affect the position of the scapula.

______________________ ______________________

7) Which two cervical vertebrae have spinous processes that protrude further posteriorly and are more distinct than the other cervical vertebrae?

______________________ ______________________

8) Which band of connective tissue lies superficial to the cervical spinous processes?

____

9) Many of the cervical transverse processes are deep to which neck muscle? ______________________

10) Your partner is supine and you passively rotate the head 45° away from the side you are palpating. This position places the cervical transverse processes in a line running between which two bony landmarks?

______________________ ______________________

11) The lamina groove is located between which two bony landmarks of the vertebrae?

______________________ ______________________

Spine and Thorax

p. 182-187

Bones and Bony Landmarks #2

Please answer the following questions.

1) The thoracic transverse processes are located deep to the ____________________ muscles and superficial to the ____________________.

2) To avoid the thick erector spinae muscles overlying the lumbar transverse processes, it is best to slide your fingers roughly how far laterally from the spinous processes? ____________________

3) Which rib attaches to the sternum at the level of the sternal angle? ____________________

4) What is the structure that extends off the ribs and attaches to the sternum? ____________________

5) Which muscles are located between the ribs? ____________________

6) Although the entire rib cage is deep to muscle tissue, which portion is easily accessed?

____

7) The first rib is deep to which bone along the anterior thorax? ____________________

8) Exploring just posterior to the clavicle, through which muscle group must you palpate to access the first rib?

____

9) What action could you ask your partner to perform to feel the first rib move? ____________________

10) In which three directions are the ribs ideally designed to move? ____________________

____________________ ____________________

11) The eleventh and twelfth ribs lie at approximately what angle on the body? ____________________

12) As you palpate medially toward the spine, you may lose contact with the twelfth rib because it is deep to which muscle group?

____

Let's Palpate!

Remember—there are no right or wrong answers here

Locate and explore the **spinous processes of the vertebral column** on three individuals. Then write three words that describe what you feel. (See p. 176 in *Trail Guide*.)

Person #1 ________	Person #2 ________	Person #3 ________
________	________	________
________	________	________
________	________	________

Spine and Thorax

Bones of the Spine and Thorax

Please identify the following structures.

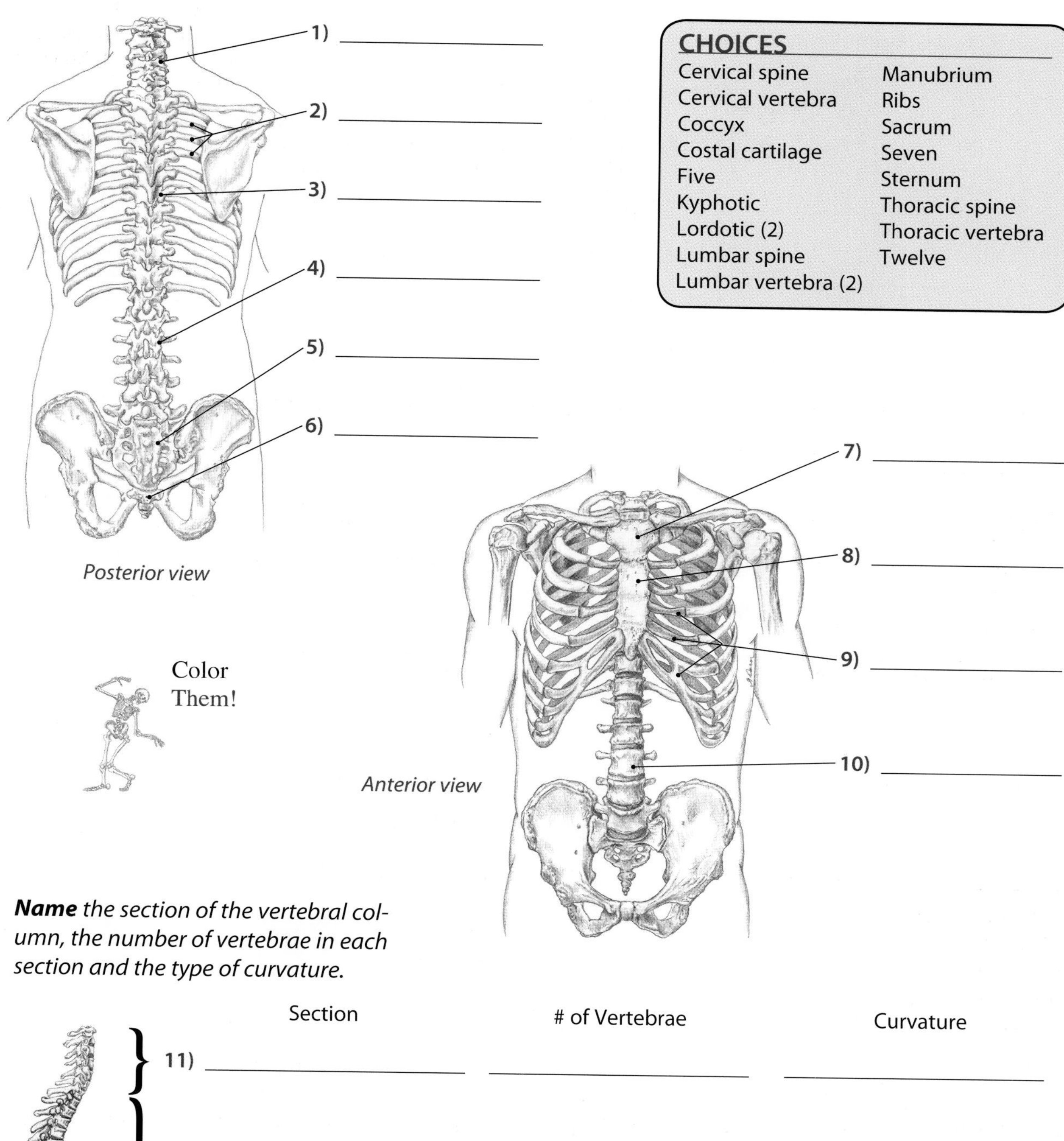

Posterior view

Anterior view

CHOICES

Cervical spine	Manubrium
Cervical vertebra	Ribs
Coccyx	Sacrum
Costal cartilage	Seven
Five	Sternum
Kyphotic	Thoracic spine
Lordotic (2)	Thoracic vertebra
Lumbar spine	Twelve
Lumbar vertebra (2)	

***Name** the section of the vertebral column, the number of vertebrae in each section and the type of curvature.*

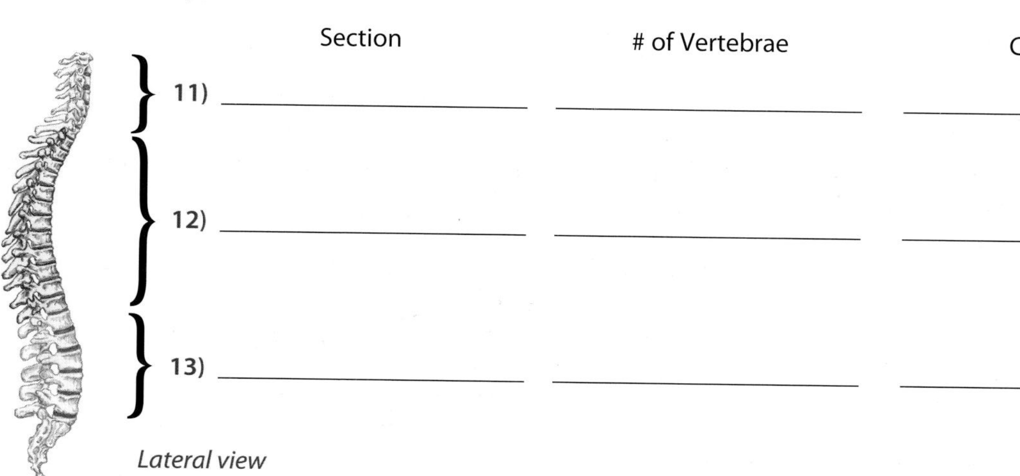

	Section	# of Vertebrae	Curvature
11)	____________	____________	____________
12)	____________	____________	____________
13)	____________	____________	____________

Lateral view

First and Second Cervical Vertebrae

p. 171

Please identify the following structures.

CHOICES

- Articular facet for odontoid process
- Groove for vertebral artery
- Lamina (4)
- Odontoid process (2)
- Posterior tubercle
- Spinous process (2)
- Superior facets (2)
- Transverse foramen (2)
- Transverse process (4)
- Vertebral foramen (3)

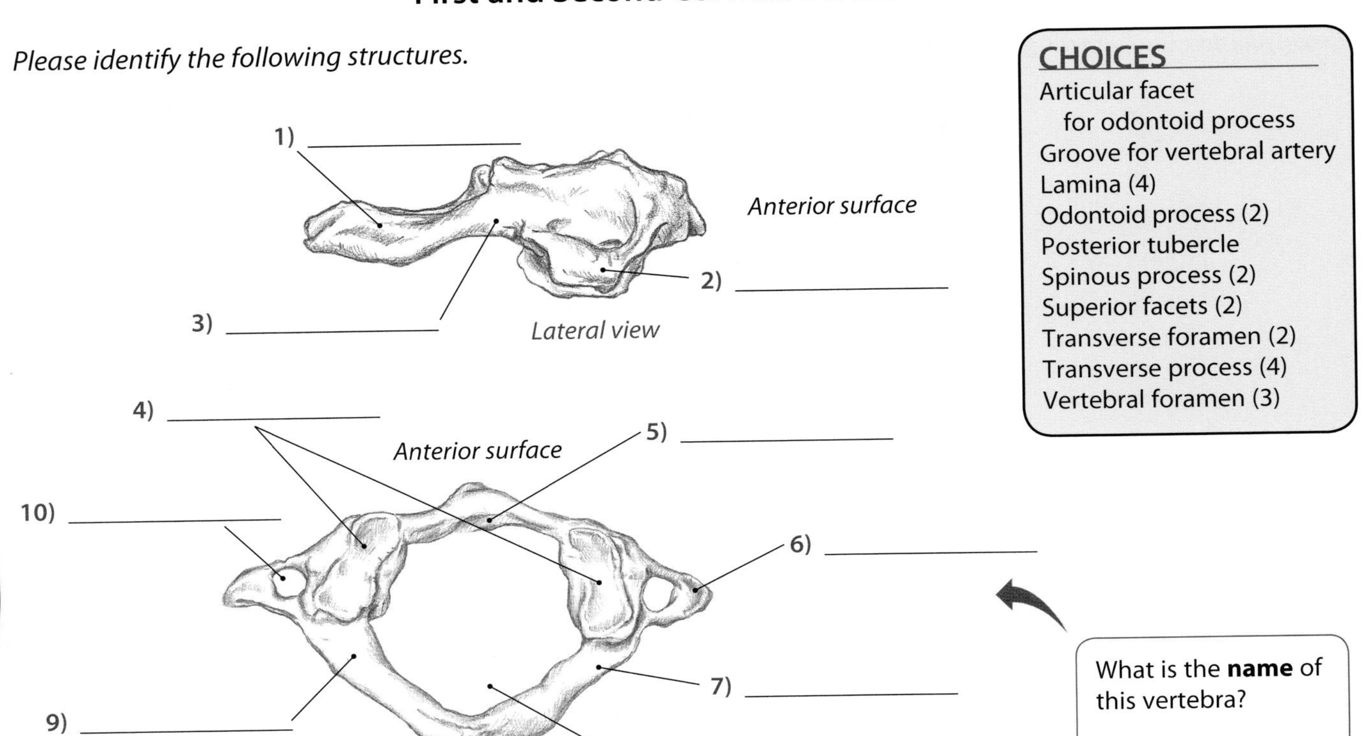

What is the **name** of this vertebra?

11) ____

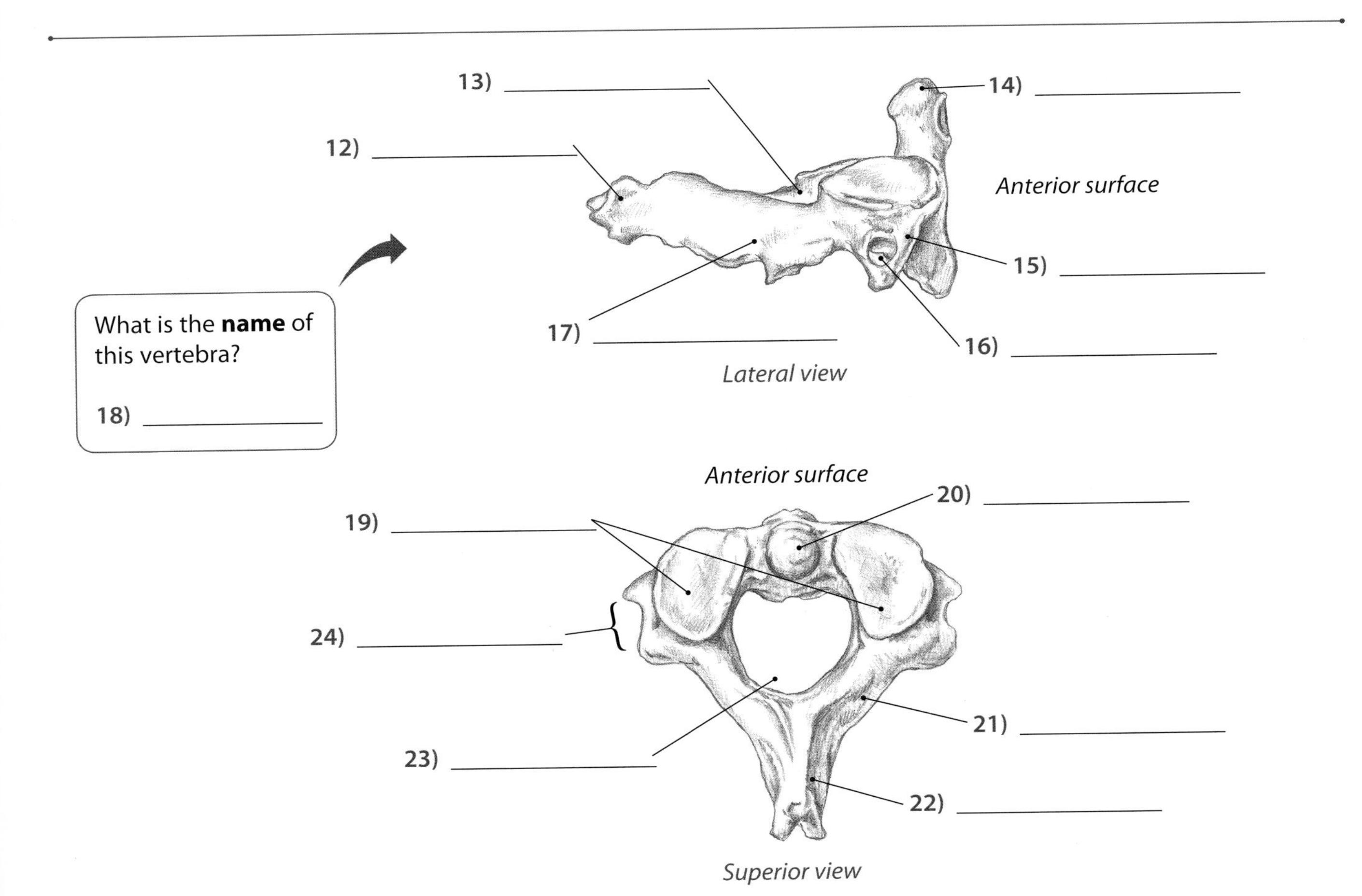

What is the **name** of this vertebra?

18) ____

Spine and Thorax

Cervical Vertebrae

Please identify the following structures.

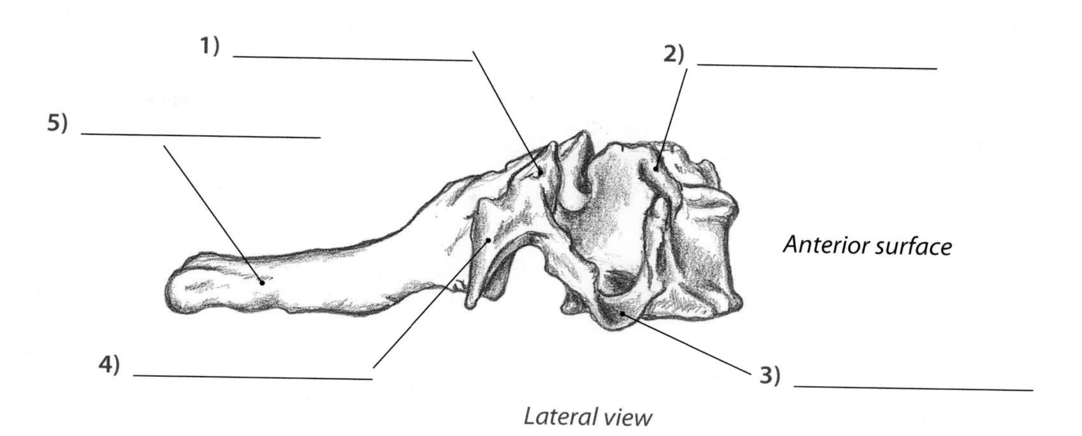

Lateral view

CHOICES

Anterior tubercle (2)	Posterior tubercle (2)
Body	Spinous process (2)
Canal for spinal nerve (2)	Superior facet
Lamina	Transverse foramen
Lamina groove	Transverse process (2)

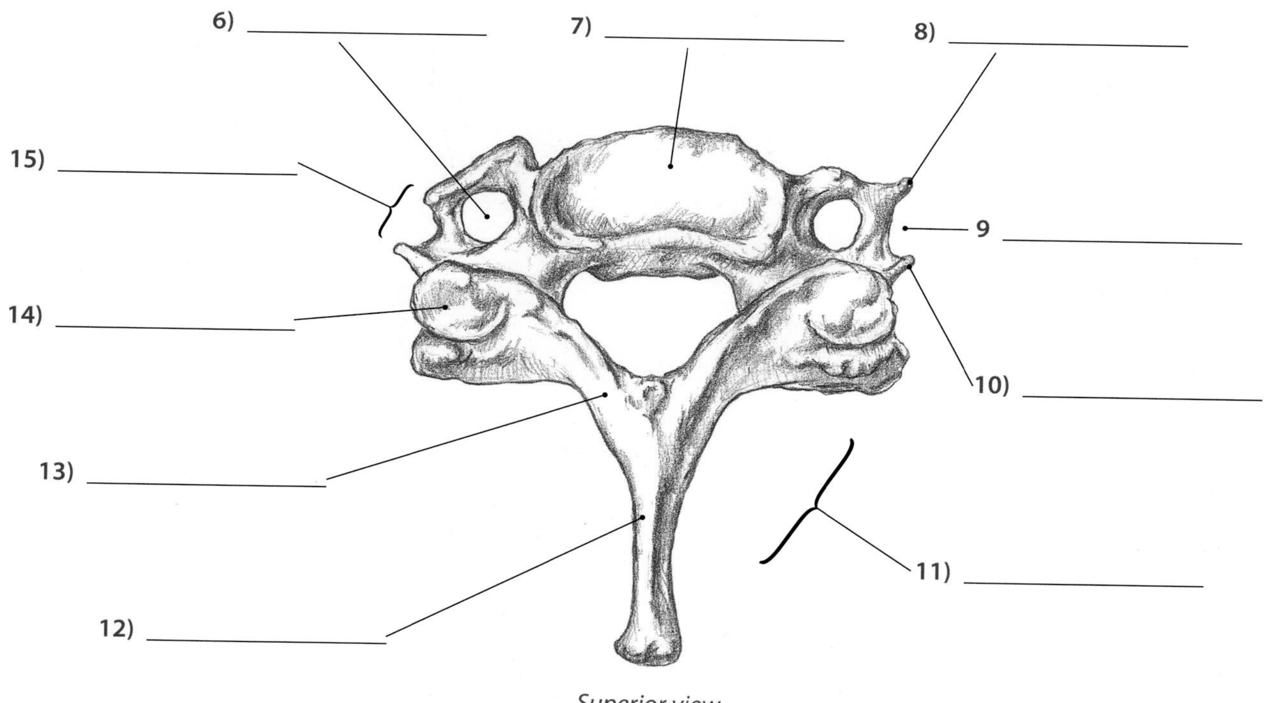

Superior view

Thoracic and Lumbar Vertebrae

Please identify the following structures.

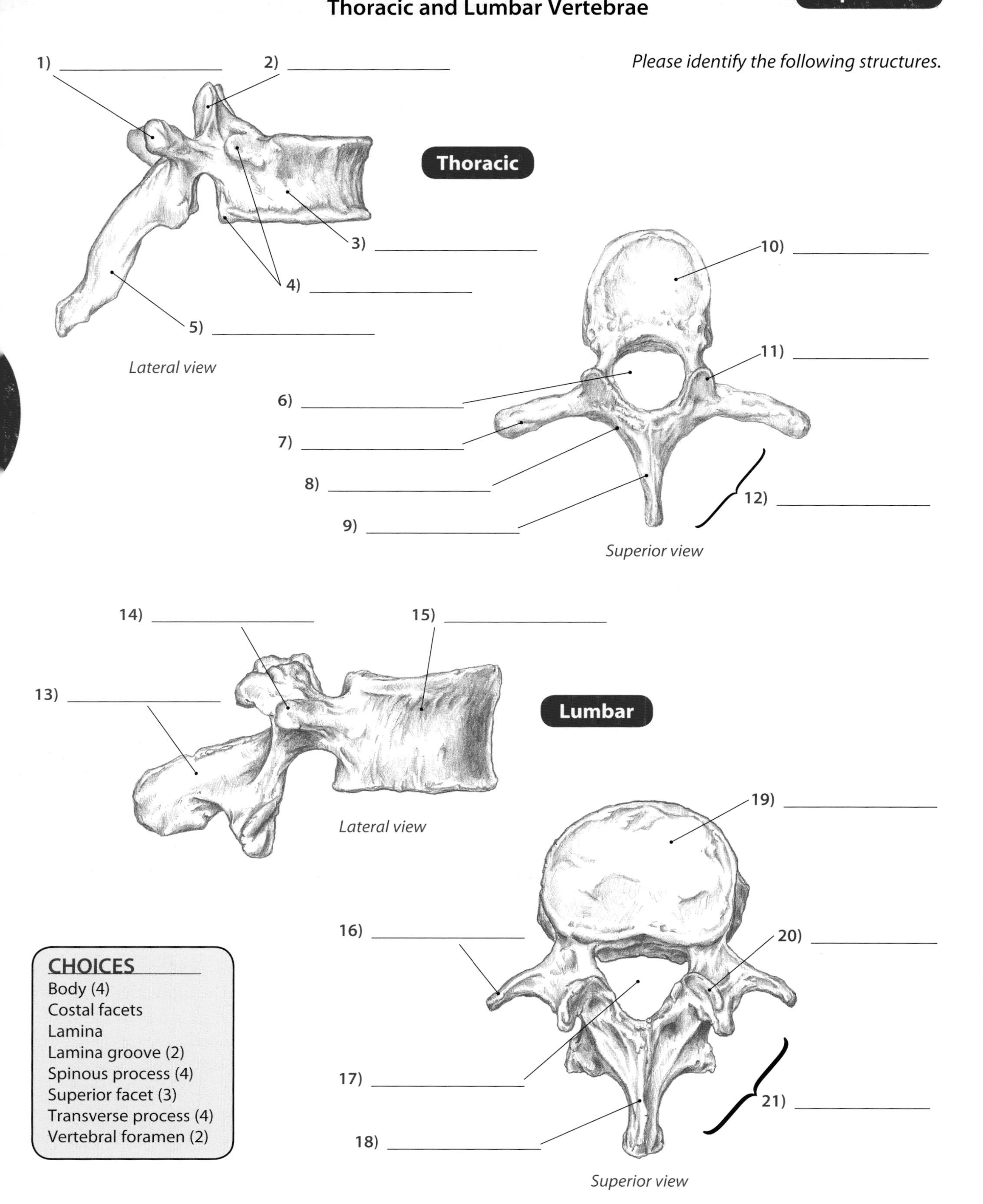

CHOICES
Body (4)
Costal facets
Lamina
Lamina groove (2)
Spinous process (4)
Superior facet (3)
Transverse process (4)
Vertebral foramen (2)

Spine and Thorax
Rib Cage and Sternum

Please identify the following structures.

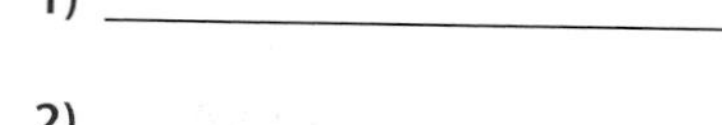
1) ____________________

2) ____________________

3) ____________________

4) ____________________

5) ____________________

6) ____________________

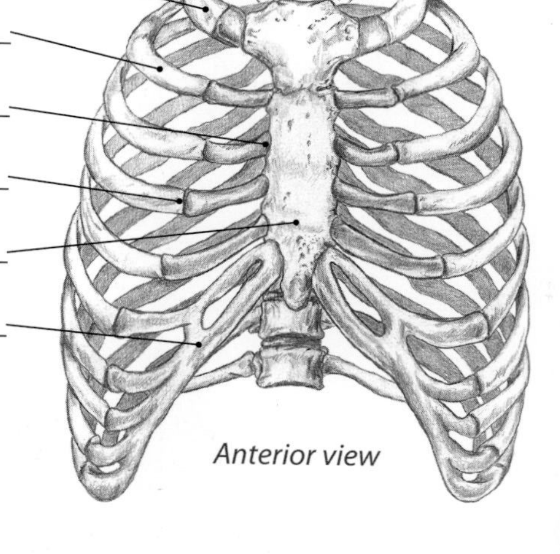

Anterior view

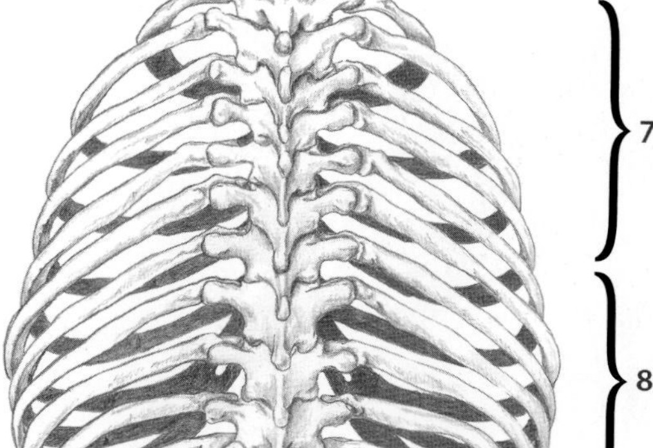

7) ____________________

8) ____________________

9) ____________________

Posterior view

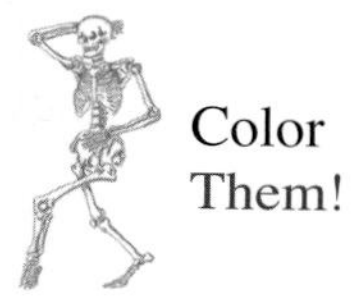

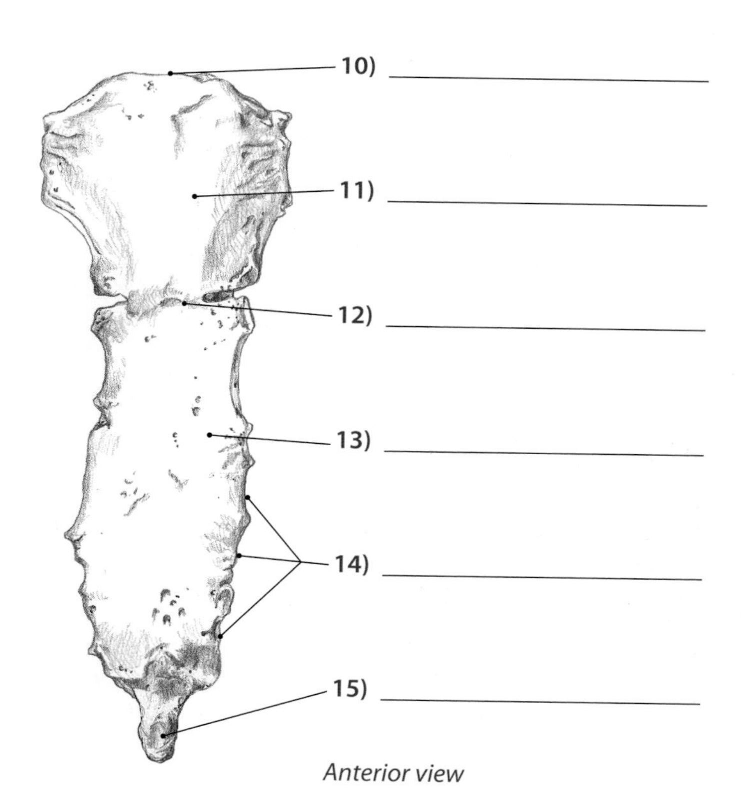

10) ____________________

11) ____________________

12) ____________________

13) ____________________

14) ____________________

15) ____________________

Anterior view

CHOICES

Articulations with ribs	Jugular notch
Body of sternum (2)	Manubrium
Costal cartilage	Second rib
Costochondral joint	Sternal angle
False ribs	Sternocostal joint
First rib	True ribs
Floating ribs	Xiphoid process

Muscles of the Spine and Thorax #1

Please identify the following structures.

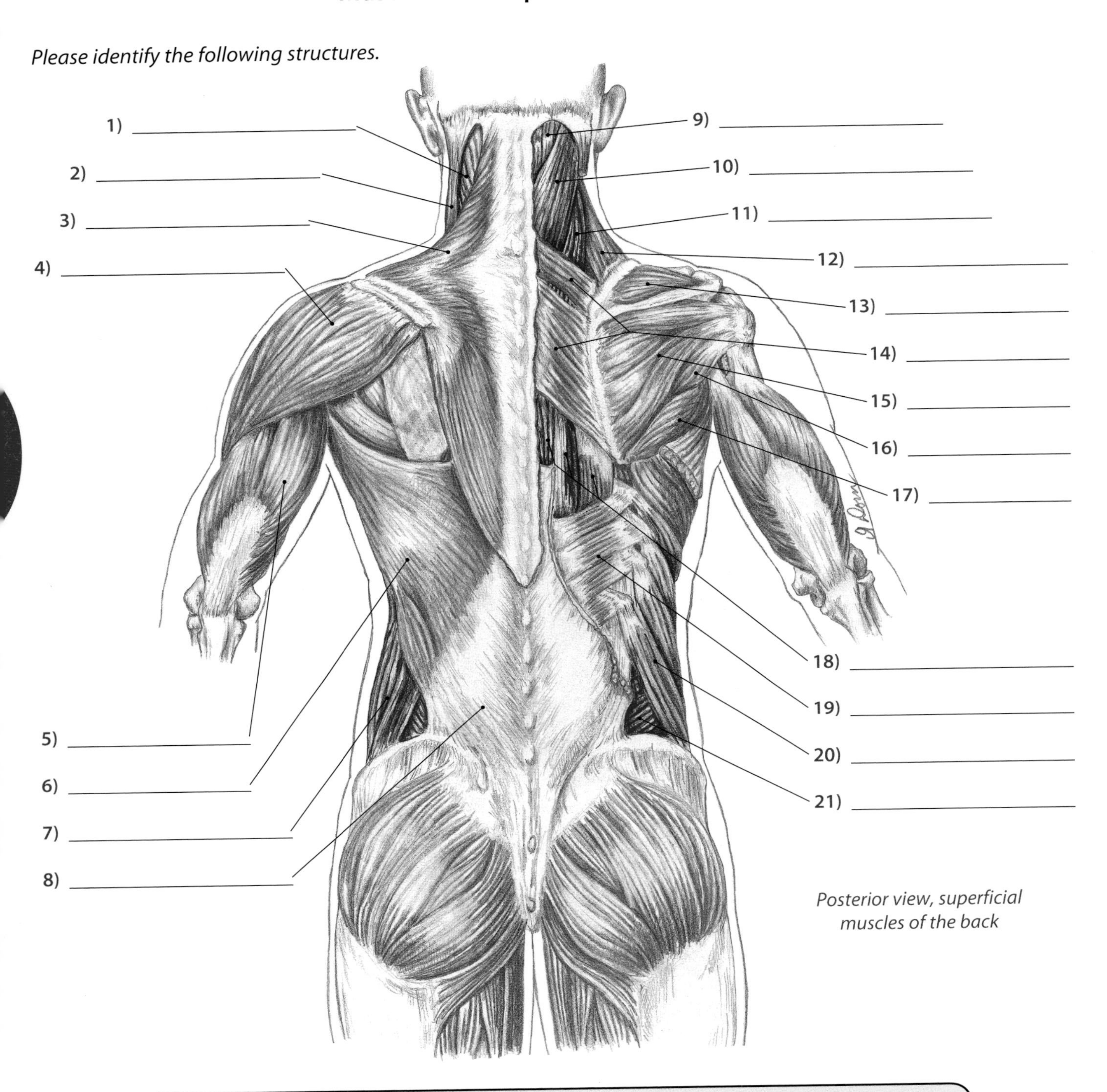

Posterior view, superficial muscles of the back

CHOICES

Deltoid
Erector spinae group
External oblique (2)
Infraspinatus
Internal oblique
Latissimus dorsi
Levator scapula
Rhomboids
Semispinalis capitis
Serratus posterior inferior
Splenius capitis (2)
Splenius cervicis
Sternocleidomastoid
Supraspinatus
Teres major
Teres minor
Thoracolumbar aponeurosis
Trapezius
Triceps brachii

Spine and Thorax

Muscles of the Spine and Thorax #2

Please identify the following structures.

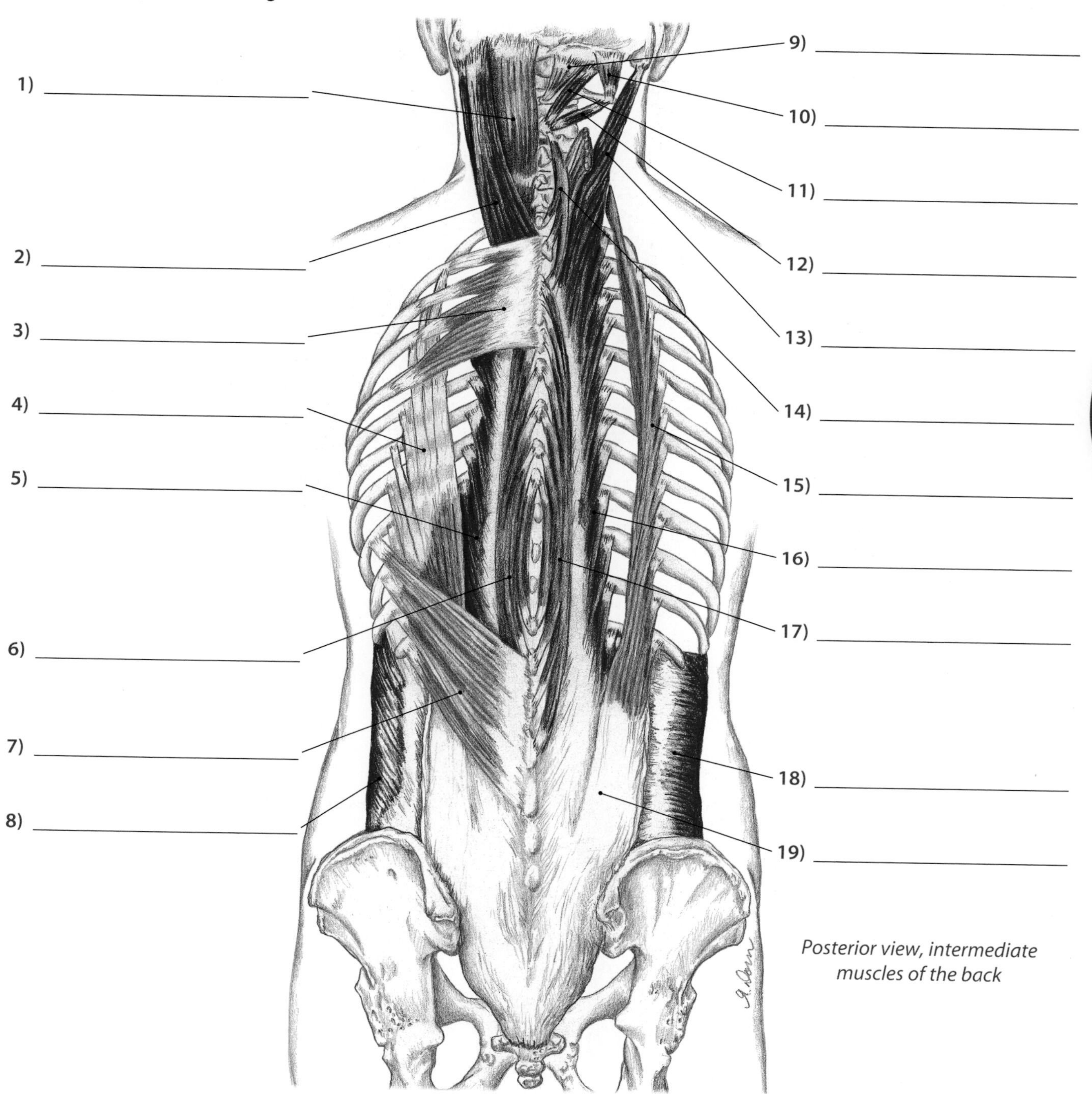

Posterior view, intermediate muscles of the back

CHOICES

Iliocostalis (2)
Internal oblique
Longissimus capitis
Longissimus thoracis (2)
Oblique capitis inferior
Oblique capitis superior
Rectus capitis posterior major
Rectus capitis posterior minor
Semispinalis capitis
Serratus posterior inferior
Serratus posterior superior
Spinalis cervicis
Spinalis thoracis (2)
Splenius capitis
Thoracolumbar aponeurosis
Transverse abdominis

Muscles of the Spine and Thorax #3

Please identify the following structures.

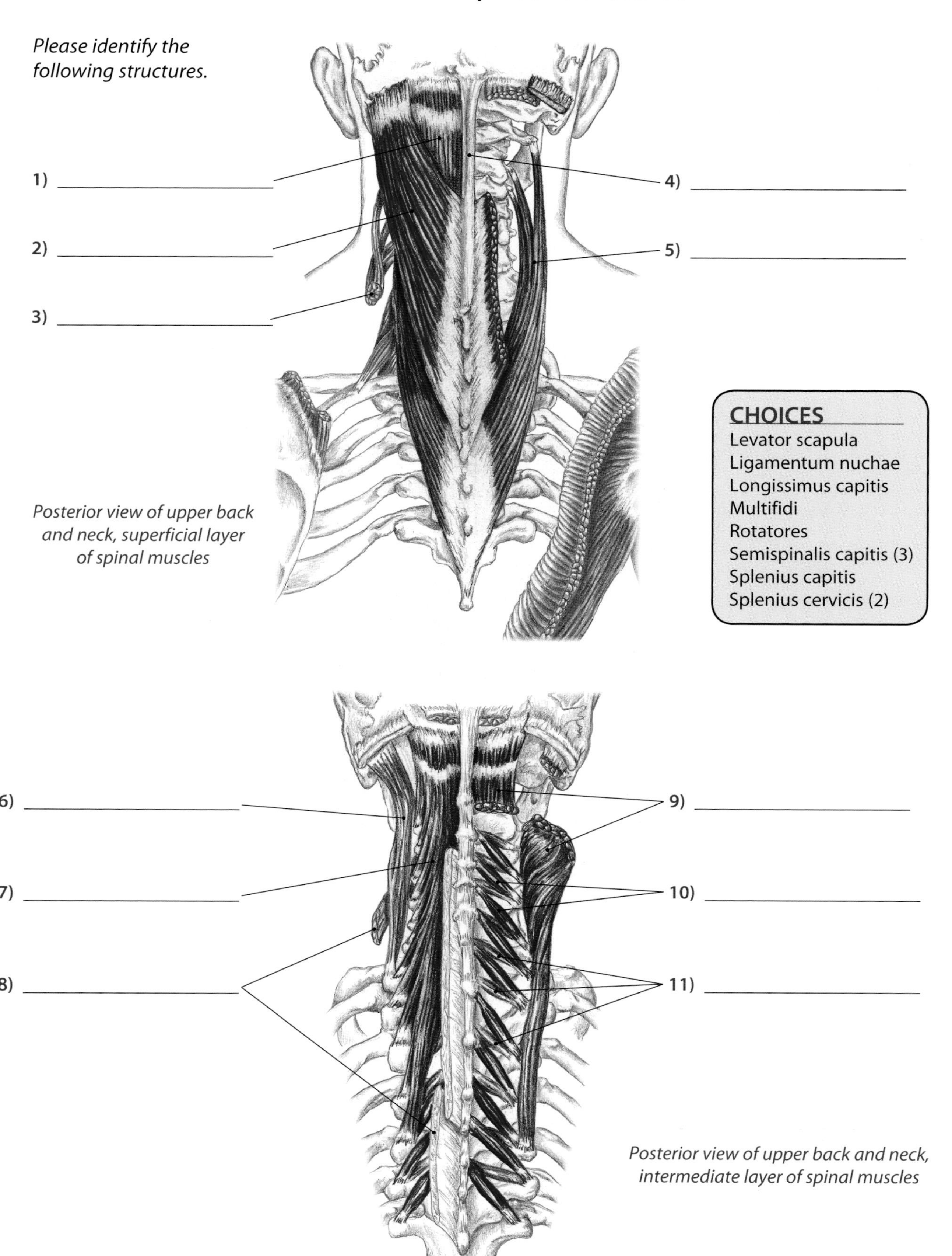

Posterior view of upper back and neck, superficial layer of spinal muscles

CHOICES

- Levator scapula
- Ligamentum nuchae
- Longissimus capitis
- Multifidi
- Rotatores
- Semispinalis capitis (3)
- Splenius capitis
- Splenius cervicis (2)

Posterior view of upper back and neck, intermediate layer of spinal muscles

Spine and Thorax
Cross Section of the Neck

Please identify the following structures.

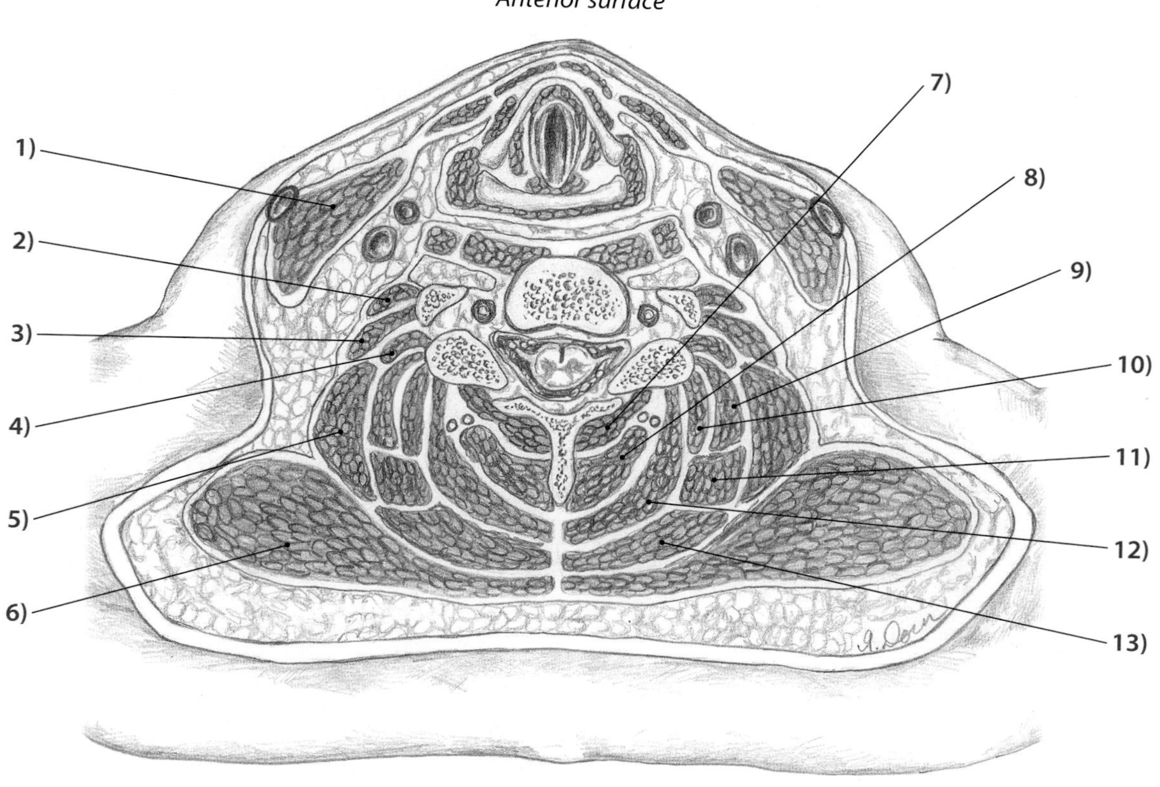

Cross section of the neck at the level of the fifth cervical vertebra

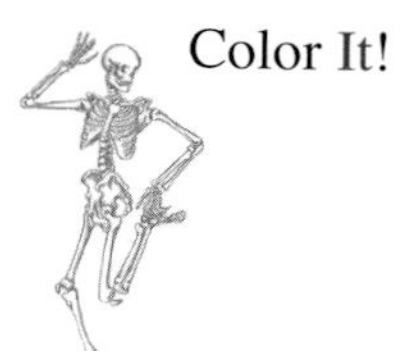

Color It!

CHOICES

Anterior scalene
Levator scapula
Longissimus capitis
Longissimus cervicis
Middle scalene
Multifidi and spinalis cervicis
Posterior scalene
Semispinalis capitis
Semispinalis cervicis
Splenius capitis
Splenius cervicis
Sternocleidomastoid
Trapezius

1) ______________________
2) ______________________
3) ______________________
4) ______________________
5) ______________________
6) ______________________
7) ______________________
8) ______________________
9) ______________________
10) ______________________
11) ______________________
12) ______________________
13) ______________________

p. 193

Cross Section of the Thorax #1

Please identify the following structures.

Color It!

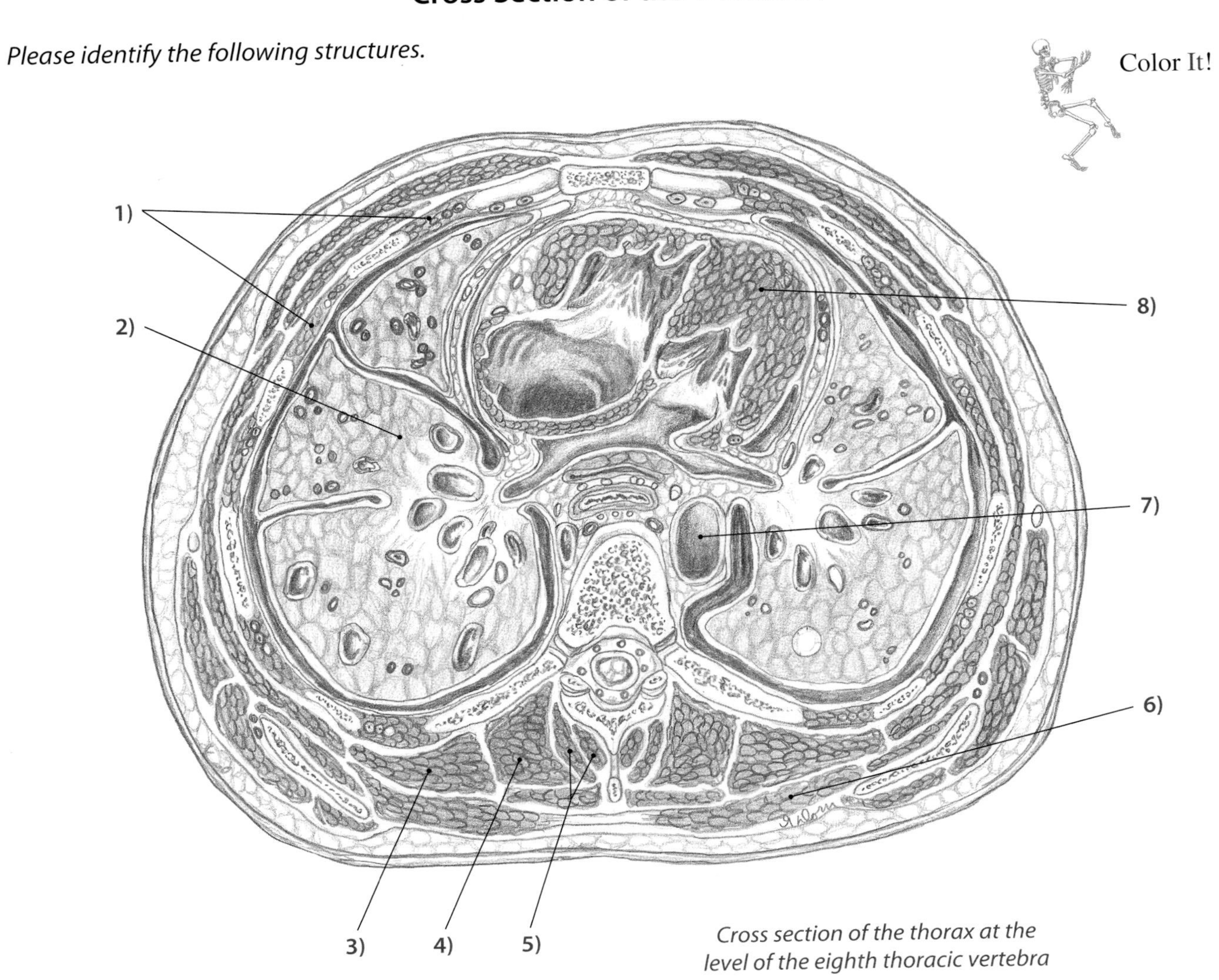

Cross section of the thorax at the level of the eighth thoracic vertebra

CHOICES

- Abdominal aorta
- Heart
- Iliocostalis
- Intercostals
- Longissimus
- Lung
- Multifidi and rotatores
- Trapezius

1) ____________________

2) ____________________

3) ____________________

4) ____________________

5) ____________________

6) ____________________

7) ____________________

8) ____________________

Cross Section of the Thorax #2

Please identify the following structures.

Color It!

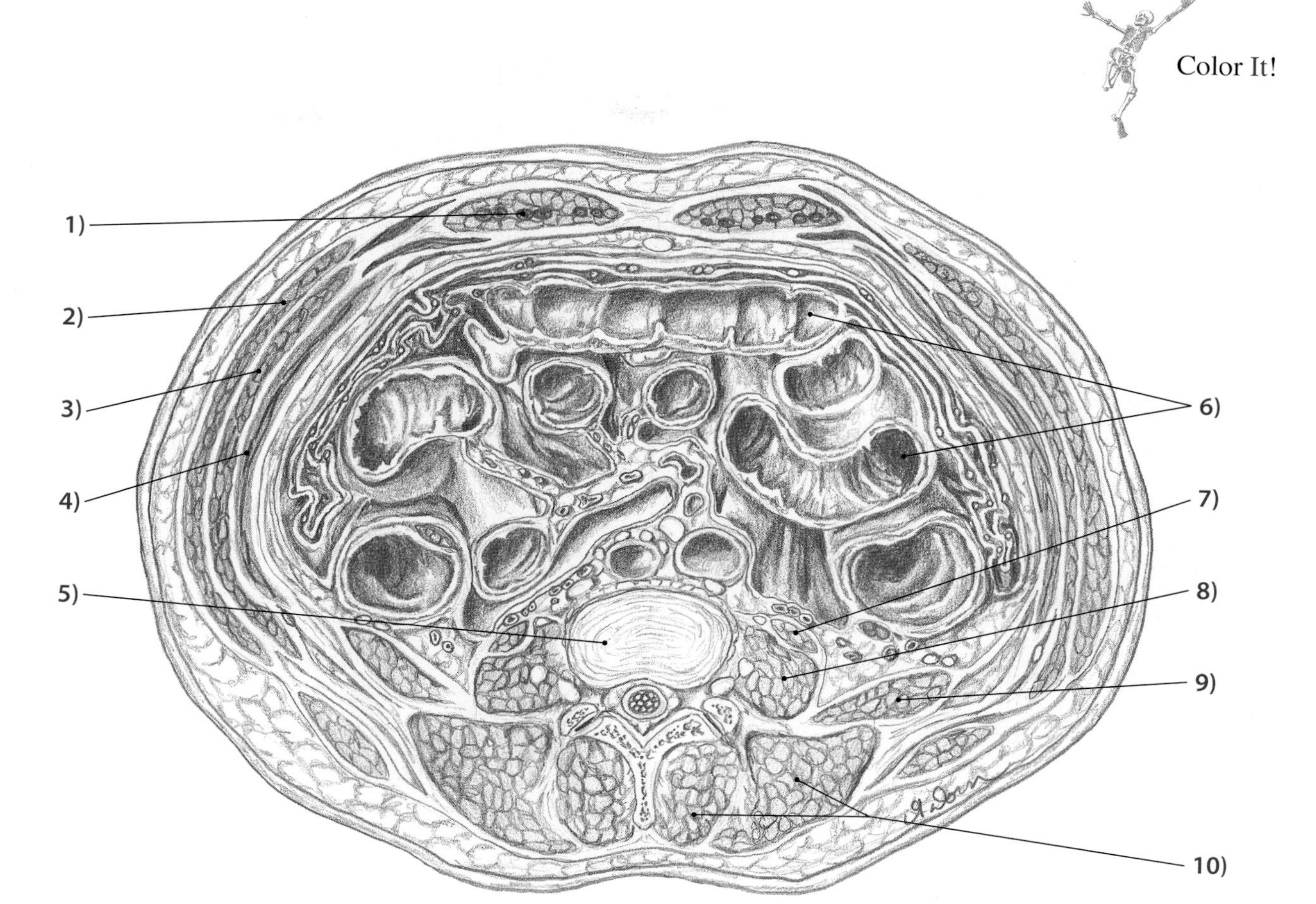

Cross section of the thorax at the level of the third lumbar vertebra

CHOICES

Body of L-3	Psoas major
Erector spinae group	Psoas minor
External oblique	Quadratus lumborum
Internal oblique	Rectus abdominis
Intestines	Transverse abdominis

1) ______________________ 6) ______________________

2) ______________________ 7) ______________________

3) ______________________ 8) ______________________

4) ______________________ 9) ______________________

5) ______________________ 10) ______________________

p. 188

Color the Muscles #1

Using different colors, please fill in and label the muscles and other structures listed below.

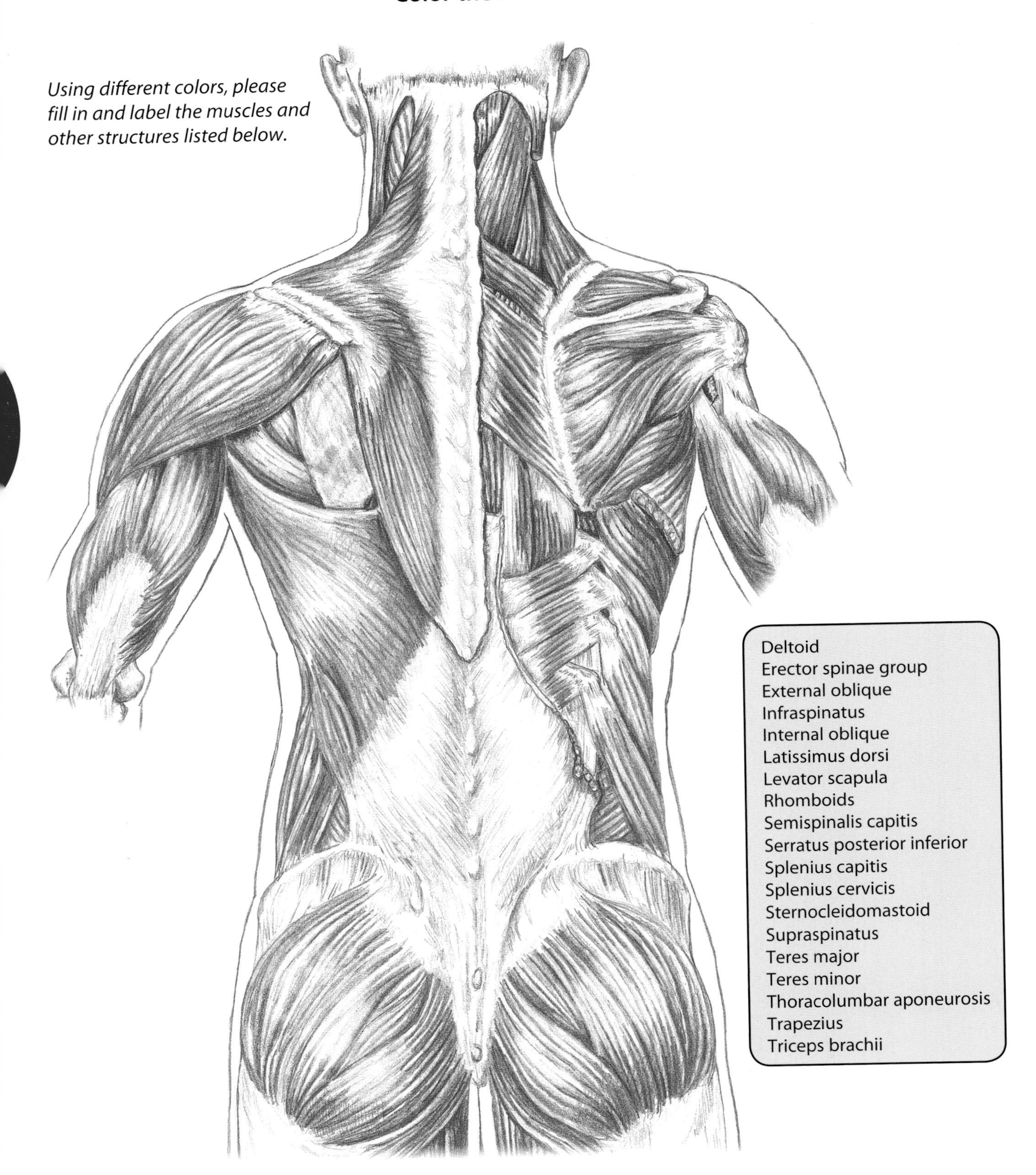

- Deltoid
- Erector spinae group
- External oblique
- Infraspinatus
- Internal oblique
- Latissimus dorsi
- Levator scapula
- Rhomboids
- Semispinalis capitis
- Serratus posterior inferior
- Splenius capitis
- Splenius cervicis
- Sternocleidomastoid
- Supraspinatus
- Teres major
- Teres minor
- Thoracolumbar aponeurosis
- Trapezius
- Triceps brachii

Posterior view, superficial muscles of the back

Spine and Thorax

Color the Muscles #2

Using different colors, please fill in and label the muscles and other structures listed below.

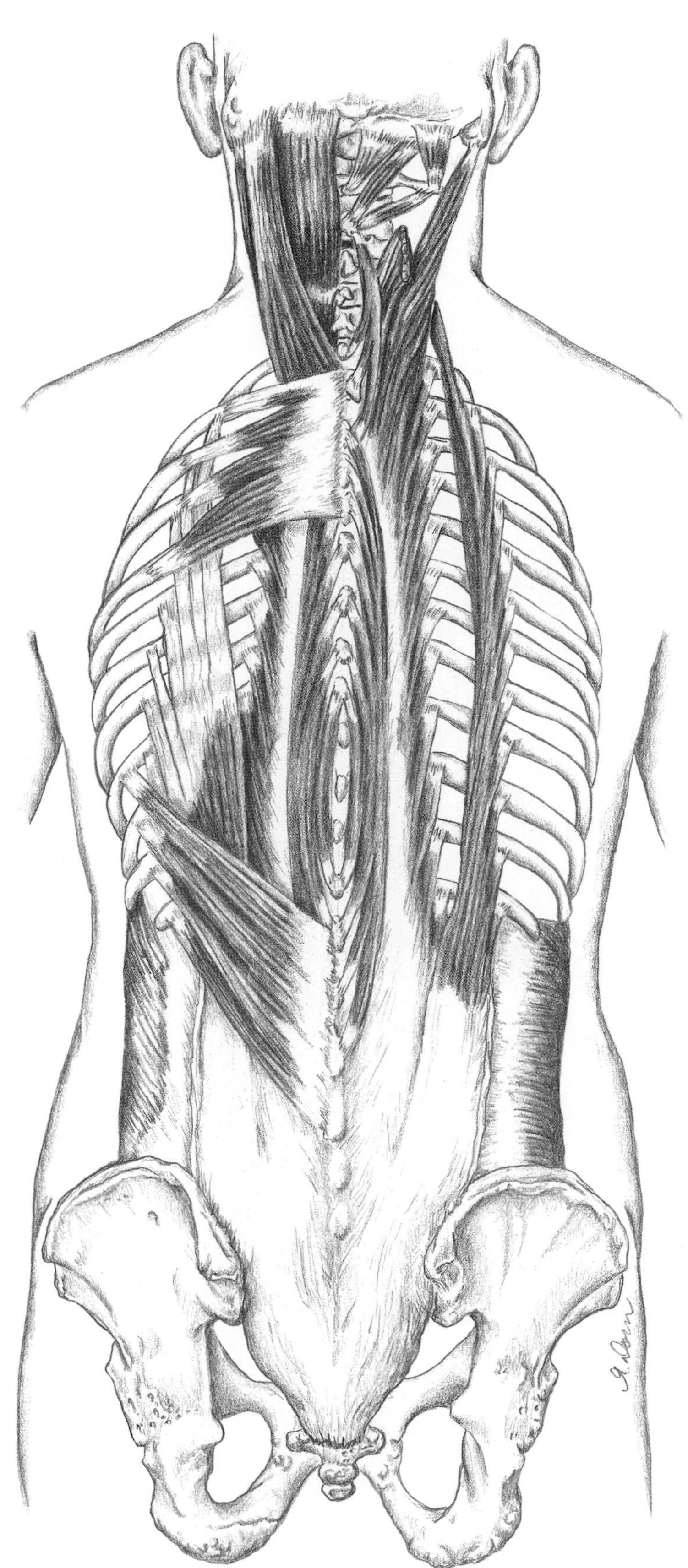

- Iliocostalis
- Internal oblique
- Longissimus
- Longissimus capitis
- Oblique capitis inferior
- Oblique capitis superior
- Rectus capitis posterior major
- Rectus capitis posterior minor
- Semispinalis capitis
- Serratus posterior inferior
- Serratus posterior superior
- Spinalis
- Spinalis cervicis
- Splenius capitis
- Thoracolumbar aponeurosis
- Transverse abdominis

Posterior view, intermediate muscles of the back

Color the Muscles #3

Using different colors, please fill in and label the muscles and other structures listed below.

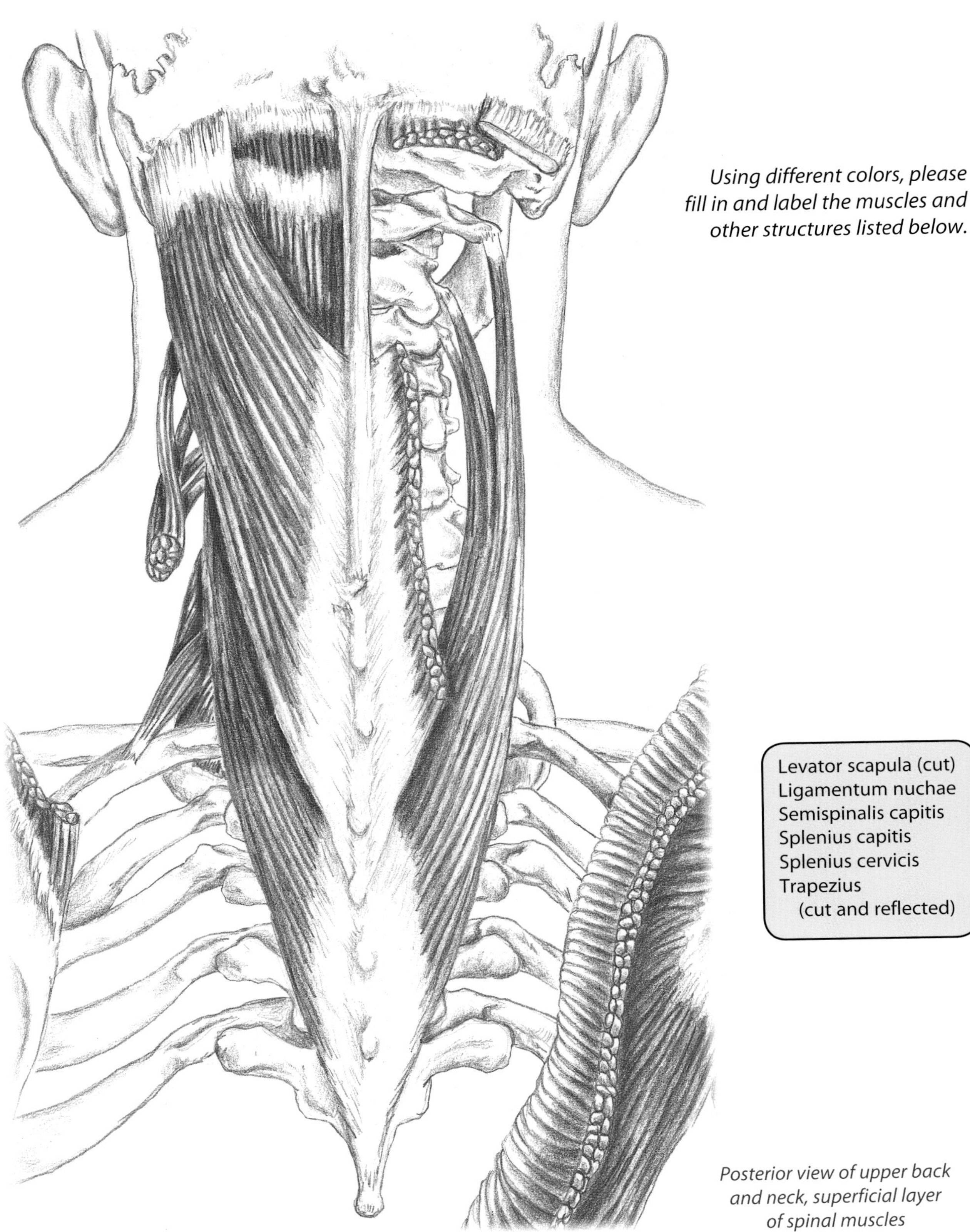

Levator scapula (cut)
Ligamentum nuchae
Semispinalis capitis
Splenius capitis
Splenius cervicis
Trapezius
(cut and reflected)

Posterior view of upper back and neck, superficial layer of spinal muscles

Color the Muscles #4

Using different colors, please fill in and label the muscles and other structures listed below.

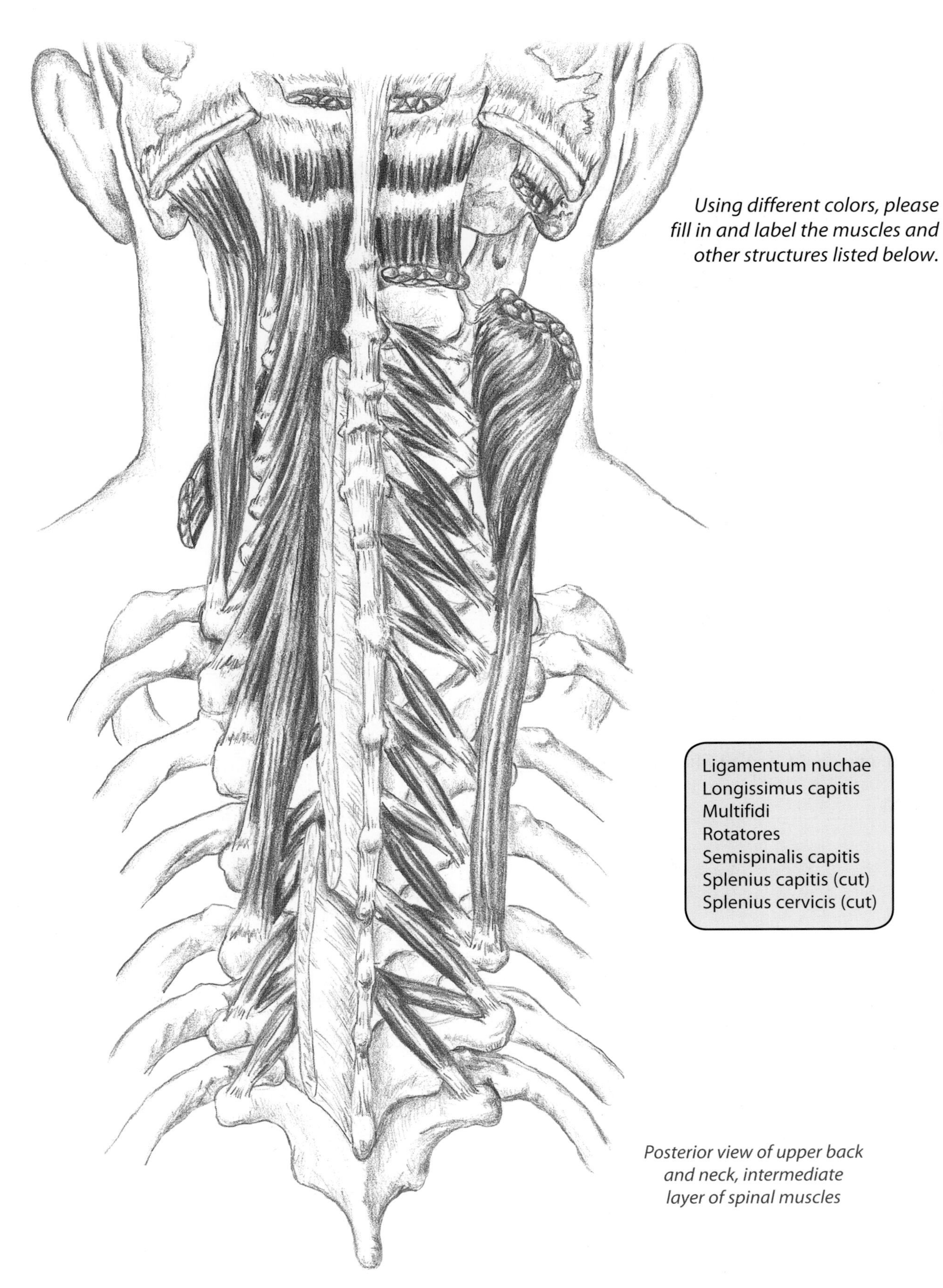

Ligamentum nuchae
Longissimus capitis
Multifidi
Rotatores
Semispinalis capitis
Splenius capitis (cut)
Splenius cervicis (cut)

Posterior view of upper back and neck, intermediate layer of spinal muscles

Please list the action demonstrated, two synergists and one antagonist.
The first letter of the muscles has been provided.

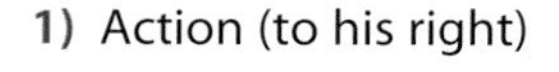

1) Action (to his right)

____

2) Synergists (and on what side—his left or right?)

E ______________________________

I ______________________________

3) Antagonist (and on what side—his left or right?)

M ______________________________

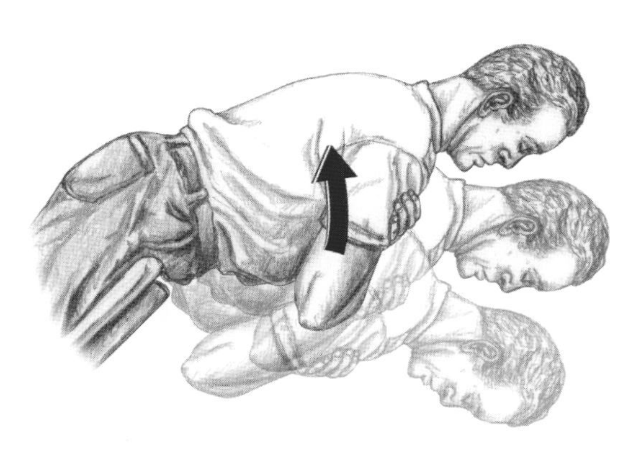

4) Action

____

5) Synergists

S ______________________________

I ______________________________

6) Antagonist

R ______________________________

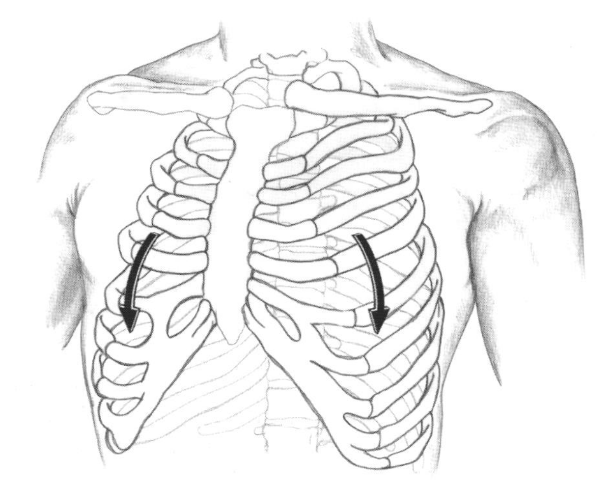

7) Action

____

8) Synergists

I ______________________________

S ______________________________

9) Antagonist

E ______________________________

Spine and Thorax

Muscles and Movements #2

Please list the action demonstrated, two synergists and one antagonist.
The first letter of the muscles has been provided.

1) Action

____

2) Synergists

E ______________________

I ______________________

3) Antagonist

Q ______________________

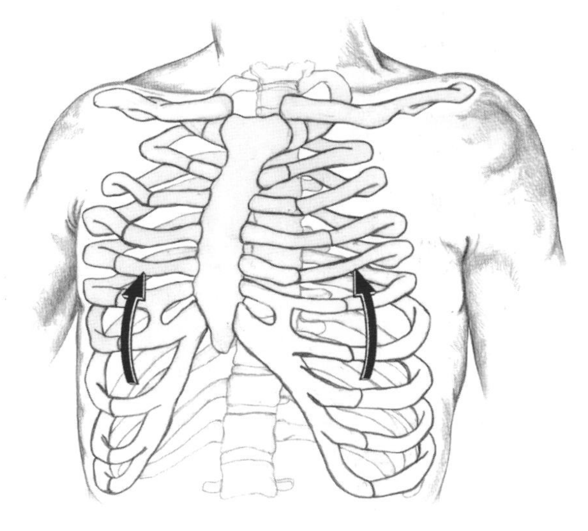

4) Action

____

5) Synergists

S ______________________

S ______________________

6) Antagonist

I ______________________

7) Action

____

8) Synergists (and on what side—his left or right?)

Q ______________________

E ______________________

9) Antagonist (and on what side—his left or right?)

S ______________________

What's the Muscle? #1

Please identify the following muscles.

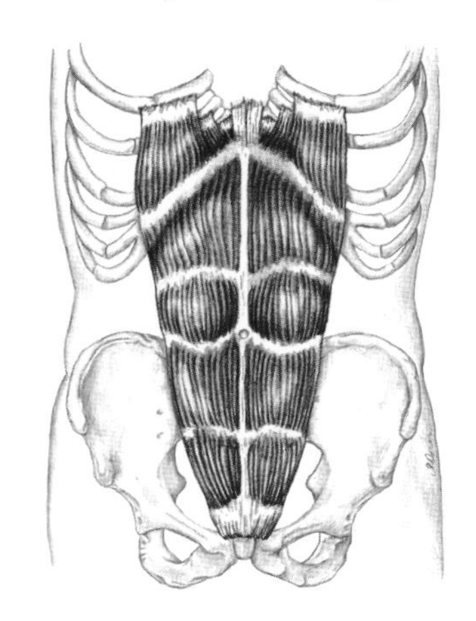

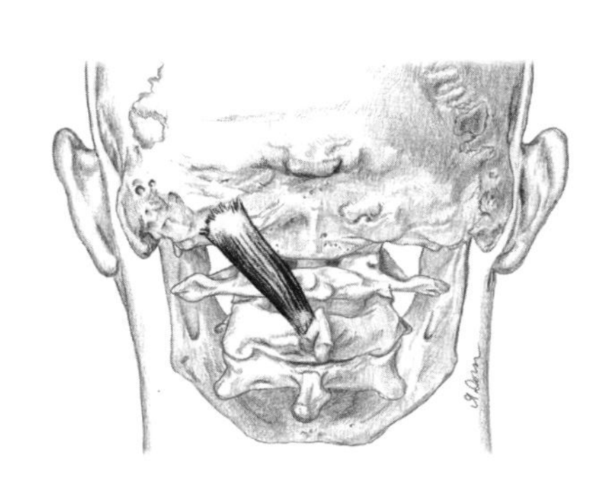

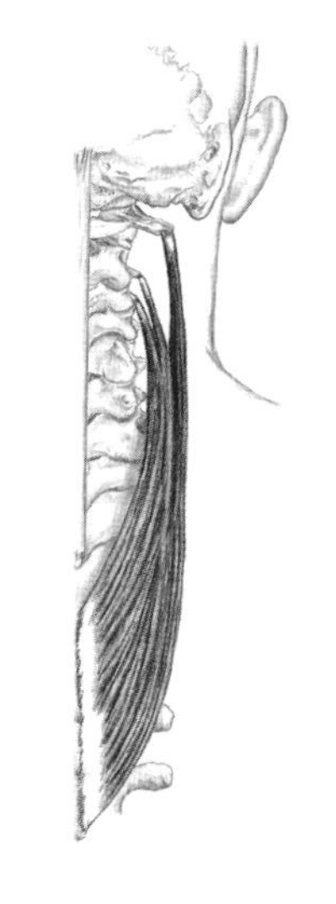

1) ______________________

2) ______________________

3) ______________________

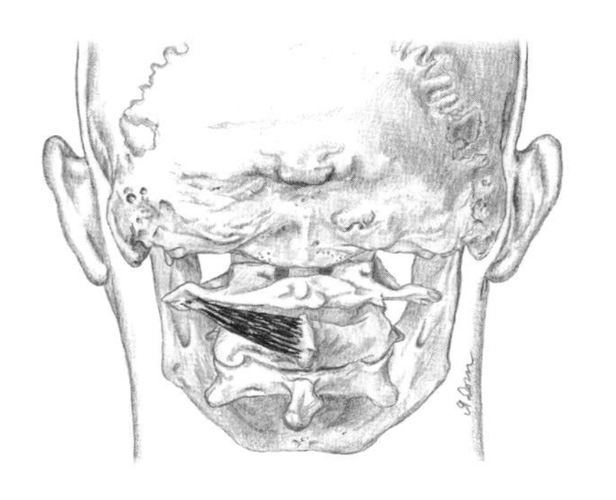

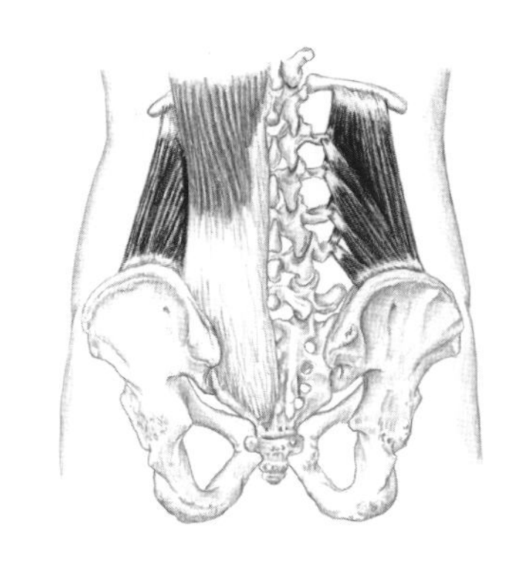

4) ______________________

5) ______________________

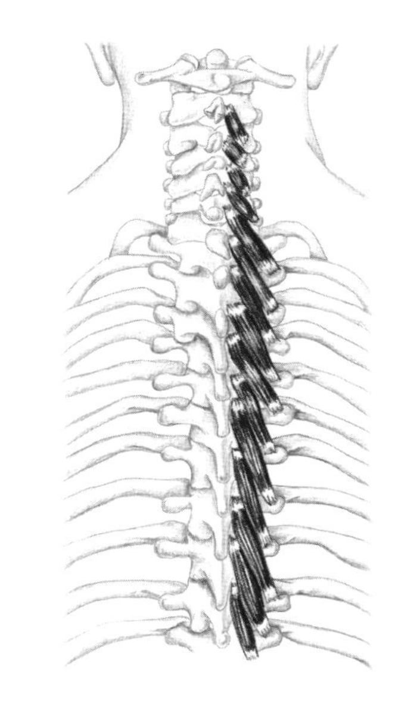

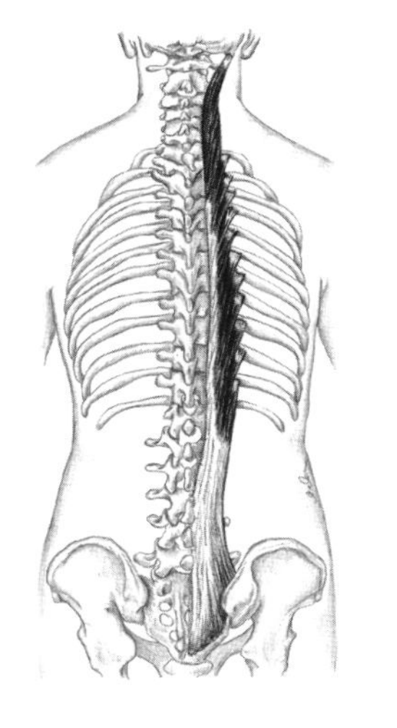

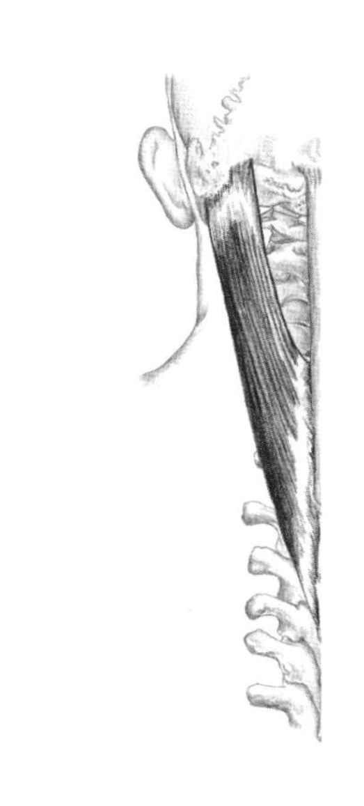

7) ______________________

6) ______________________

8) ______________________

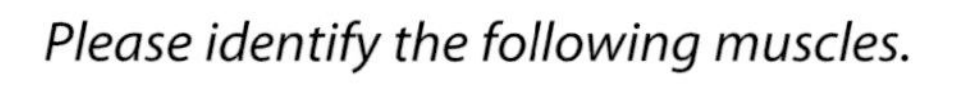

What's the Muscle? #2

Please identify the following muscles.

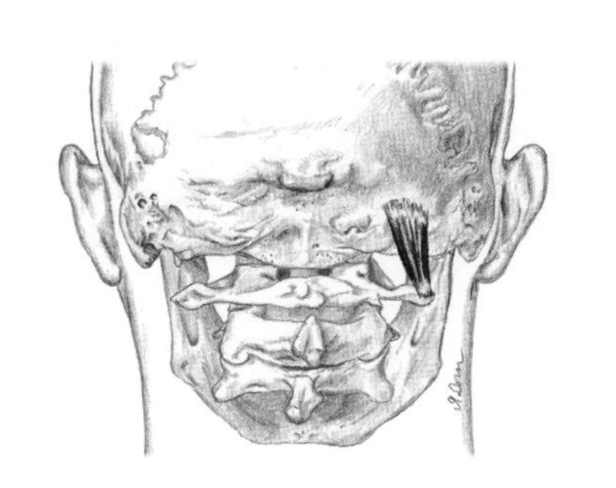

1) ____________________

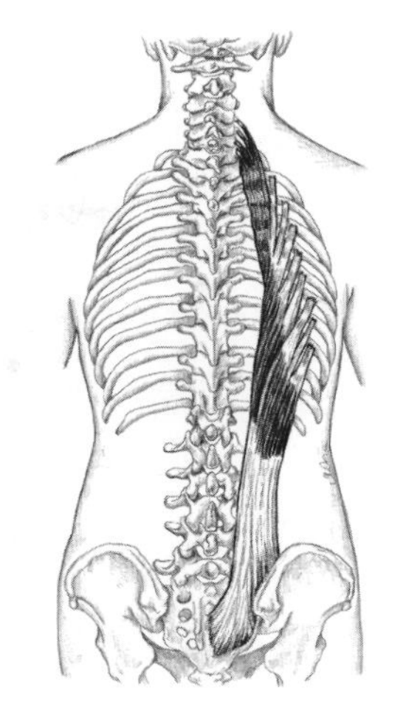

2) ____________________

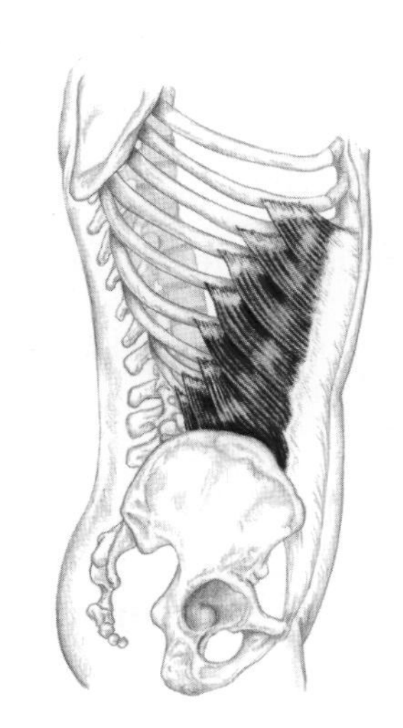

3) ____________________

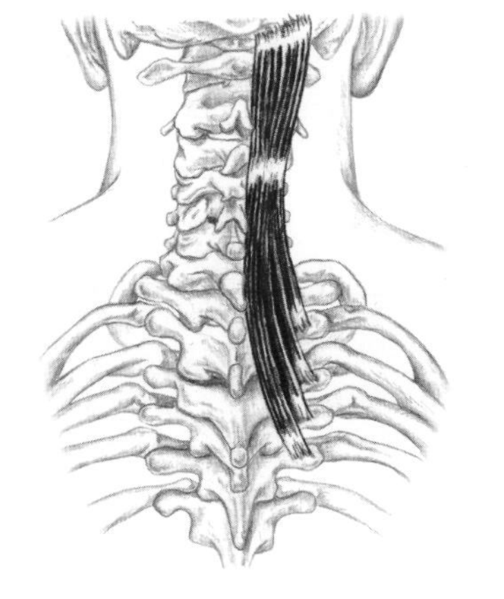

4) ____________________

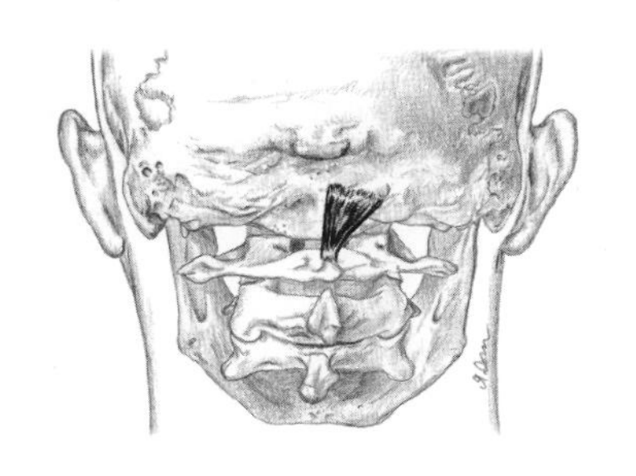

5) ____________________

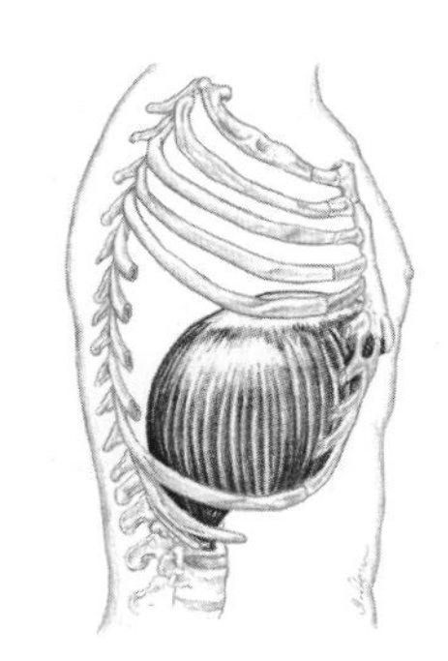

6) ____________________

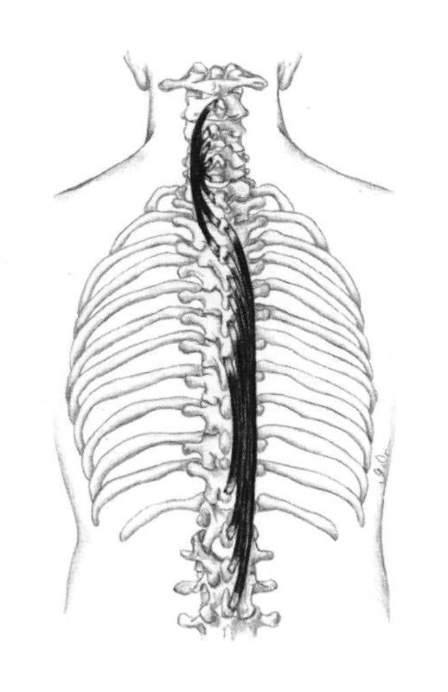

7) ____________________

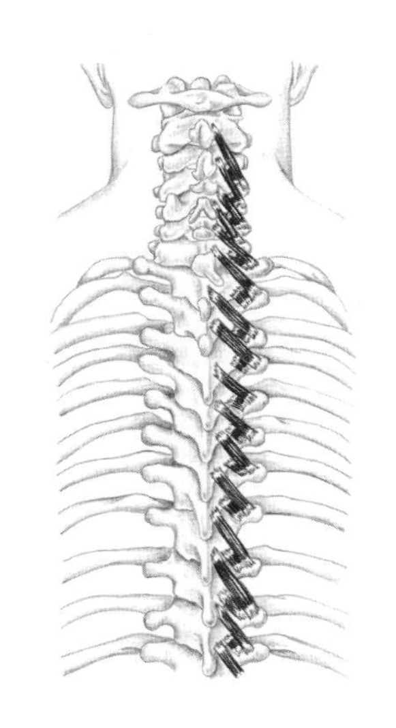

8) ____________________

Spine and Thorax, Muscle Group #1

Erector Spinae and Transversospinalis Groups

p. 196-202

Please answer the following questions.

1) The most medial branch of the erector spinae group is the ______________________________, while the most lateral is the ______________________________.

2) In the lumbar region, the erectors lie deep to what connective tissue structure? ______________________________

3) To contract the lower fibers of the erector spinae group in a prone position, you could ask your partner to perform what action?

____

4) When exploring between the scapulae, can you name two muscles through which you will have to palpate to access the deeper erector spinae fibers?

______________________________ ______________________________

5) Unlike the long, vertical erector fibers, the branches of the transversospinalis group consist of many ______________________________ fibers.

6) As a group, the transversospinalis muscles can be easily located along the ______________________________ of the thoracic and lumbar vertebrae.

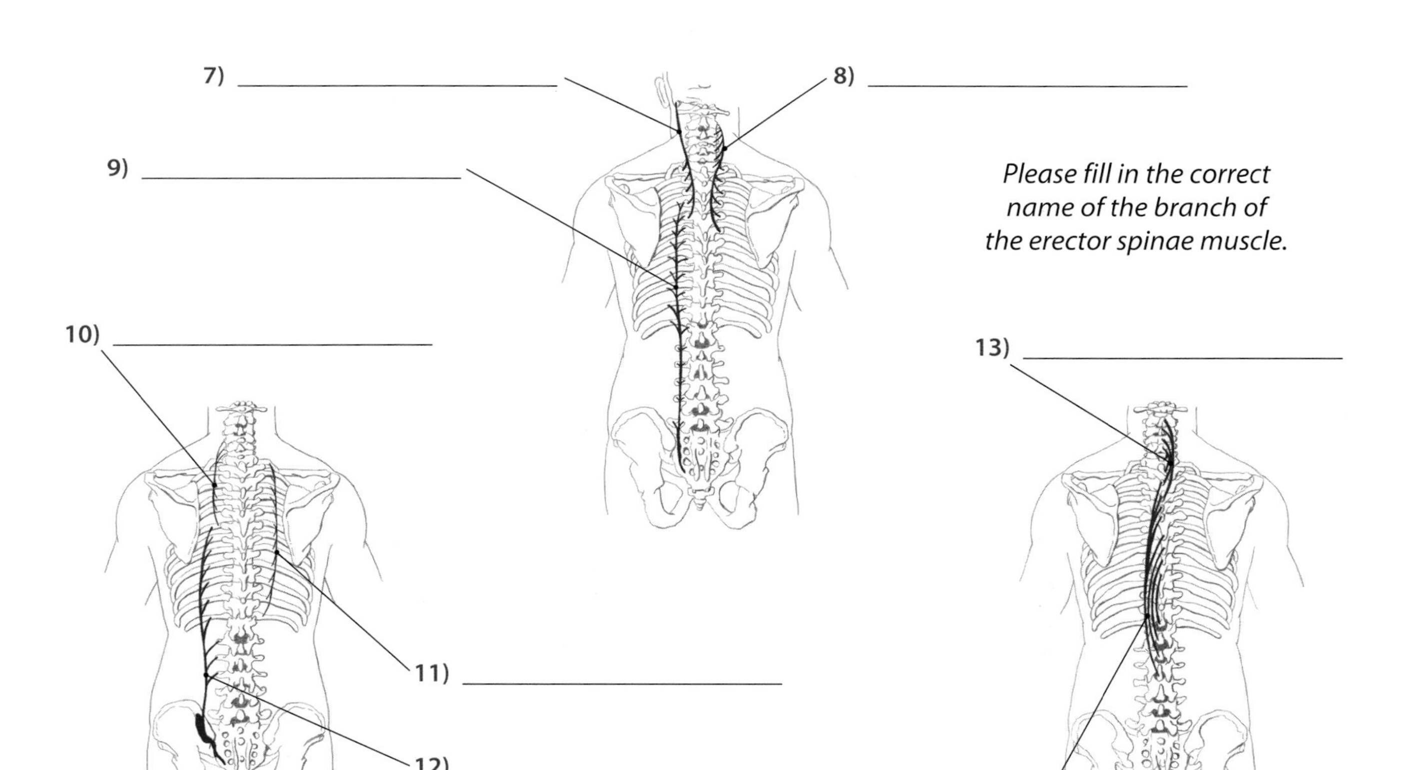

p. 196-202

Spine and Thorax, Muscle Group #1
Erector Spinae and Transversospinalis Groups

Matching
Match the origin and insertion to the correct muscle.

Origins
1) Common tendon (lumborum), posterior surface of ribs 1-12 (thoracis and cervicis)
2) Common tendon (thoracis), TVPs of upper five thoracic vertebrae (cervicis and capitis)
3) Sacrum and TVPs of lumbar through cervical vertebrae
4) Spinous processes of the upper lumbar and lower thoracic vertebrae (thoracis), ligamentum nuchae, spinous process of C-7 (cervicis)
5) TVPs of lumbar through cervical vertebrae
6) Transverse processes of C-4 to T-5

Muscle	O	I
Iliocostalis	______	______
Longissimus	______	______
Multifidi	______	______
Rotatores	______	______
Semispinalis capitis	______	______
Spinalis	______	______

Insertions
7) Lower 9 ribs and TVPs of thoracic vertebrae (thoracis), TVPs of cervical vertebrae (cervicis), mastoid process (capitis)
8) Spinous processes of lumbar vertebrae through second cervical vertebra (each belly spanning 1-2 vertebrae)
9) Spinous processes of lumbar vertebrae through second cervical vertebra (each belly spanning 2-4 vertebrae)
10) Spinous processes of upper thoracic (thoracis), spinous processes of cervicals, except C-1 (cervicis)
11) Between the superior and inferior nuchal lines of the occiput
12) TVPs of lumbar vertebrae 1-3 and posterior surface of ribs 6-12 (lumborum), posterior surface of ribs 1-6 (thoracis), TVPs of lower cervicals (cervicis)

Shorten or Lengthen?

13) Passive flexion of the spine would ____________________ the iliocostalis.

14) Passive rotation of the spine to the opposite side would ____________________ the rotatores.

15) Passive lateral flexion of the spine to the same side would ____________________ the longissimus.

16) Passive rotation of the spine to the same side would ____________________ the multifidi.

Let's Palpate!
Remember—there are no right or wrong answers here

Locate and explore the **erector spinae group** on three individuals. Then write three words that describe what you feel. (See p. 196-199 in *Trail Guide.*)

Person #1 ____________	Person #2 ____________	Person #3 ____________
____________________	____________________	____________________
____________________	____________________	____________________
____________________	____________________	____________________

Splenii and Suboccipitals

Please answer the following questions.

1) Rotating the head to the left demands the contraction of which splenius capitis—the left or right?

____

2) The splenius capitis is a deep muscle except on the lateral side of the neck where it is superficial between which two muscles?

_______________________________ _______________________________

3) To distinguish the trapezius fibers from the splenius capitis fibers, you could ask your partner to perform what action?

____

4) How can the upper fibers of the trapezius be helpful in locating the suboccipitals?

____

5) What two bony landmarks and one region can be helpful to isolate the location of the suboccipitals?

_______________________________ _______________________________ _______________________________

Please draw the four suboccipital muscles reflecting their correct origins and insertions.

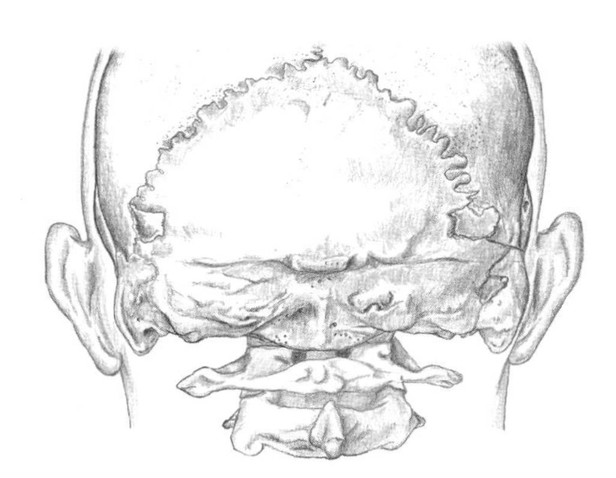

1) Rectus capitis posterior major

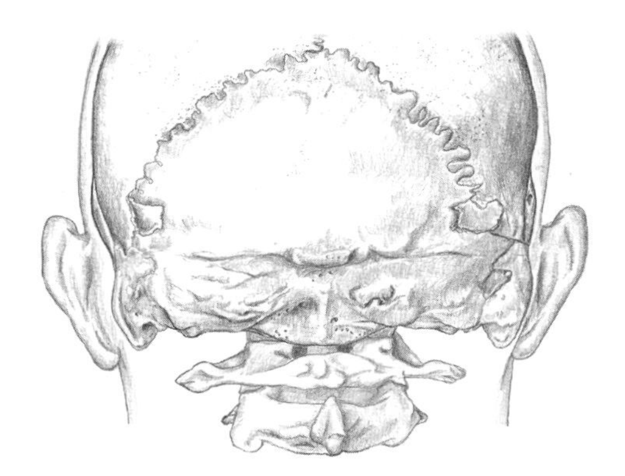

3) Oblique capitis superior

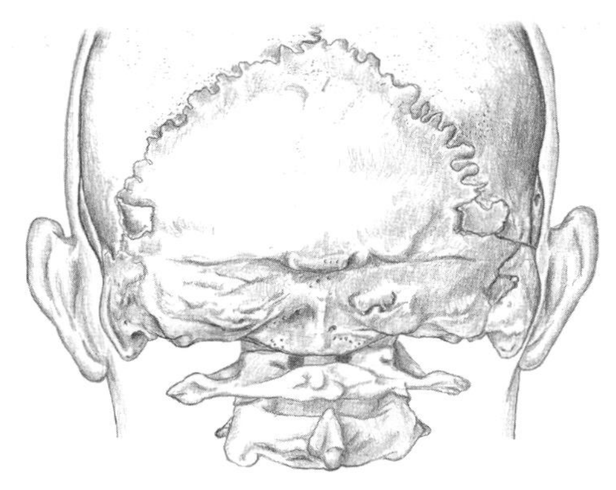

2) Rectus capitis posterior minor

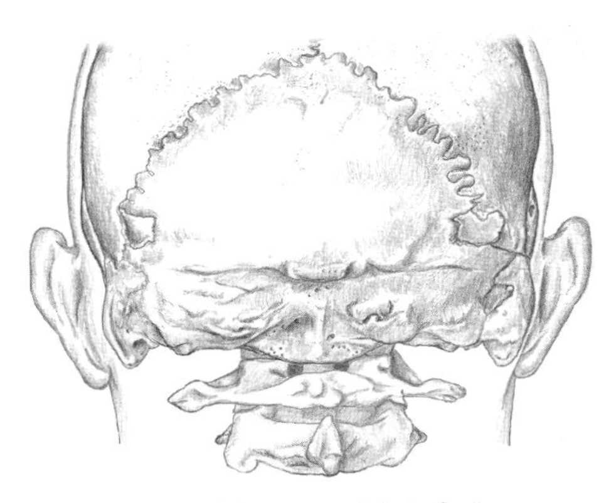

4) Oblique capitis inferior

p. 203-206

Spine and Thorax, Muscle Group #2

Splenii and Suboccipitals

Matching

Match the origin and insertion to the correct muscle.

Origins

1) Inferior one-half of ligamentum nuchae and spinous processes of C-7 to T-4
2) Spinous process of the axis (C-2) (2)
3) Spinous processes of T-3 to T-6
4) Transverse process of the atlas (C-1)
5) Tubercle of the posterior arch of the atlas (C-1)

Insertions

6) Between the nuchal lines of the occiput
7) Inferior nuchal line of the occiput (2)
8) Mastoid process and lateral portion of superior nuchal line
9) Transverse process of the atlas (C-1)
10) Transverse processes of C-1 to C-3

Muscle	O	I
Oblique capitis inferior	______	______
Oblique capitis superior	______	______
Rectus capitis posterior major	______	______
Rectus capitis posterior minor	______	______
Splenius capitis	______	______
Splenius cervicis	______	______

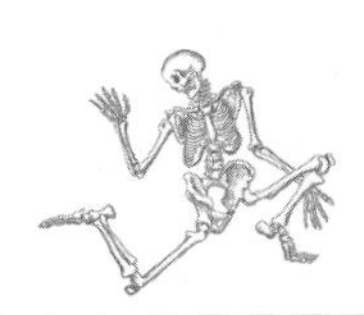

Let's Palpate!

Remember—there are no right or wrong answers here

Locate and explore the **splenius capitis** on three individuals. Then write three words that describe what you feel. (See p. 203-204 in *Trail Guide*.)

Person #1 ____________

____

____

____

Person #2 ____________

____

____

____

Person #3 ____________

____

____

____

Shorten or Lengthen?

11) Passive rotation of the head to the same side would ________________ the rectus capitis posterior major.
12) Passive lateral flexion of the head and neck would ________________ the splenius cervicis.
13) Passive extension of the head and neck would ________________ the splenius capitis.
14) Passive rotation of the head to the opposite side would ________________ the oblique capitis inferior.
15) Passive rotation of the head and neck to the opposite side would ________________ the splenius capitis.

Spine and Thorax, Muscle Group #3

QL, Abdominals, Diaphragm and Intercostals

p. 207-215

Please answer the following questions.

1) Which edge of the quadratus lumborum is accessible from the side of the torso? ______________________

2) Which three bony landmarks can help you to isolate the borders of the quadratus lumborum?

______________________ ______________________ ______________________

3) What action could you ask your partner to perform to feel the quadratus lumborum contract?

____

4) Which abdominal muscle runs vertically from the rib cage to the pubic crest? ______________________

5) Rotating your trunk to the right would engage your left or right internal oblique muscle? ______________________

6) You are palpating lateral to the edge of rectus abdominis and the fibers you feel are superficial and running at an angle. Which muscle is this? ______________________

7) What is the primary muscle of respiration? ______________________

8) When the diaphragm muscle fibers contract, what connective tissue structure is pulled inferiorly?

____

9) When is it best to move your fingers as you curl them underneath the rib cage to feel the diaphragm?

____

10) Name two large muscles through which you would have to palpate to access various portions of the intercostals.

______________________ ______________________

Please identify the following structures.

11) ______________________

12) ______________________

13) ______________________

14) ______________________

15) ______________________

16) ______________________

17) ______________________

18) ______________________

Color It!

Spine and Thorax, Muscle Group #3
QL, Abdominals, Diaphragm and Intercostals

Matching

Match the origin and insertion to the correct muscle.

Origins

1) Inner surface of lower six ribs, upper two or three lumbar vertebrae and inner part of xiphoid process
2) Lateral inguinal ligament, iliac crest and thoracolumbar fascia
3) Lateral inguinal ligament, iliac crest, thoracolumbar fascia and internal surface of lower six ribs
4) External surfaces of fifth to twelfth ribs
5) Posterior iliac crest
6) Pubic crest, pubic symphysis

Insertions

7) Abdominal aponeurosis to linea alba
8) Anterior part of the iliac crest, abdominal aponeurosis to linea alba
9) Cartilage of fifth, sixth and seventh ribs and xiphoid process
10) Central tendon
11) Internal surface of lower three ribs, abdominal aponeurosis to linea alba
12) Last rib and transverse processes of first through fourth lumbar vertebrae

Muscle	O	I
Diaphragm	_______	_______
External oblique	_______	_______
Internal oblique	_______	_______
Quadratus lumborum	_______	_______
Rectus abdominis	_______	_______
Transverse abdominis	_______	_______

Shorten or Lengthen?

13) Increasing the volume of the thoracic cavity would ____________________ the diaphragm's fibers.
14) Passive rotation of the vertebral column to the same side would ____________________ the external oblique.
15) Compression of the abdominal contents would ____________________ the transverse abdominis.
16) Drawing the ventral part of the ribs upward would ____________________ the external intercostals.
17) Passive rotation of the vertebral column to the opposite side would ____________________ the internal oblique.

Other Structures

p. 218-220

Please answer the following questions.

1) The ligamentum nuchae spans between which two bony landmarks?

_______________ _______________

2) To feel the ligamentum nuchae change tension underneath your fingers, what two passive movements can you perform at the head? _______________

3) What superficial ligament can be felt between the spinous processes of the thoracic and lumbar vertebrae?

____

4) The abdominal aorta is located where in relationship to the psoas major? _______________

5) The thoracolumbar aponeurosis serves as an attachment site for which two muscles?

_______________ _______________

Let's Palpate!

Remember—there are no right or wrong answers here

Locate and explore the **quadratus lumborum** on three individuals. Then write three words that describe what you feel. (See p. 207-208 in *Trail Guide*.)

Person #1 ________	Person #2 ________	Person #3 ________
________	________	________
________	________	________
________	________	________

Craniovertebral Joints #1

Please identify the following structures.

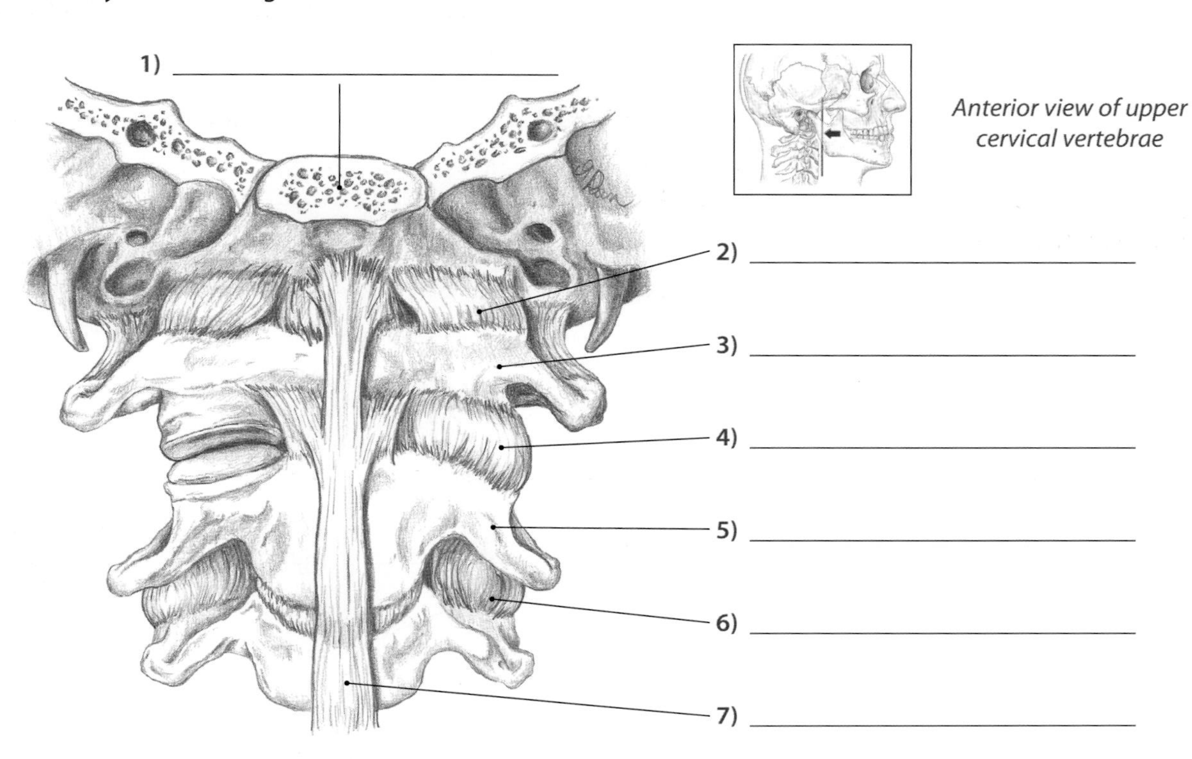

Anterior view of upper cervical vertebrae

CHOICES

Alar ligaments	Capsule of lateral atlantoaxial joint
Anterior longitudinal ligament	Capsule of zygapophyseal (lateral) joint
Atlas (2)	Cruciform ligament
Axis (2)	Inferior longitudinal fibers
Basilar portion of occiput	Superior longitudinal fibers
Capsule of atlantooccipital joint	Transverse ligament of atlas

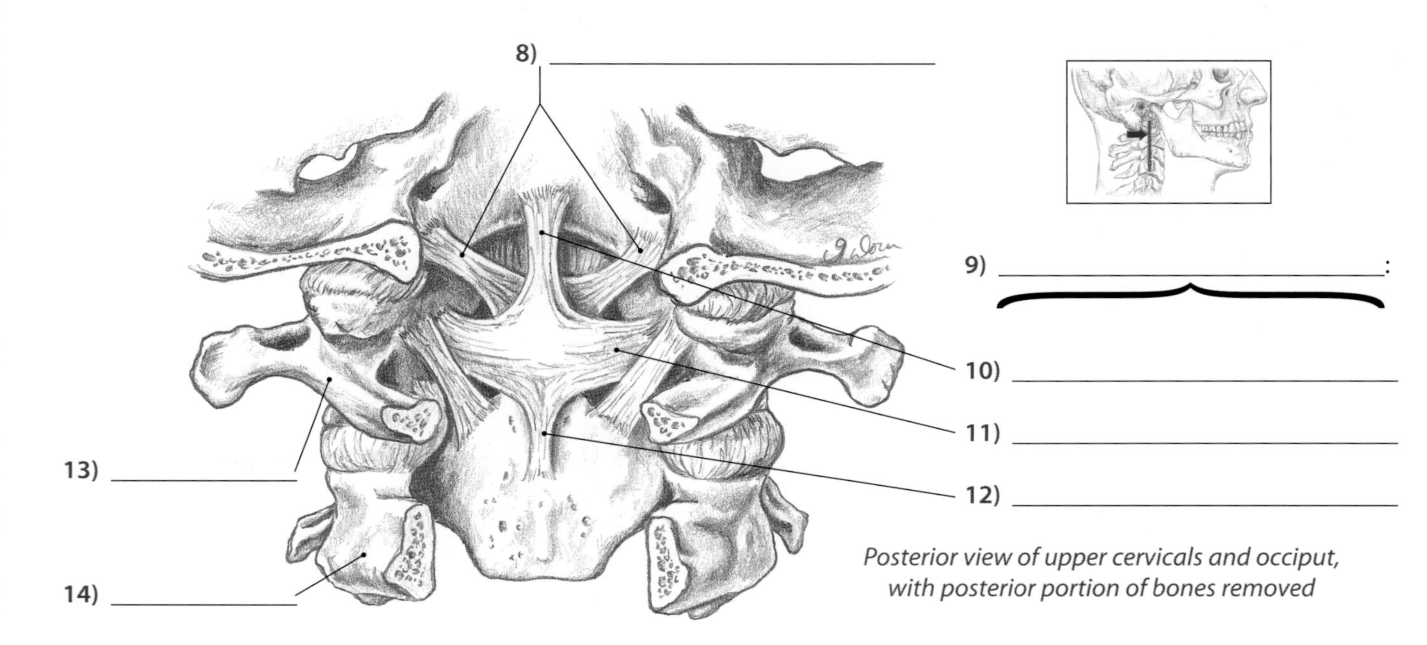

Posterior view of upper cervicals and occiput, with posterior portion of bones removed

Craniovertebral Joints #2

Please identify the following structures.

Color Them!

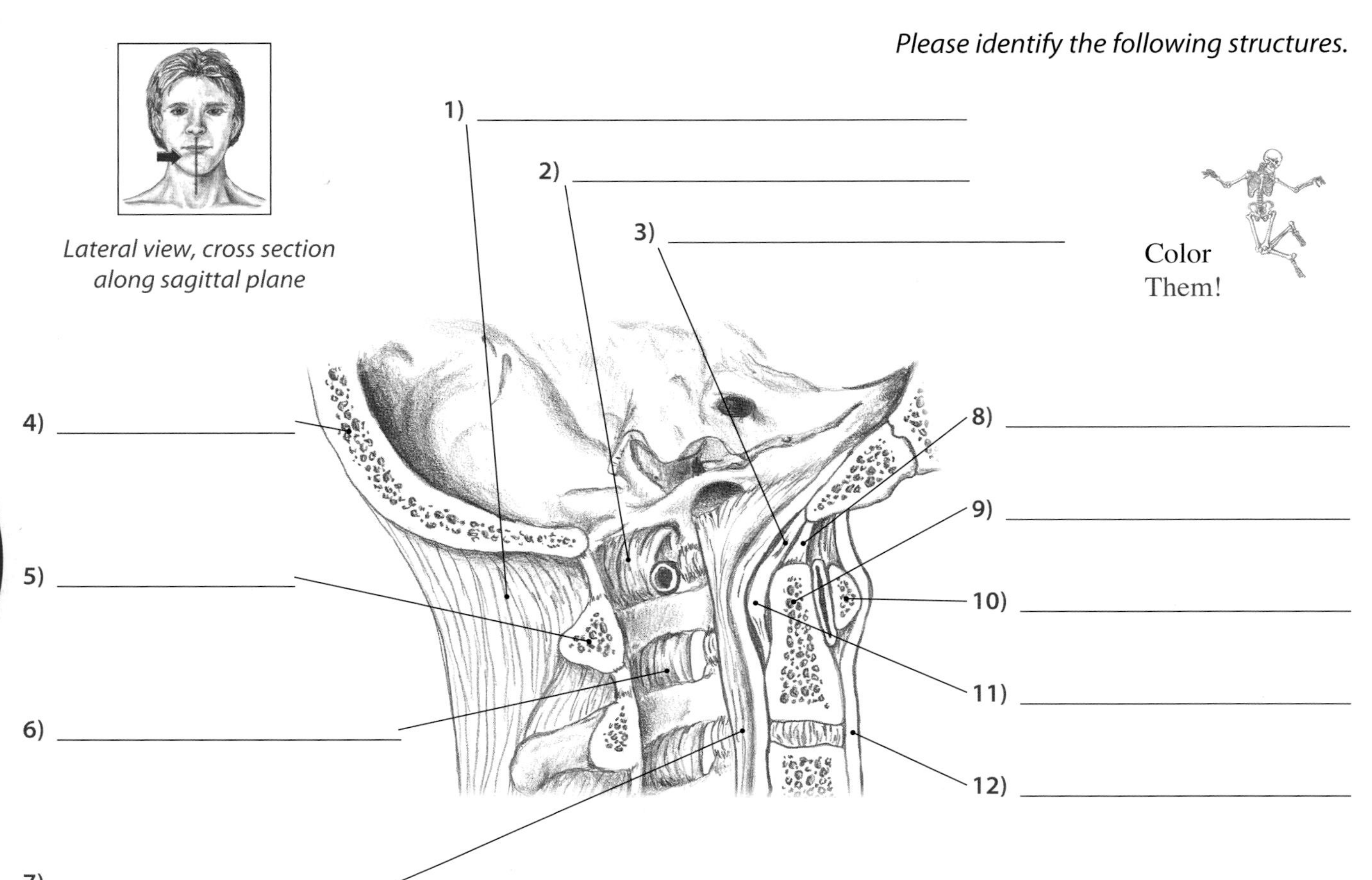

Lateral view, cross section along sagittal plane

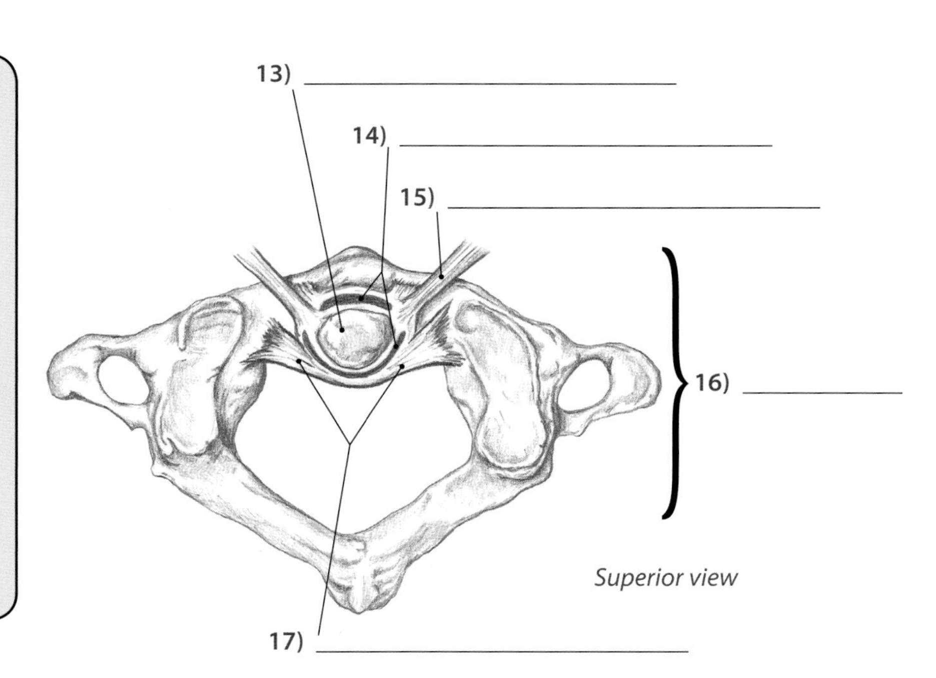

Superior view

CHOICES

- Alar ligament
- Anterior longitudinal ligament
- Anterior tubercle of atlas
- Apical ligament
- Atlas
- Ligamentum nuchae
- Occiput
- Odontoid process of axis (2)
- Posterior atlantoaxial membrane
- Posterior atlantooccipital membrane
- Posterior longitudinal ligament
- Posterior tubercle of atlas
- Superior longitudinal fibers of cruciform ligament
- Synovial cavities
- Transverse ligament of atlas (2)

Intervertebral Joints

Please identify the following structures.

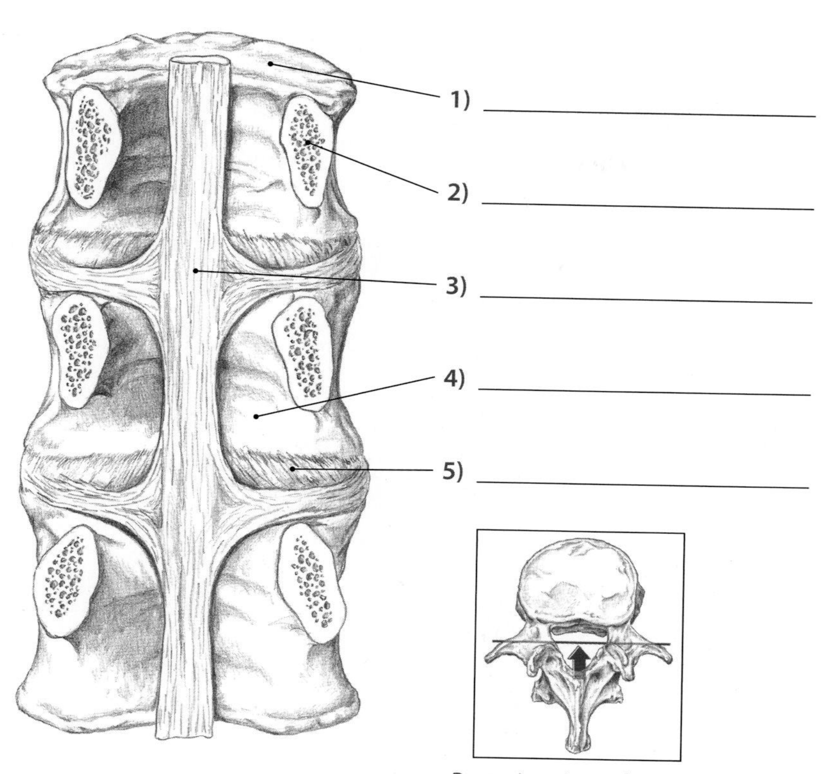

Posterior view of anterior portions of vertebrae

CHOICES
- Body of vertebra
- Inferior articular facet
- Intervertebral disc
- Lamina
- Ligamentum flavum
- Pedicle (cut) (2)
- Posterior longitudinal ligament
- Posterior surface of vertebral body
- Superior articular process
- Transverse process

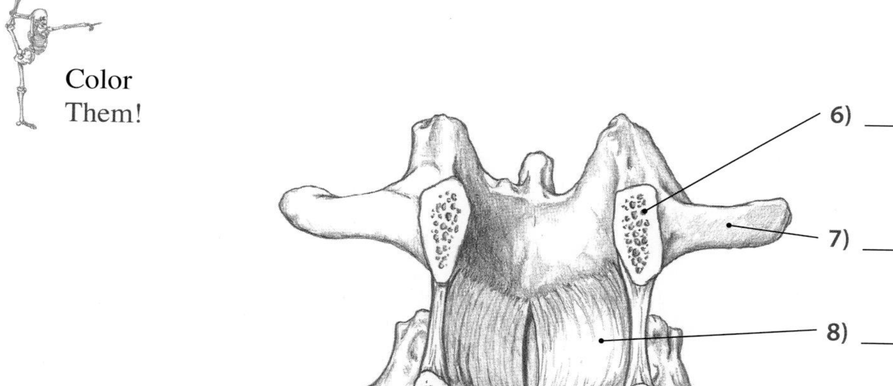

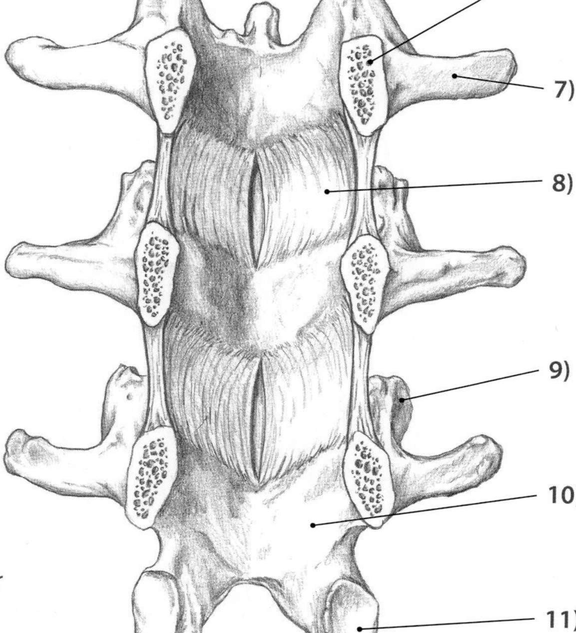

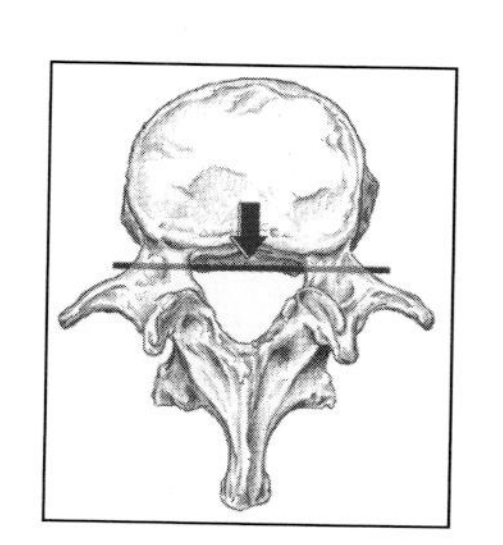

Anterior view of posterior portion of vertebrae

Spine and Thorax

Costovertebral and Intervertebral Joints

Please identify the following structures.

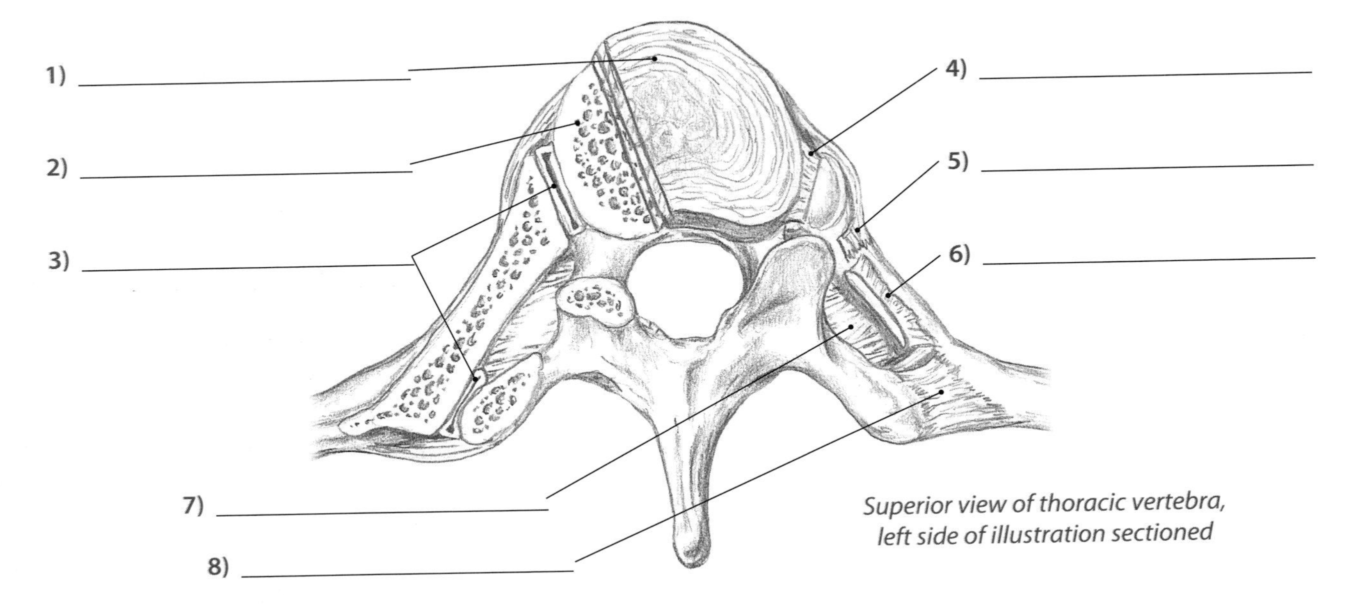

Superior view of thoracic vertebra, left side of illustration sectioned

CHOICES

- Anterior longitudinal ligament
- Body of vertebra (2)
- Costotransverse ligament
- Interarticular ligament
- Interspinous ligament
- Intervertebral disc (2)
- Intervertebral foramen
- Lateral costotransverse ligament
- Ligamentum flavum
- Posterior longitudinal ligament
- Radiate ligament
- Spinous process
- Superior costotransverse ligament (cut)
- Supraspinous ligament
- Synovial cavities
- Transverse process

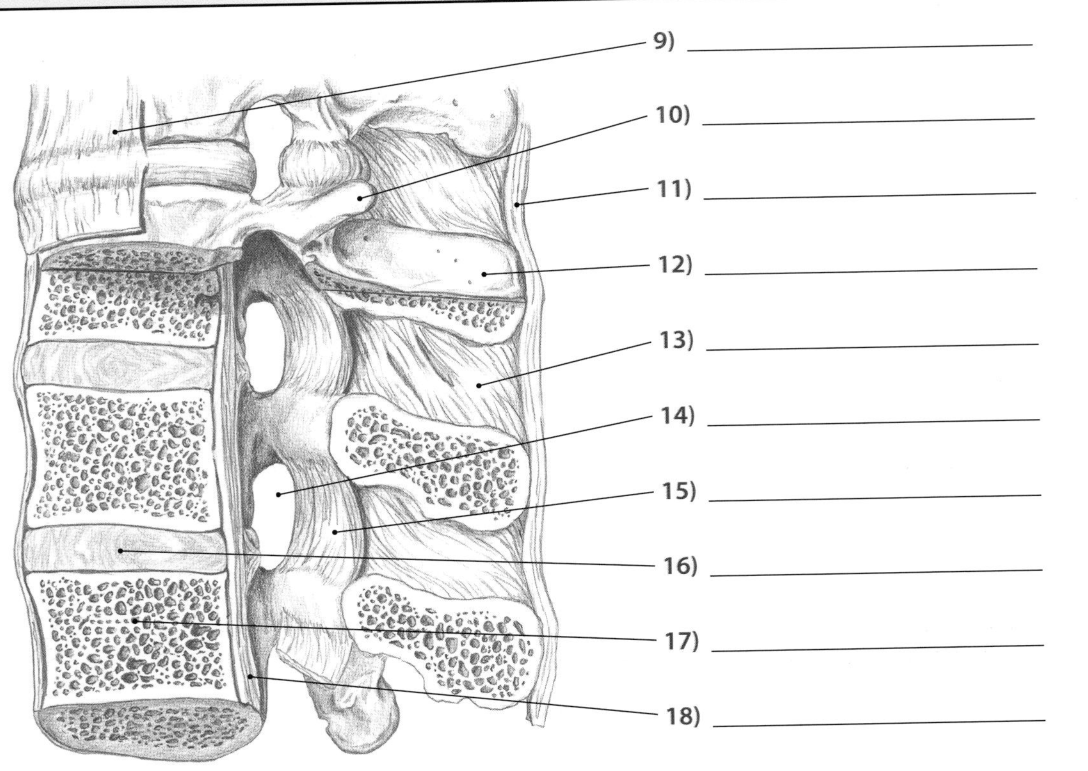

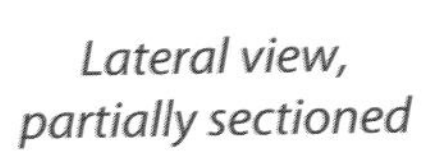

Lateral view, partially sectioned

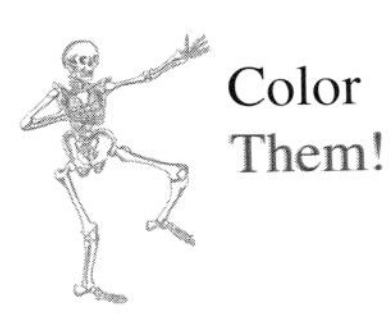

Spine and Thorax

Costovertebral and Sternocostal Joints

Please identify the following structures.

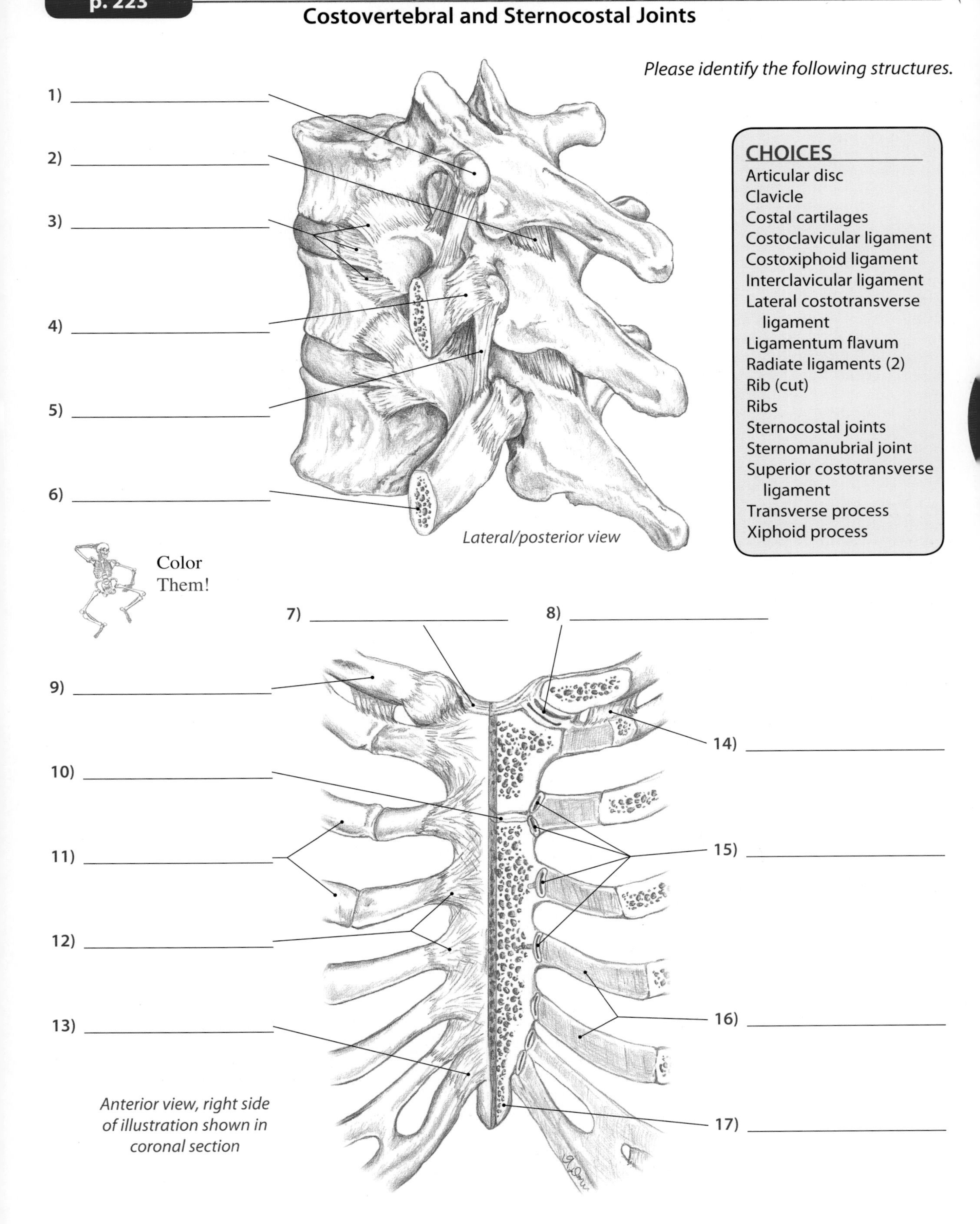

Anterior view, right side of illustration shown in coronal section

p. 226

Head, Neck and Face

Topographical View

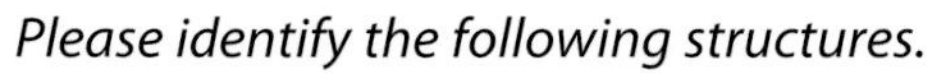

Please identify the following structures.

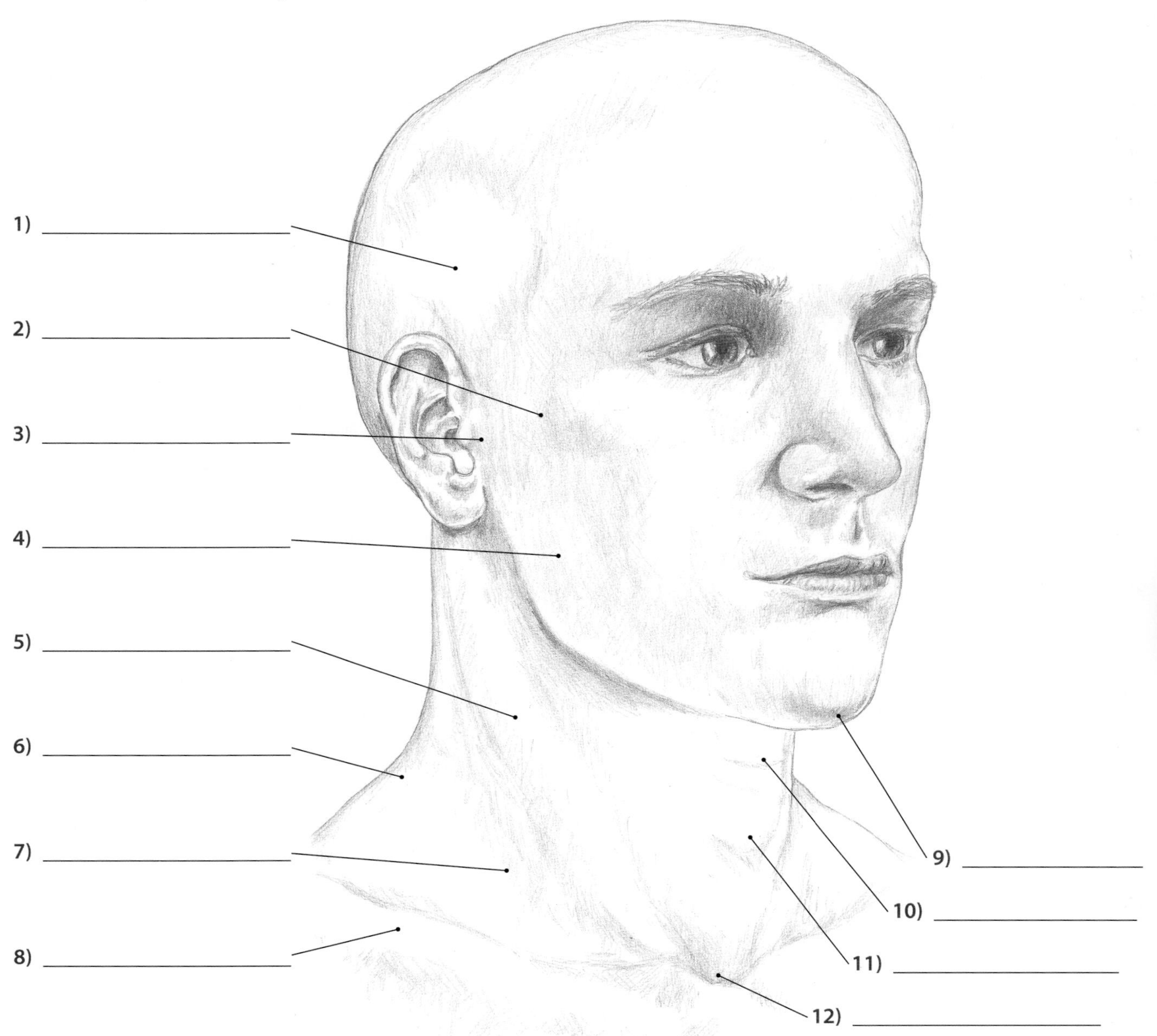

CHOICES

Base of the mandible	Scalenes
Clavicle	Sternocleidomastoid
Condyle of the mandible	Temporalis
Hyoid bone	Thyroid cartilage
Jugular notch	Trapezius
Masseter	Zygomatic arch

Head, Neck and Face
Bones and Bony Landmarks

Please answer the following questions.

1) What three landmarks create the borders of the neck's anterior triangle? ____________

____________ ____________

2) The sternocleidomastoid, clavicle and trapezius form the ____________ of the neck.

3) How many bones compose the skull? ____________

4) The cranial bones are connected by ____________ joints which form tight-fitting sutures.

5) The ____________ is located at the posterior and inferior aspects of the cranium.

6) Located at the center of the occiput, the ____________ is the superior attachment site for the ligamentum nuchae.

7) Which bony landmark of the occiput serves as an attachment site for several neck muscles?

____

8) The ____________ bones merge at the body's midline to form the sagittal suture.

9) Which bony landmark is located directly behind the earlobe and serves as an attachment site for the sternocleidomastoid? ____________

10) The space between the zygomatic arch and the cranium is filled by the ____________ muscle.

11) The ____________ bone forms the forehead and upper rim of the eye sockets.

12) Which bony landmark is located on the underside of the mandible and acts as an attachment site for the suprahyoid muscles? ____________

13) While palpating the mandible, in which area should one use extra sensitivity? ____________

Let's Palpate!

Remember—there are no right or wrong answers here

Locate and explore the **external occipital protuberance and superior nuchal lines** on three individuals. Then write three words that describe what you feel. (See p. 231-232 in *Trail Guide*.)

Person #1 ________	Person #2 ________	Person #3 ________
________	________	________
________	________	________
________	________	________

Head, Neck and Face

Skull #1

Please identify the following structures.

1) ______________________
2) ______________________
3) ______________________
4) ______________________
5) ______________________
6) ______________________
7) ______________________
8) ______________________
9) ______________________
10) ______________________
11) ______________________

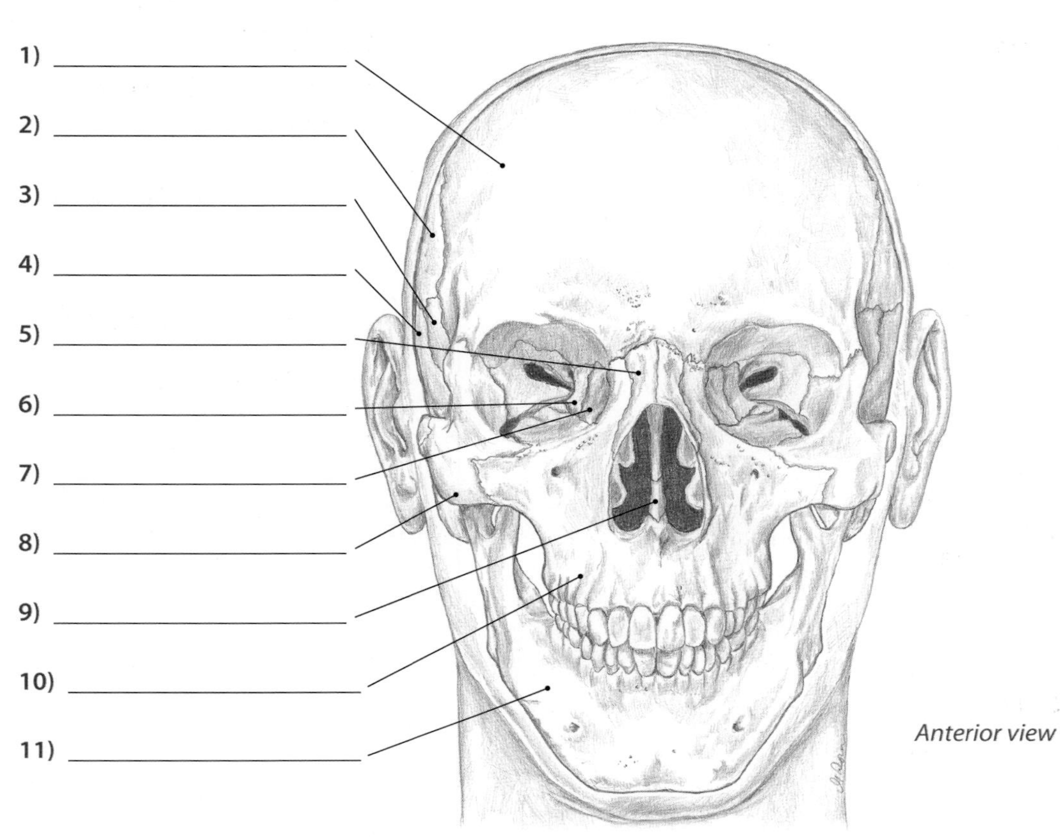

Anterior view

CHOICES
- Ethmoid
- External occipital protuberance
- Frontal
- Lacrimal
- Lambdoid suture
- Mandible (2)
- Mastoid process
- Maxilla (2)
- Nasal
- Occiput
- Parietal (2)
- Sagittal suture
- Sphenoid
- Superior nuchal line
- Temporal
- Vomer
- Zygomatic

Color Them

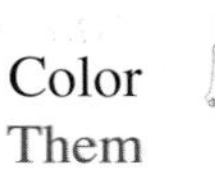

12) ______________________
13) ______________________
14) ______________________
15) ______________________
16) ______________________
17) ______________________
18) ______________________
19) ______________________
20) ______________________

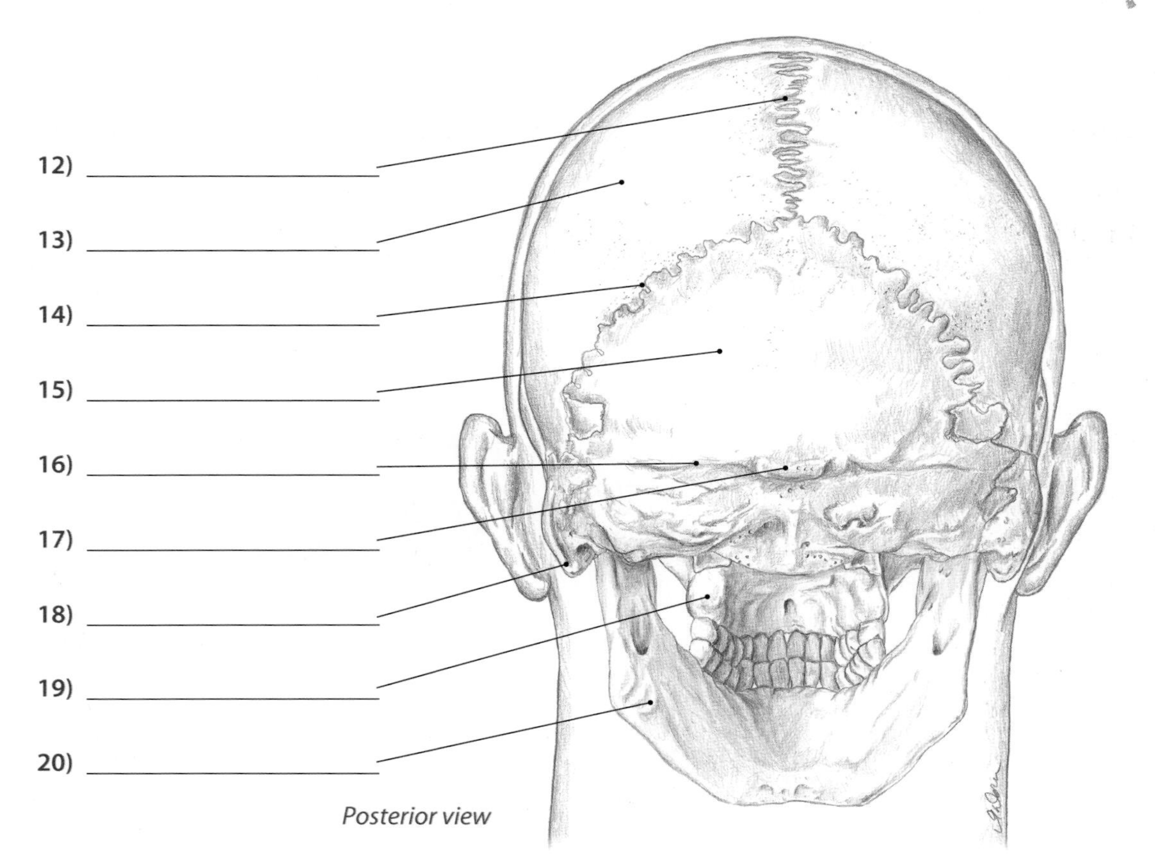

Posterior view

Skull #2

*Please identify the following structures. Numbers in **black** indicate bones, numbers in **red** are bony landmarks.*

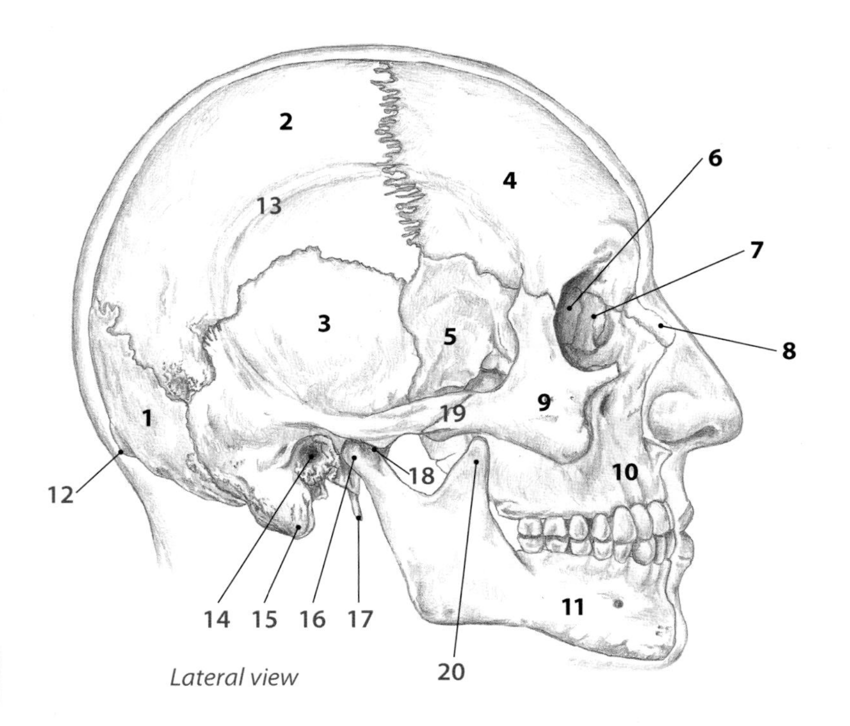

Lateral view

CHOICES 1-20

- ____ Condyle of the mandible
- ____ Coronoid process
- ____ Ethmoid
- ____ External auditory meatus
- ____ External occipital protuberance
- ____ Frontal
- ____ Lacrimal
- ____ Mandible
- ____ Mastoid process
- ____ Maxilla
- ____ Nasal
- ____ Occiput
- ____ Parietal
- ____ Sphenoid
- ____ Styloid process
- ____ Temporal
- ____ Temporal lines
- ____ Temporomandibular joint
- ____ Zygomatic
- ____ Zygomatic arch

Color Them!

CHOICES 21-32

- ____ External occipital protuberance
- ____ Foramen magnum
- ____ Inferior nuchal line
- ____ Mastoid process
- ____ Maxilla
- ____ Occiput
- ____ Palatine
- ____ Sphenoid
- ____ Superior nuchal line
- ____ Temporal
- ____ Vomer
- ____ Zygomatic

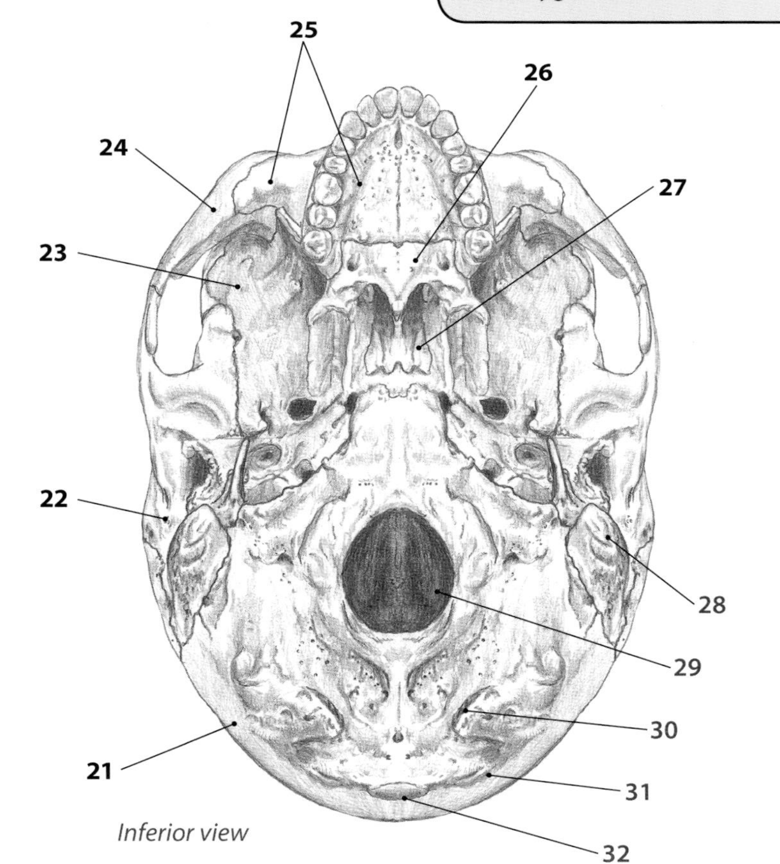

Inferior view

p. 235 & 239

Head, Neck and Face

Mandible and Hyoid Bone

Please identify the following structures.

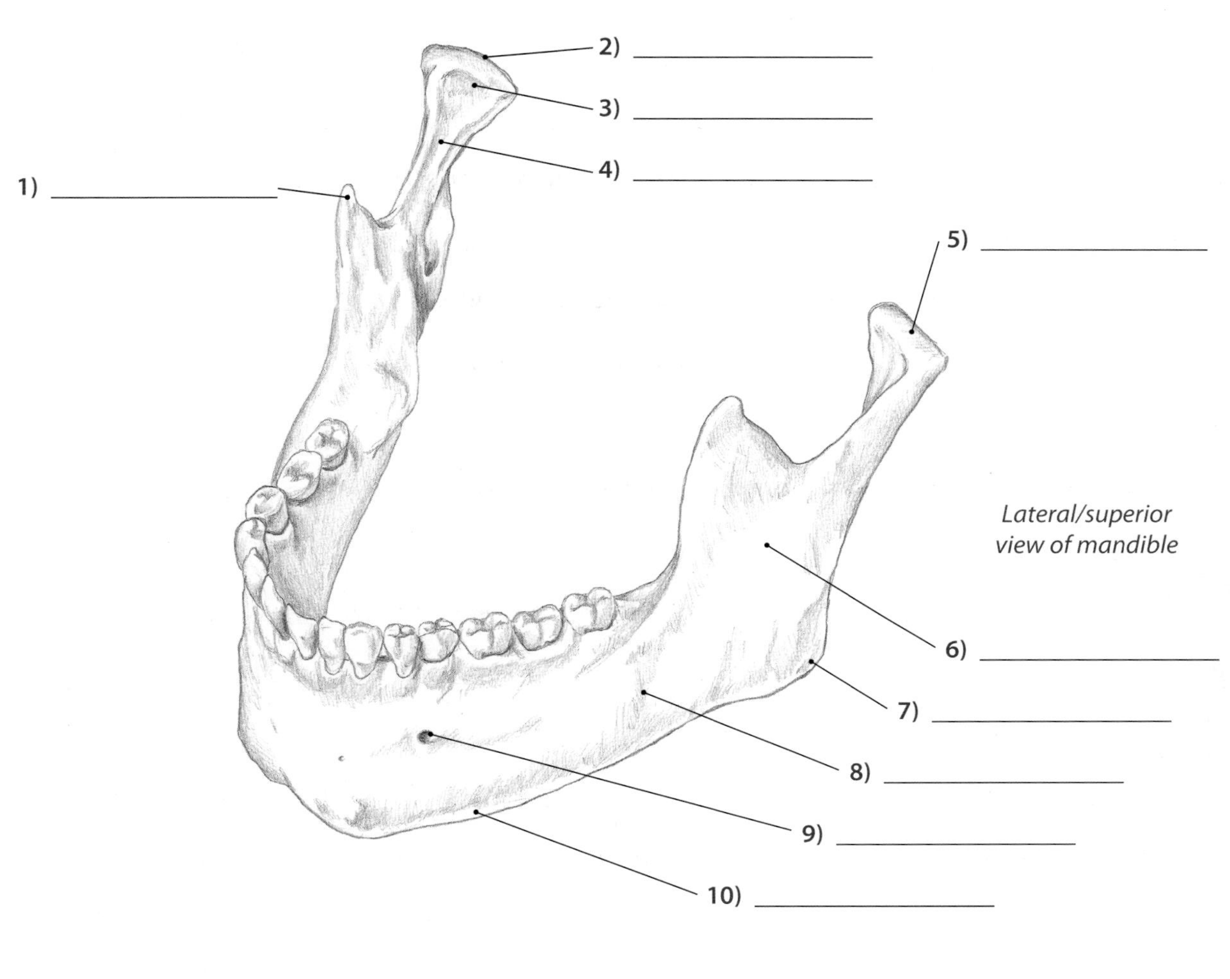

Lateral/superior view of mandible

CHOICES
- Angle
- Base
- Body (2)
- Condyle
- Coronoid process
- Greater horn
- Head
- Lesser horn
- Mental foramen
- Neck
- Pterygoid fossa
- Ramus

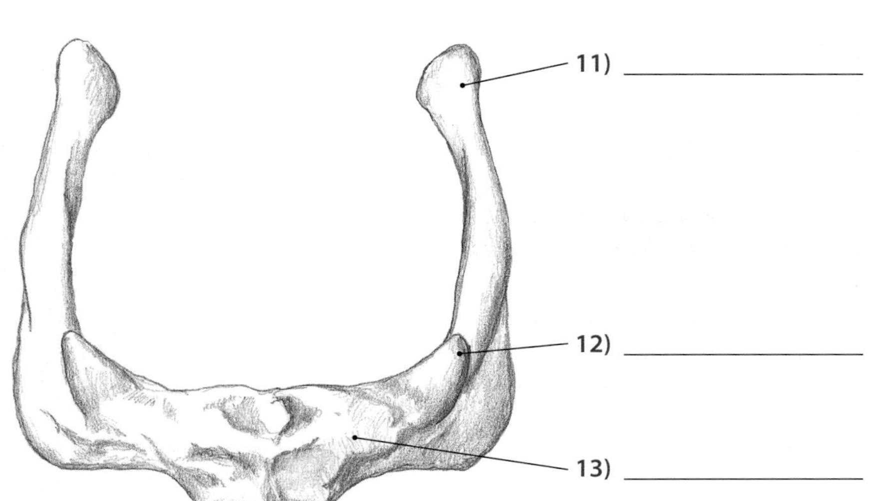

Superior view of hyoid bone

Temporomandibular Joint

Please identify the following structures.

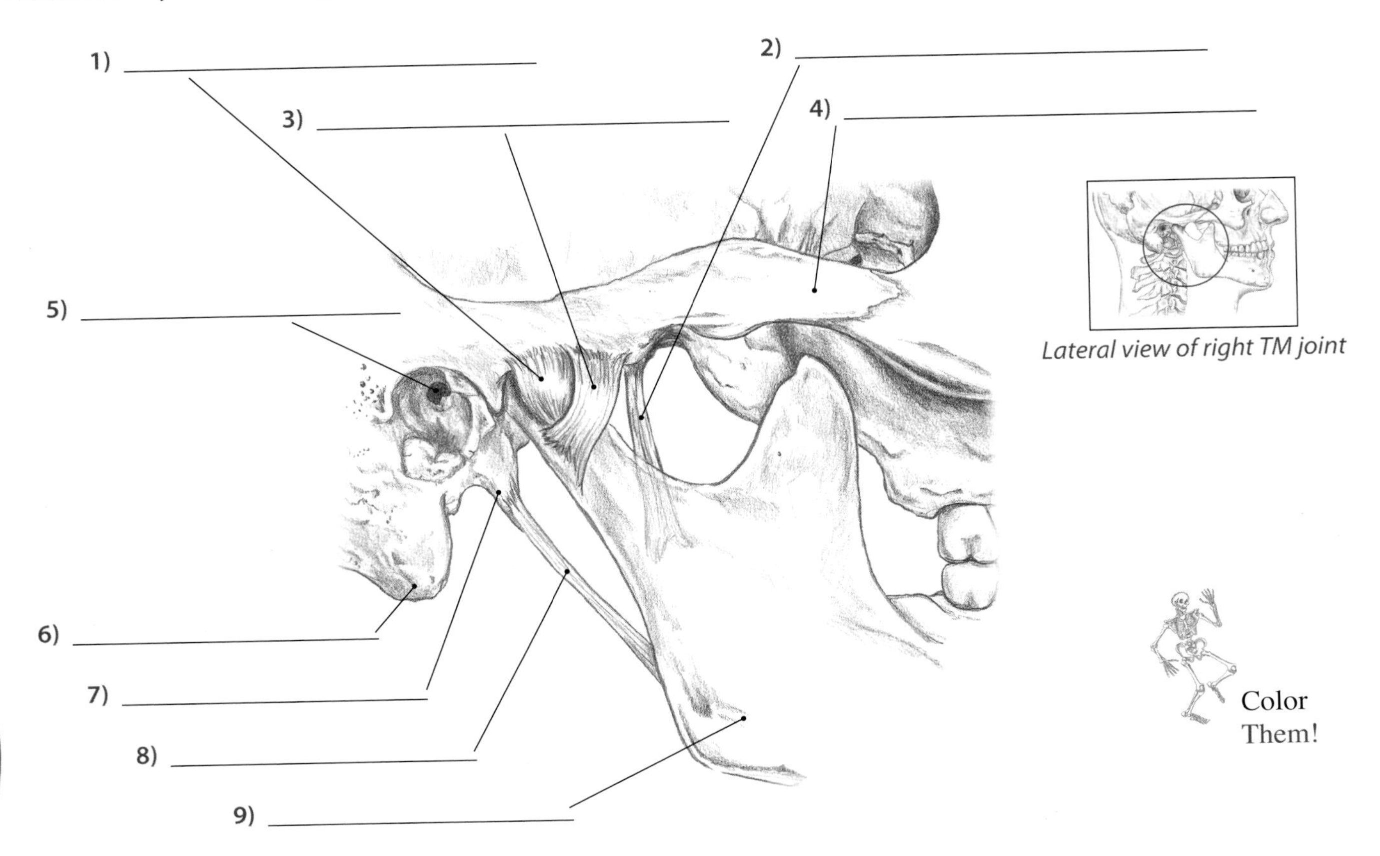

Lateral view of right TM joint

Color Them!

CHOICES

- Articular disc of temporomandibular joint
- Condyle of mandible (cut)
- External auditory meatus
- Joint capsule (2)
- Lateral pterygoid
- Lateral temporomandibular ligament
- Mandible (2)
- Mastoid process
- Sphenomandibular ligament (2)
- Styloid process
- Stylomandibular ligament
- Zygomatic arch

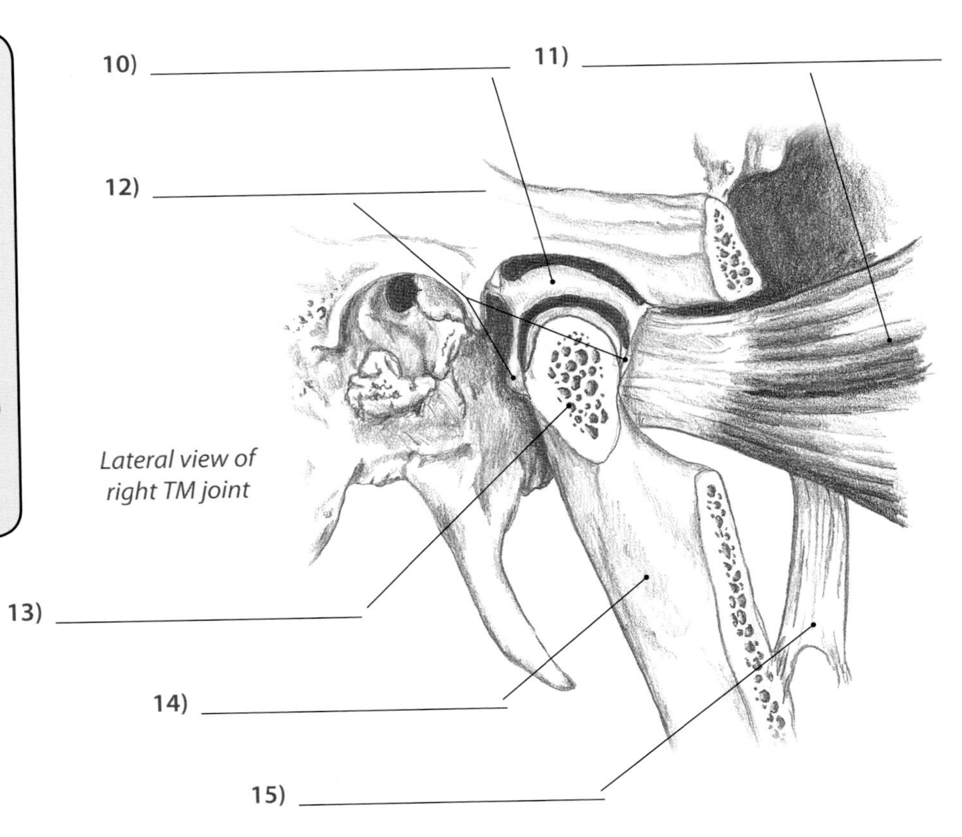

Lateral view of right TM joint

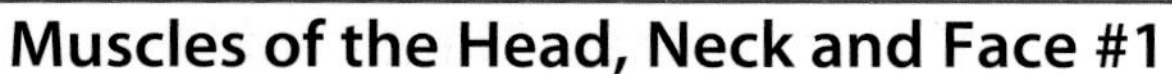

Muscles of the Head, Neck and Face #1

Please identify the following structures.

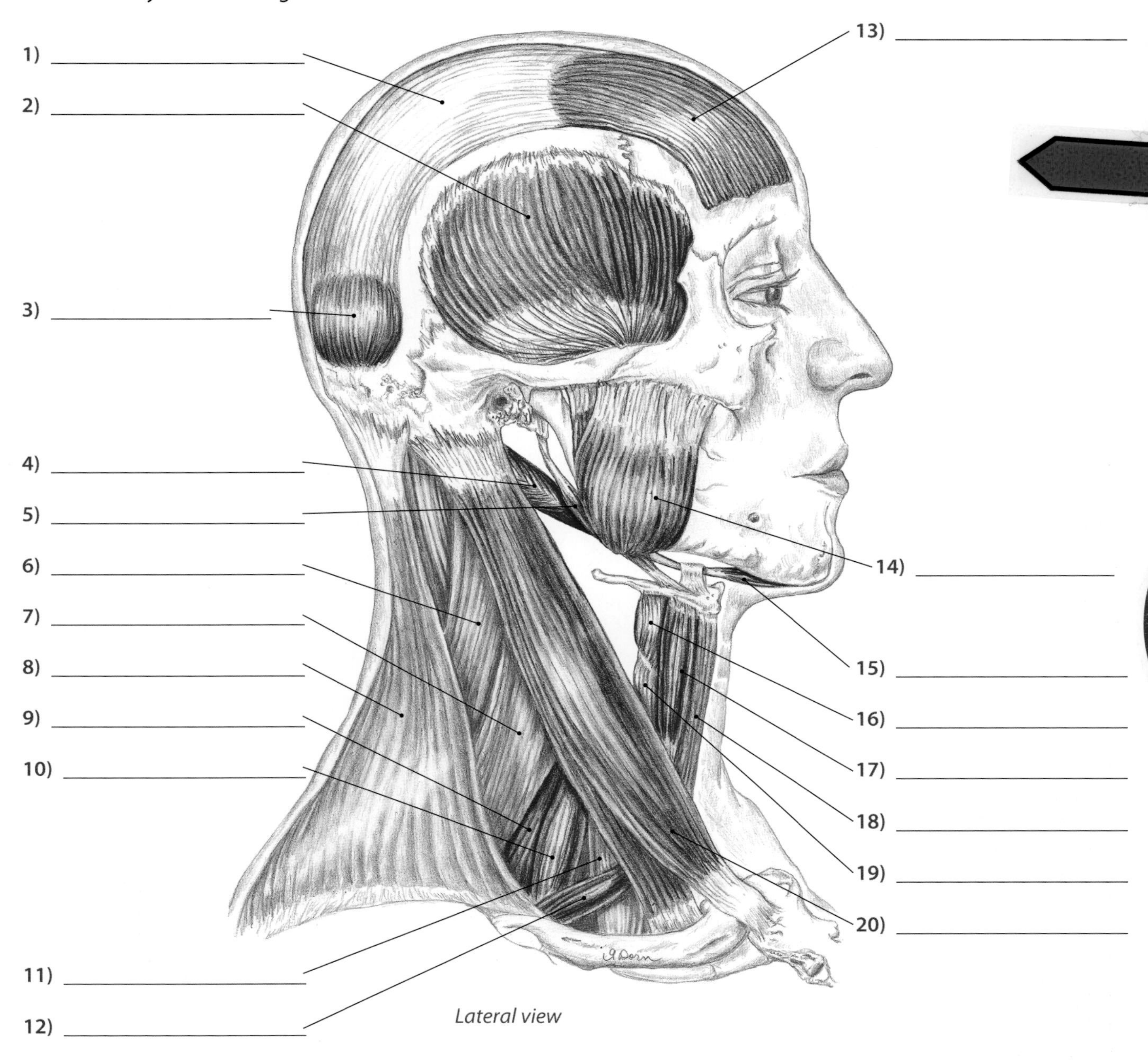

Lateral view

CHOICES

Anterior scalene	Middle scalene	Sternohyoid
Digastric (anterior belly)	Occipitalis	Sternothyroid
Digastric (posterior belly)	Omohyoid (inferior belly)	Stylohyoid
Frontalis	Omohyoid (superior belly)	Temporalis
Galea aponeurotica	Posterior scalene	Thyrohyoid
Levator scapula	Splenius capitis	Trapezius
Masseter	Sternocleidomastoid	

Muscles of the Head, Neck and Face #2

Please identify the following structures.

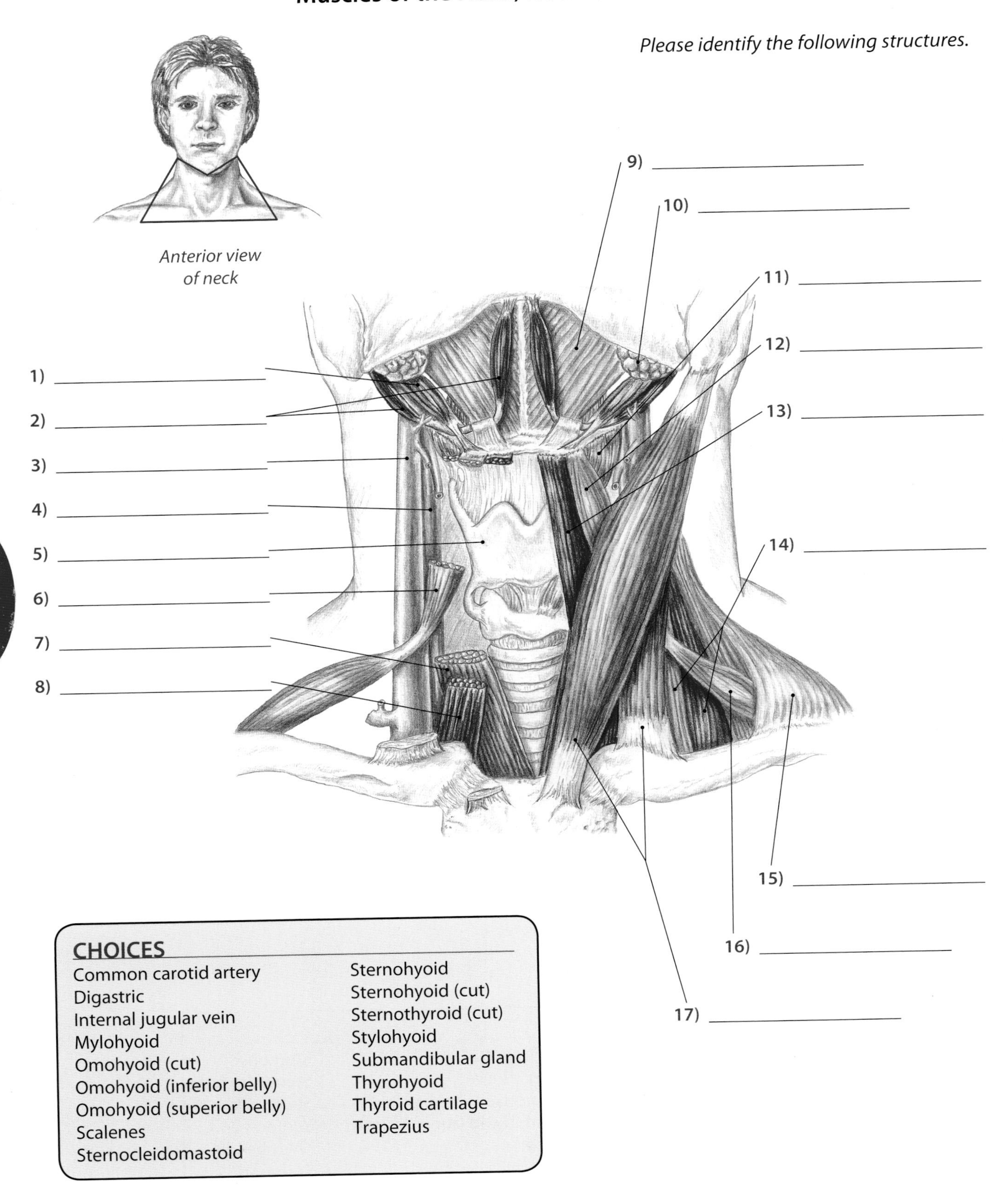

CHOICES

Common carotid artery
Digastric
Internal jugular vein
Mylohyoid
Omohyoid (cut)
Omohyoid (inferior belly)
Omohyoid (superior belly)
Scalenes
Sternocleidomastoid
Sternohyoid
Sternohyoid (cut)
Sternothyroid (cut)
Stylohyoid
Submandibular gland
Thyrohyoid
Thyroid cartilage
Trapezius

Head, Neck and Face

Muscles of Facial Expression

Please identify the following structures.

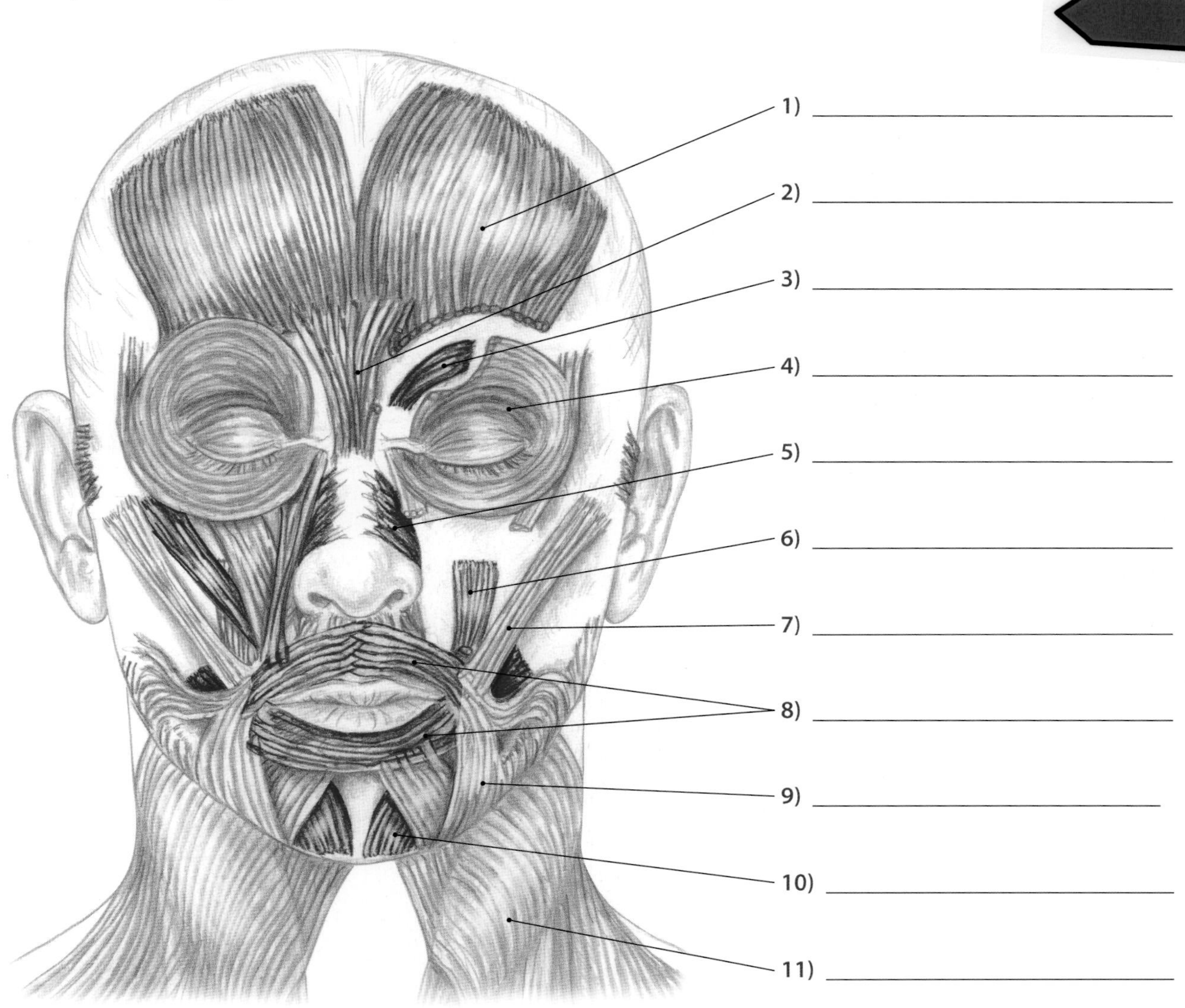

CHOICES

- Corrugator supercili
- Depressor anguli oris
- Frontalis
- Levator anguli oris
- Mentalis
- Nasalis
- Orbicularis oculi
- Orbicularis oris
- Platysma
- Procerus
- Zygomaticus major

Color the Muscles #1

Using different colors, please fill in and label the muscles and other structures listed below.

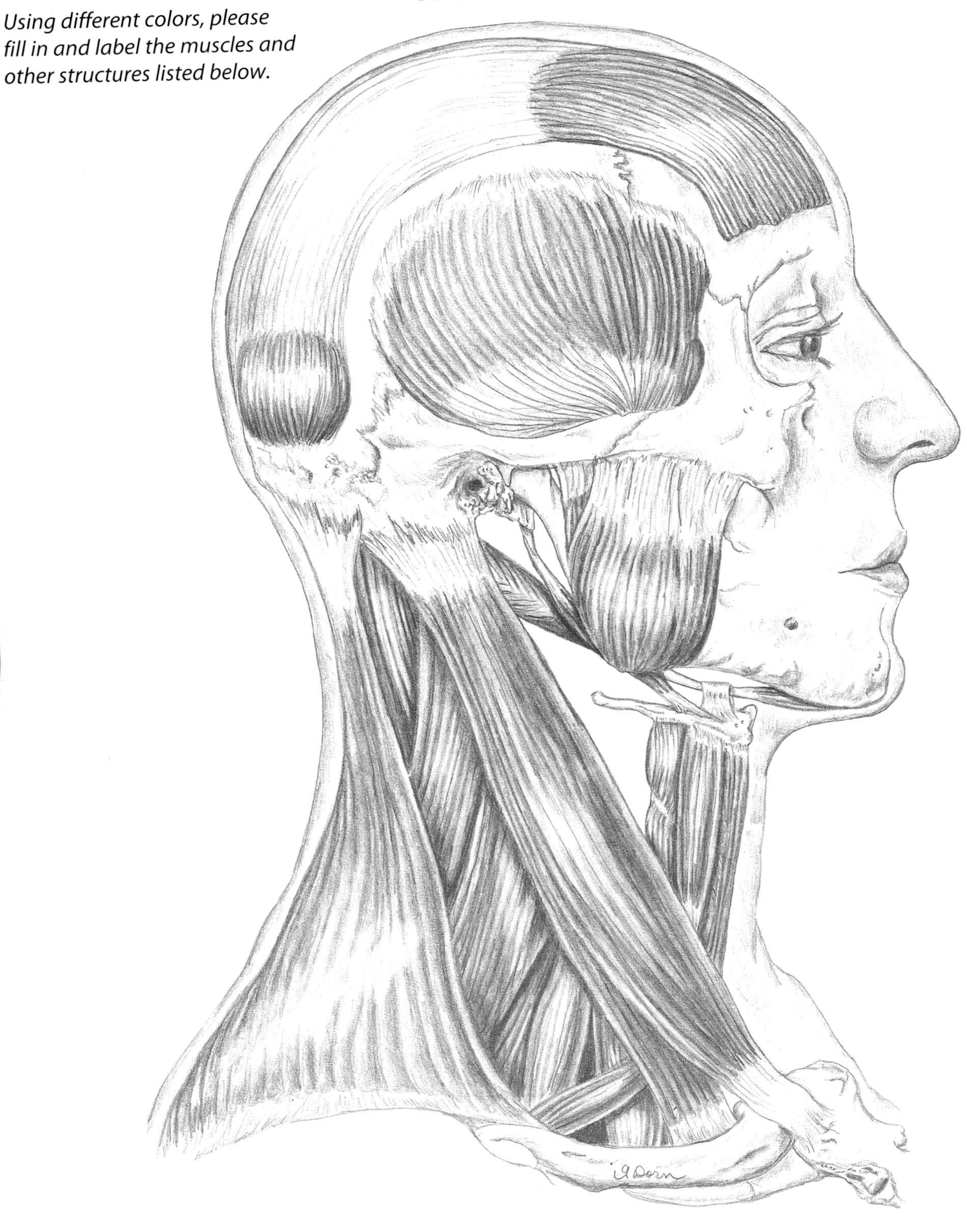

Anterior scalene	Levator scapula	Omohyoid (superior belly)	Sternothyroid
Digastric (anterior belly)	Masseter	Posterior scalene	Stylohyoid
Digastric (posterior belly)	Middle scalene	Splenius capitis	Temporalis
Frontalis	Occipitalis	Sternocleidomastoid	Thyrohyoid
Galea aponeurotica	Omohyoid (inferior belly)	Sternohyoid	Trapezius

Head, Neck and Face

Color the Muscles #2

Using different colors, please fill in and label the muscles and other structures listed below.

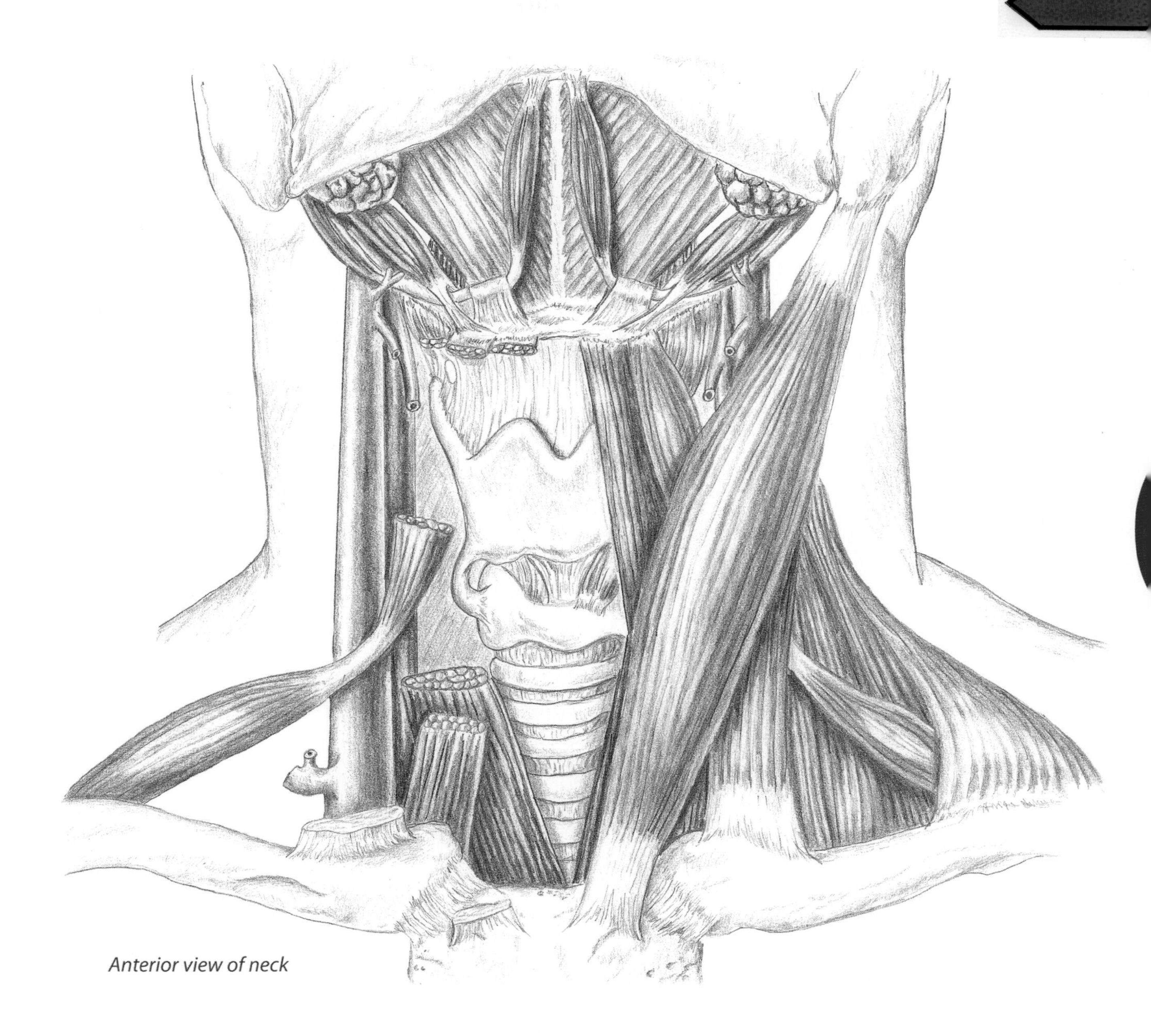

Anterior view of neck

Common carotid artery	Omohyoid (inferior belly)	Sternohyoid	Submandibular gland
Digastric	Omohyoid (superior belly)	Sternohyoid (cut)	Thyrohyoid
Internal jugular vein	Scalenes	Sternothyroid (cut)	Thyroid cartilage
Mylohyoid	Sternocleidomastoid	Stylohyoid	Trapezius
Omohyoid (cut)			

Color the Muscles #3

Using different colors, please fill in and label the muscles and other structures listed below.

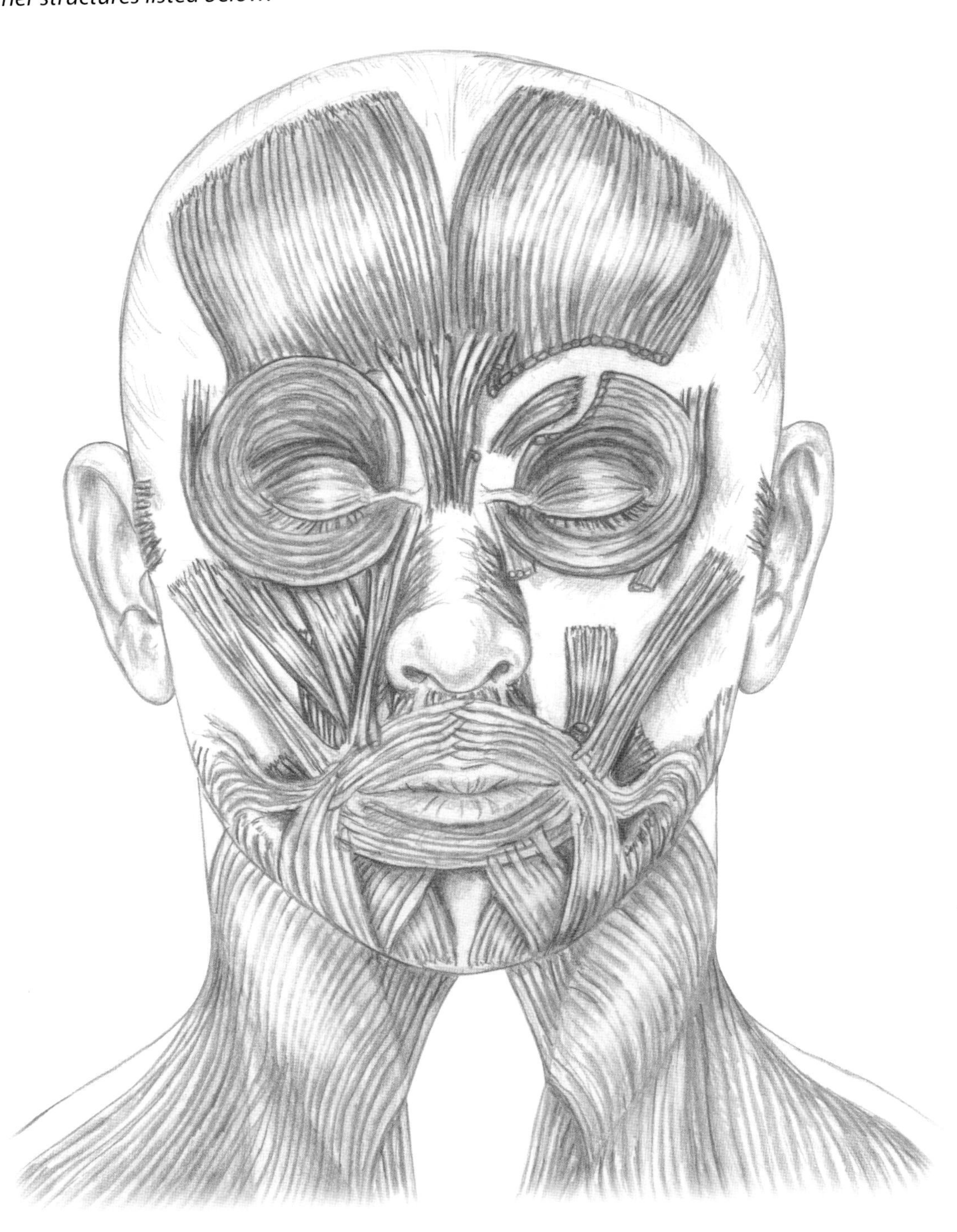

Corrugator supercili	Levator labii superioris	Orbicularis oculi	Procerus
Depressor anguli oris	Mentalis	Orbicularis oris	Zygomaticus major and minor
Frontalis	Nasalis	Platysma	

Head, Neck and Face

Muscles and Movements #1

Please list the action demonstrated, synergists and antagonist(s).

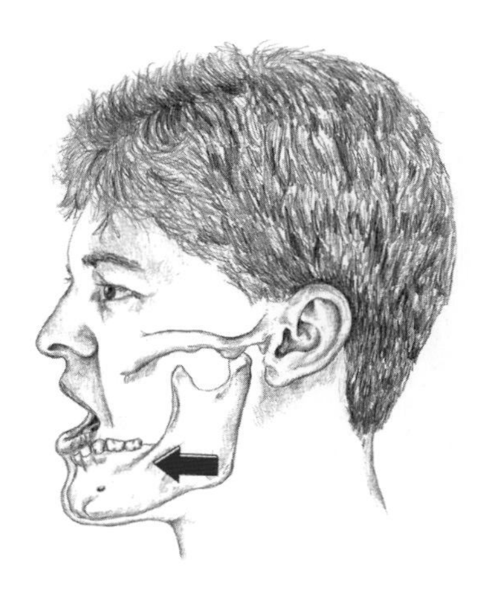

1) Action

____

2) Synergists

____

____

3) Antagonists

____

____

4) Action

____

5) Synergists

S ______________________________

S ______________________________

6) Antagonist

S ______________________________

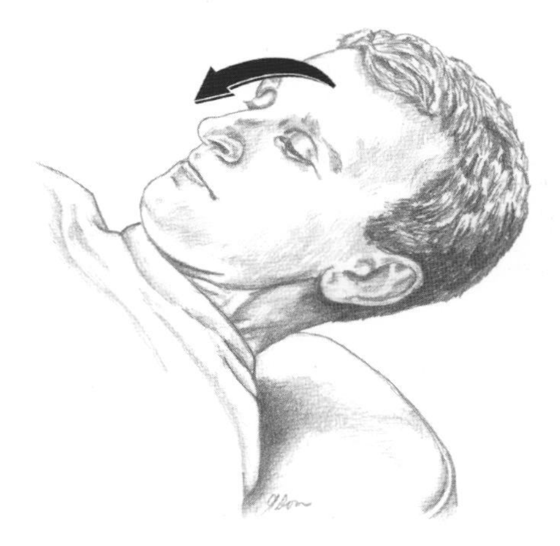

7) Action

____

8) Synergists

L ______________________________

L ______________________________

9) Antagonist

L ______________________________

Muscles and Movements #2

Please list the action demonstrated, two synergists and one antagonist. The first letter of each muscle has been provided.

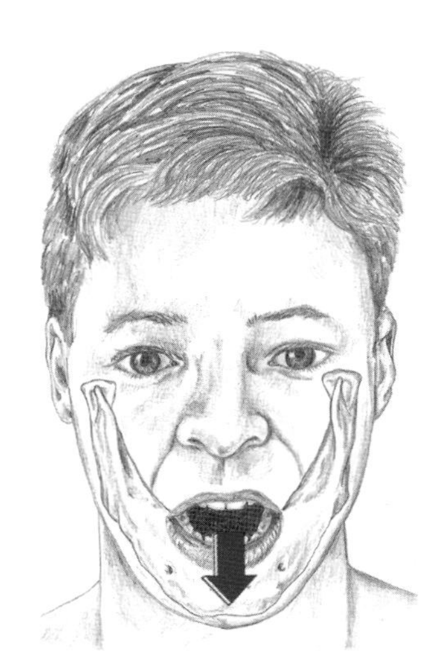

1) Action

____

2) Synergists

G ______________________________

D ______________________________

3) Antagonist

T ______________________________

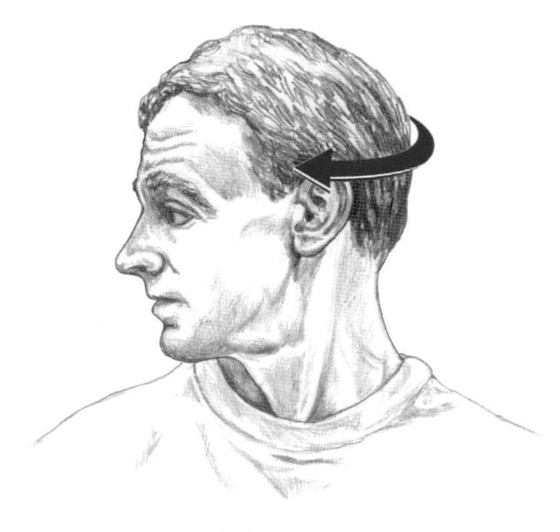

4) Action (to his right)

____

5) Synergists (and on what side—his left or right?)

M ______________________________

T ______________________________

6) Antagonist (and on what side—his left or right?)

L ______________________________

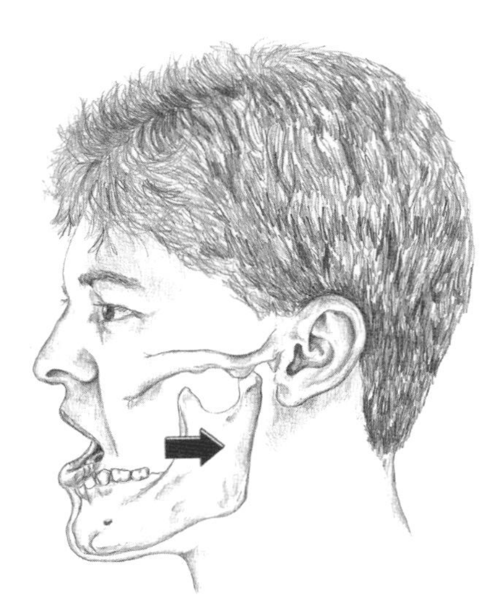

7) Action

____

8) Synergists

T ______________________________

D ______________________________

9) Antagonist

L ______________________________

Head, Neck and Face

Muscles and Movements #3

Please list the action demonstrated, synergists and antagonist.
The first letter of each muscle has been provided.

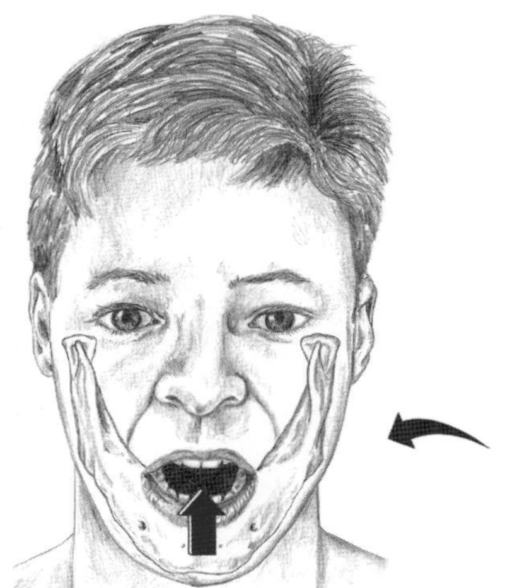

1) This action happens at which joint?

____

2) Action

____

3) Synergists

T ______________________________

M ______________________________

4) Antagonist

G ______________________________

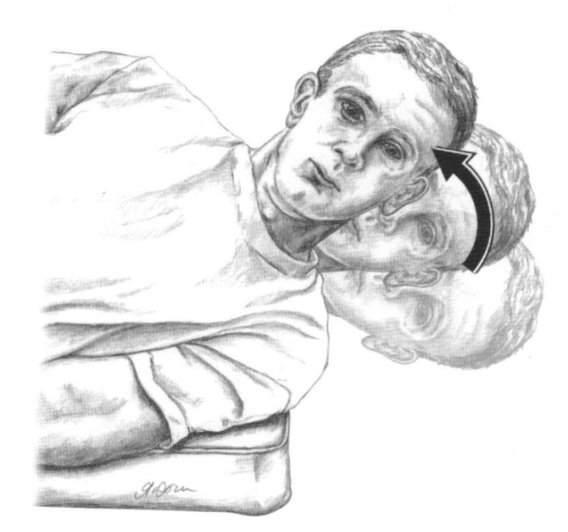

5) Action

____

6) Synergists

S ______________________________

S ______________________________

S ______________________________

What's the Muscle? #1

Please identify the following muscles.

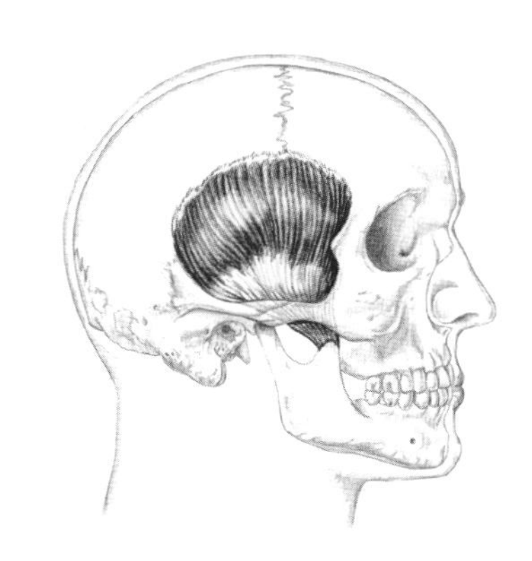

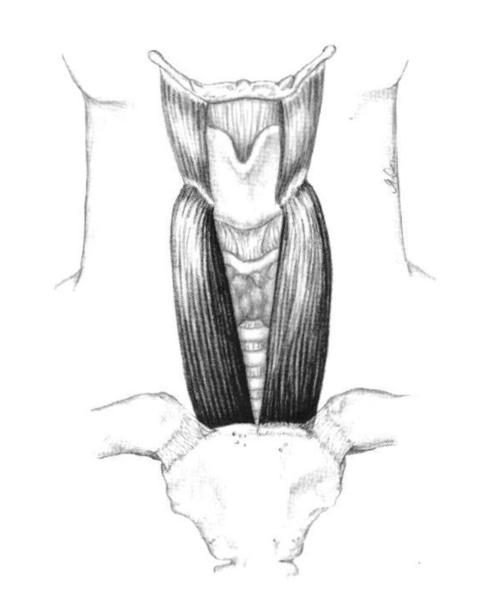

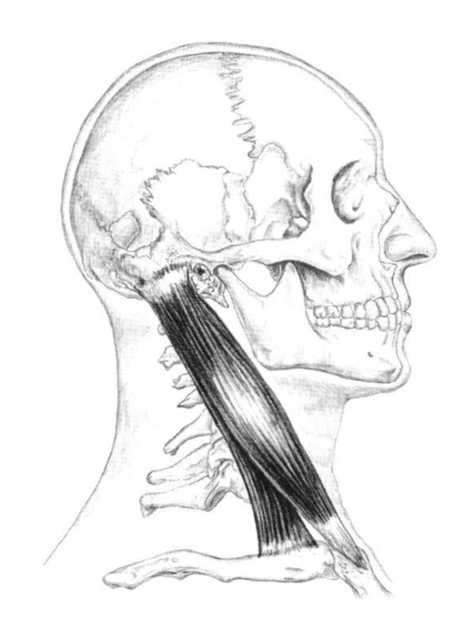

1) ______________________

2) ______________________

3) ______________________

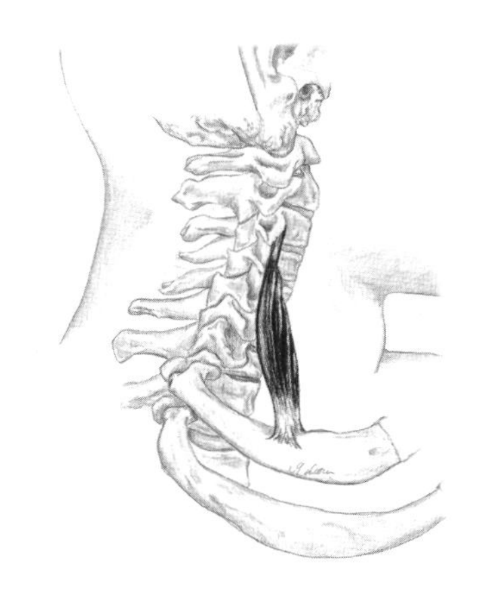

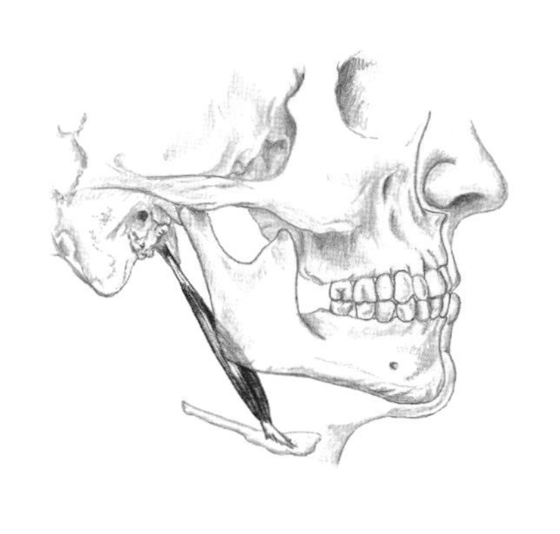

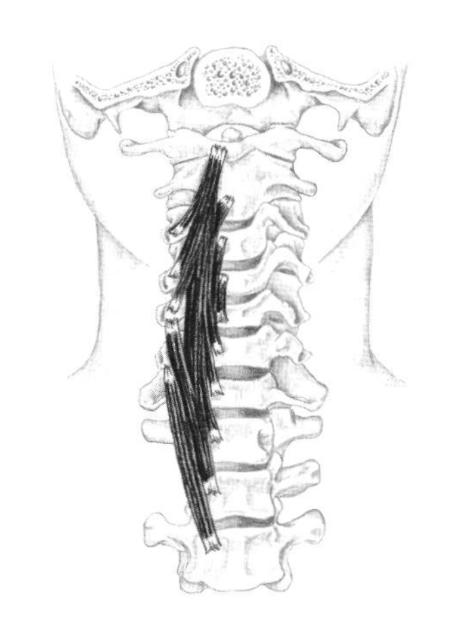

4) ______________________

5) ______________________

6) ______________________

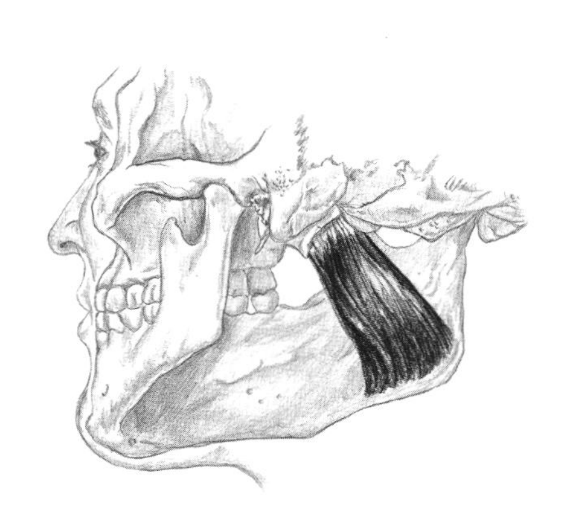

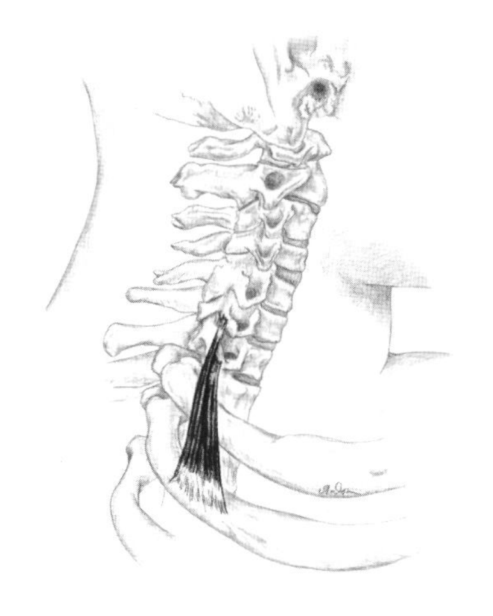

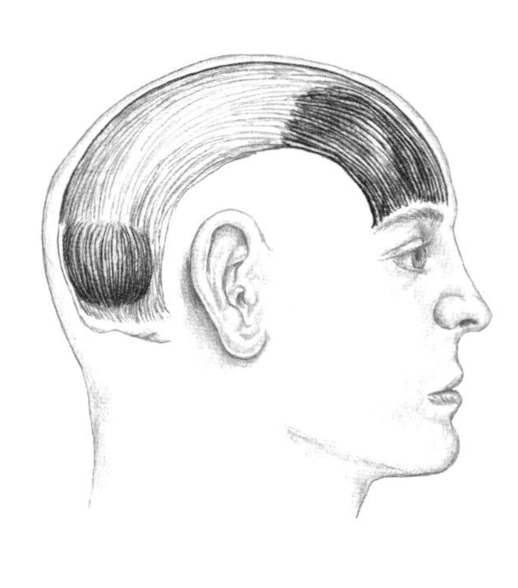

7) ______________________

8) ______________________

9) ______________________

What's the Muscle? #2

Please identify the following muscles.

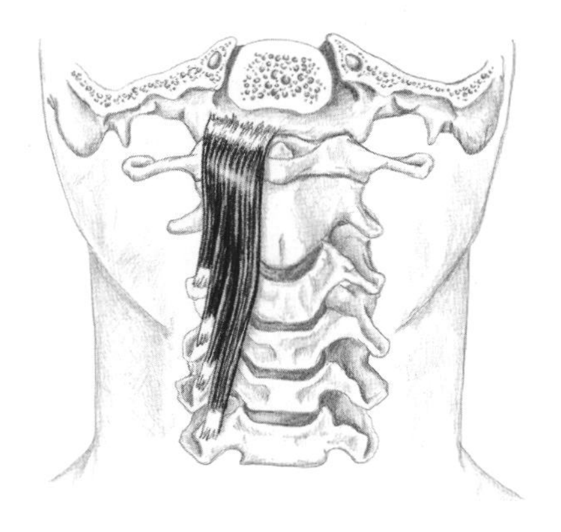

1) ______________________________

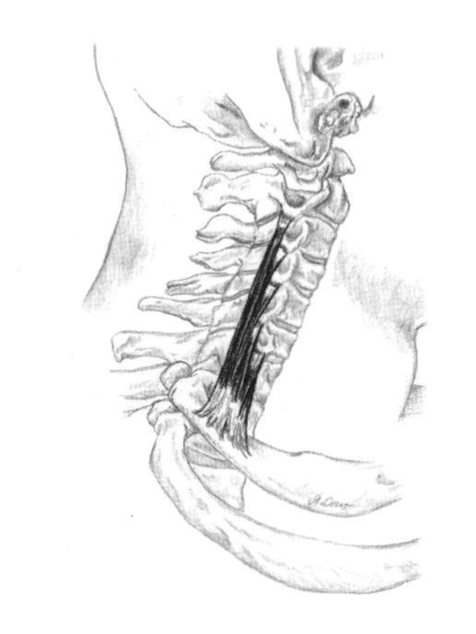

2) ______________________________

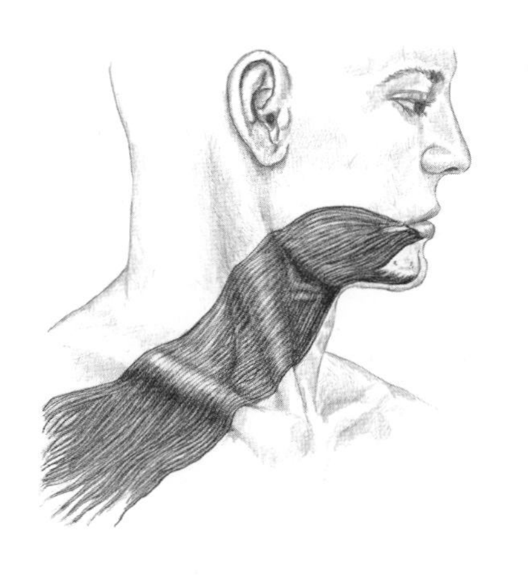

3) ______________________________

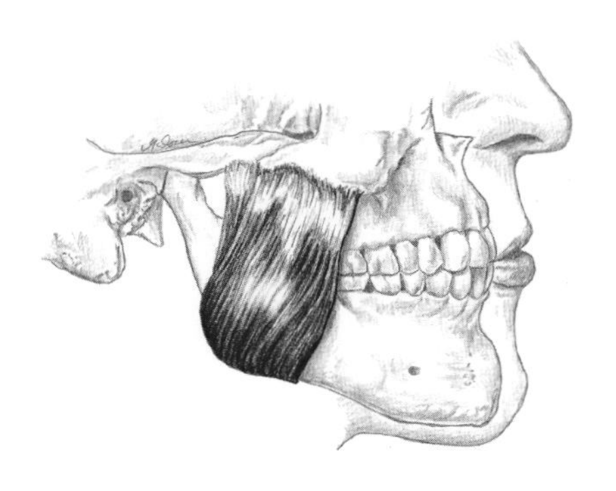

4) ______________________________

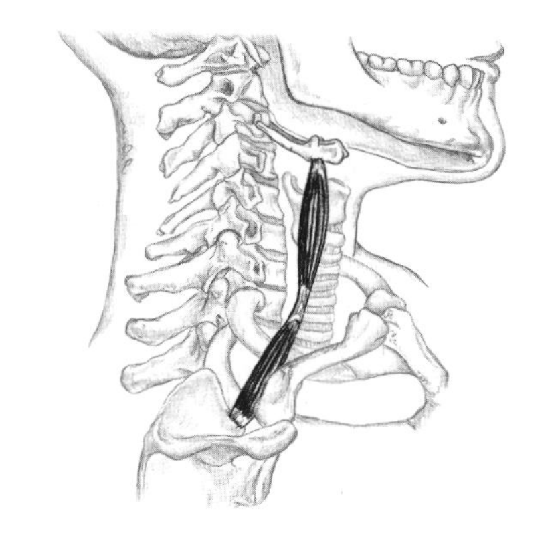

5) ______________________________

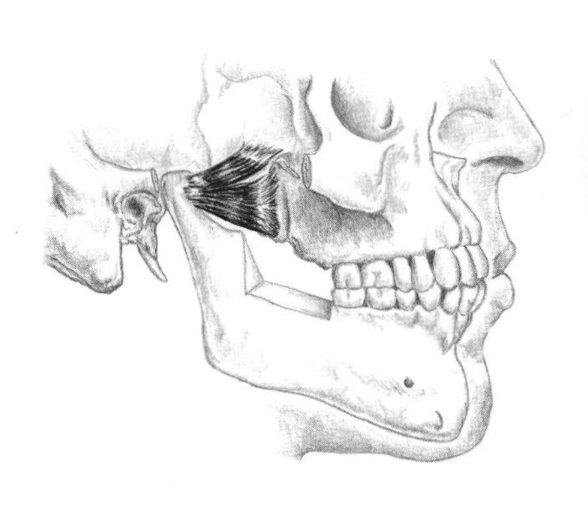

6) ______________________________

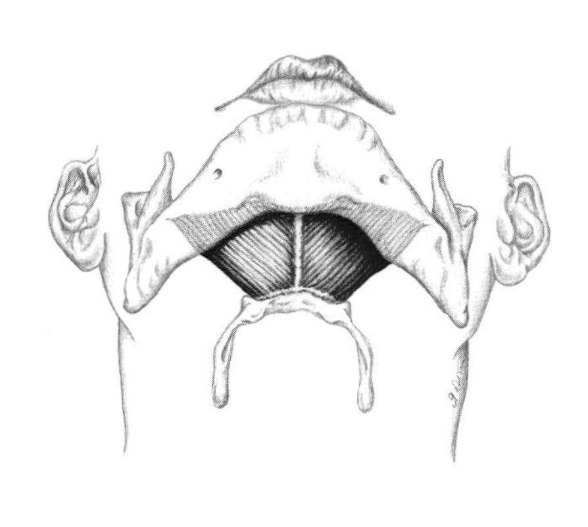

7) ______________________________

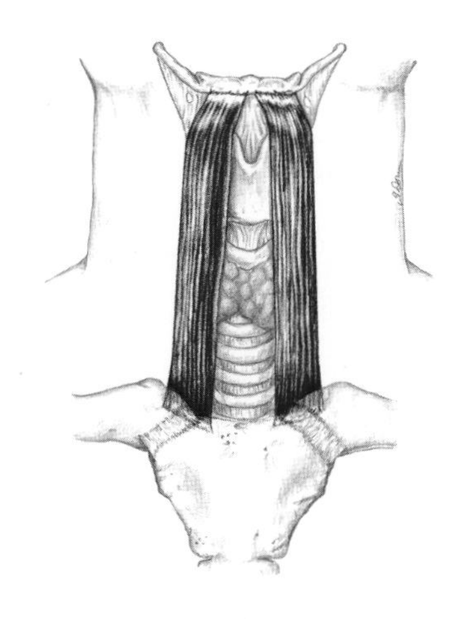

8) ______________________________

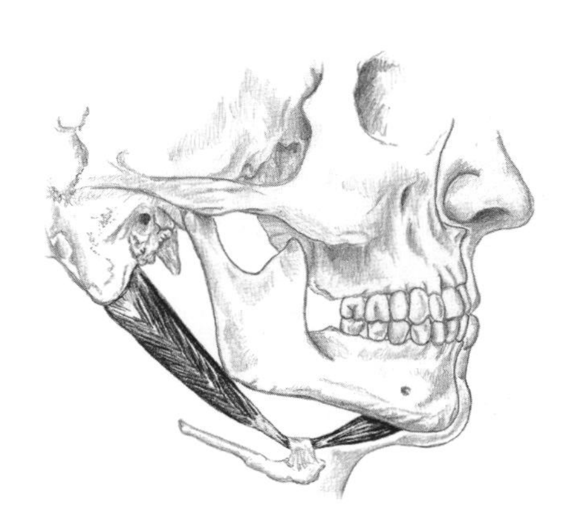

9) ______________________________

Head, Neck and Face, Muscle Group #1

p. 244-252

SCM, Scalenes, Masseter and Temporalis

Please answer the following questions.

1) The two heads of the sternocleidomastoid originate at the ______________ and the ______________.

2) To create an even more visible contraction in the sternocleidomastoid (SCM), ask your partner to flex her neck after making what adjustment? ______________

3) Which scalene is difficult to distinguish from surrounding muscle bellies? ______________

4) Which muscles are located between the SCM and the anterior flap of the trapezius? ______________

5) The brachial plexus and subclavian artery pass through a small gap between which two muscles on the anterior, lateral neck?

______________ ______________

6) You might ask your partner to "breathe deeply into your upper chest" when palpating which muscle group?

____

7) The anterior scalene lies partially deep to the lateral edge of which muscle? ______________

8) To discern the posterior scalene from the levator scapula, what action could you ask your partner to perform that would contract the levator but not the scalene? ______________

9) The ______________ is the strongest muscle in the body relative to its size.

10) The broad origin of which muscle attaches to the frontal, temporal and parietal bones? ______________

11) To access the insertion of the temporalis, you must ask your partner to perform what action?

____

Shorten or Lengthen?

12) Passive protraction of the mandible would ______________ the temporalis.

13) Passive rotation of the head and neck to the opposite side would ______________ the scalenes.

14) Passive lateral flexion of the head and neck to the same side would ______________ the sternocleidomastoid and scalenes.

15) Passive flexion of the head and neck would ______________ the anterior scalene.

16) Passive rotation of the head and neck to the same side would ______________ the sternocleidomastoid.

17) Passive elevation of the mandible would ______________ the masseter.

Head, Neck and Face, Muscle Group #1

SCM, Scalenes, Masseter and Temporalis

Matching

Match the origin and insertion to the correct muscle.

Origins

1) Temporal fossa and fascia
2) Top of manubrium, medial one-third of the clavicle
3) Transverse processes of sixth and seventh cervical vertebrae (posterior tubercles)
4) Transverse processes of second through seventh cervical vertebrae (posterior tubercles)
5) Transverse processes of third through sixth cervical vertebrae (anterior tubercles)
6) Zygomatic arch

Insertions

7) Angle and ramus of mandible
8) Coronoid process and anterior edge of ramus of the mandible
9) First rib (2)
10) Mastoid process of temporal bone and the lateral portion of superior nuchal line of occiput
11) Second rib

Muscle	O	I
Anterior scalene	_______	_______
Masseter	_______	_______
Middle scalene	_______	_______
Posterior scalene	_______	_______
Sternocleidomastoid	_______	_______
Temporalis	_______	_______

Let's Palpate!

Remember—there are no right or wrong answers here

Locate and explore the **sternocleidomastoid** on three individuals. Then write three words that describe what you feel. (See p. 244-245 in *Trail Guide*.)

Person #1 ____________	Person #2 ____________	Person #3 ____________
________________________	________________________	________________________
________________________	________________________	________________________
________________________	________________________	________________________

Head, Neck and Face, Muscle Group #2

Suprahyoids, Infrahyoids and More

p. 253-260

Please answer the following questions.

1) Name the four muscles which comprise the suprahyoids.

______________________ ______________________

______________________ ______________________

2) Which muscle originates at the mastoid process, loops through a tendinous sling at the hyoid bone and inserts to the inferior border of the mandible? ______________________

3) What direction should you give your partner in order to locate the suprahyoids? ______________________

4) Which muscle runs from the hyoid bone to the superior border of the scapula and is mostly inaccessible?

____

5) Which muscle becomes visually distinct when your partner forms a *Creature from the Black Lagoon* expression?

____

6) The galea aponeurotica forms the bridge between which two muscle bellies?

______________________ ______________________

7) The frontalis is best seen and felt by asking your partner to do what action?

____

8) What two muscles attach from the anterior surface of the cervical vertebrae to the occiput and atlas?

______________________ ______________________

Let's Palpate!

Remember—there are no right or wrong answers here

Locate and explore the **suprahyoids** on three individuals. Then write three words that describe what you feel. (See p. 253-254 in *Trail Guide.*)

Person #1 ________	Person #2 ________	Person #3 ________
________	________	________
________	________	________
________	________	________

Head, Neck and Face, Muscle Group #2

Suprahyoids, Infrahyoids and More

Matching

Match the origin and insertion to the correct muscle.

Origins

1) Top of manubrium (2)
2) Mastoid process (deep to SCM and splenius capitis)
3) Styloid process
4) Superior border of the scapula
5) Underside of mandible (2)
6) Thyroid cartilage

Insertions

7) Hyoid bone (6)
8) Inferior border of the mandible
9) Thyroid cartilage

Muscle	O	I
Digastric	_______	_______
Geniohyoid	_______	_______
Mylohyoid	_______	_______
Omohyoid	_______	_______
Sternohyoid	_______	_______
Sternothyroid	_______	_______
Stylohyoid	_______	_______
Thyrohyoid	_______	_______

Shorten or Lengthen?

10) Passively raising the eyebrows would ____________________ the frontalis fibers.

11) Tightening the fascia of the neck would ____________________ the platysma.

12) Passive protraction of the mandible would ____________________ the digastric.

13) Passive depression of the mandible would ____________________ the suprahyoids.

14) Not that you'd ever want to, but passive elevation of the hyoid bone would ____________________ the infrahyoids.

Muscles of Facial Expression #1

Please identify the following muscles.

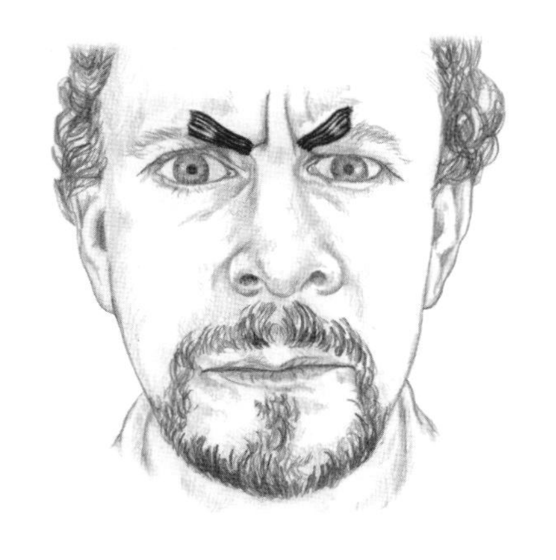

1) ______________________________

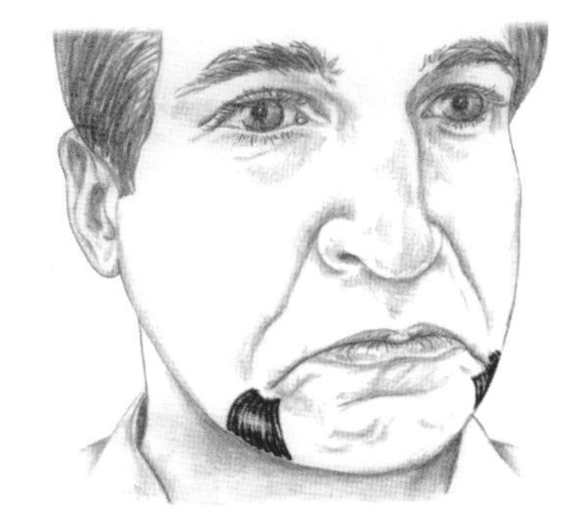

2) ______________________________

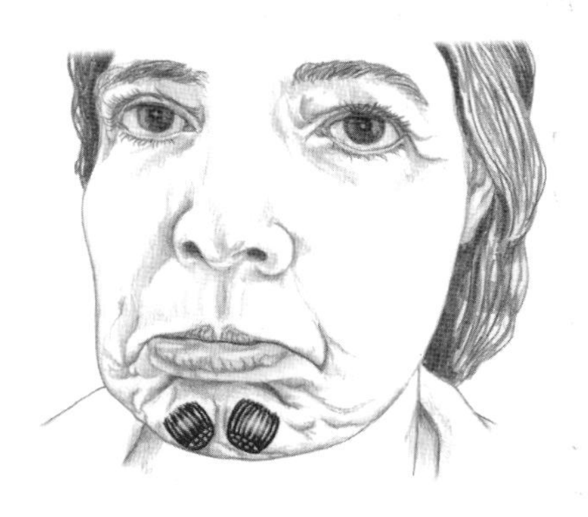

3) ______________________________

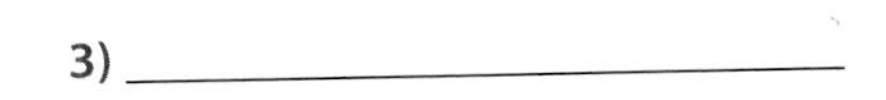

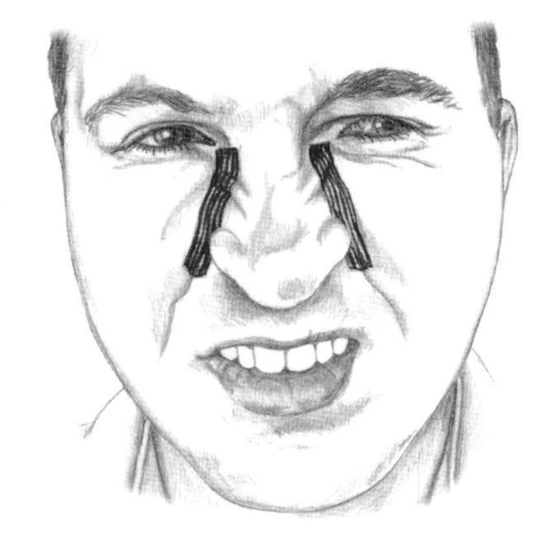

4) ______________________________

5) ______________________________

6) ______________________________

7) ______________________________

8) ______________________________

9) ______________________________

Head, Neck and Face

Muscles of Facial Expression #2

There are seven primary facial expressions. Can you match the expression to the face and then list three muscles you think might be involved in creating the expression?

1) Expression

____

Muscles

____

____

____

2) Expression

____

Muscles

____

____

____

3) Expression

____

Muscles

____

____

____

CHOICES
- Anger
- Contempt
- Disgust
- Fear
- Happiness
- Sadness
- Surprise

4) Expression

____

Muscles

____

____

____

5) Expression

____

Muscles

____

____

____

6) Expression

____

Muscles

____

____

____

7) Expression

____

Muscles

____

____

____

Head, Neck and Face
Other Structures

p. 270-274

Please identify the following structures.

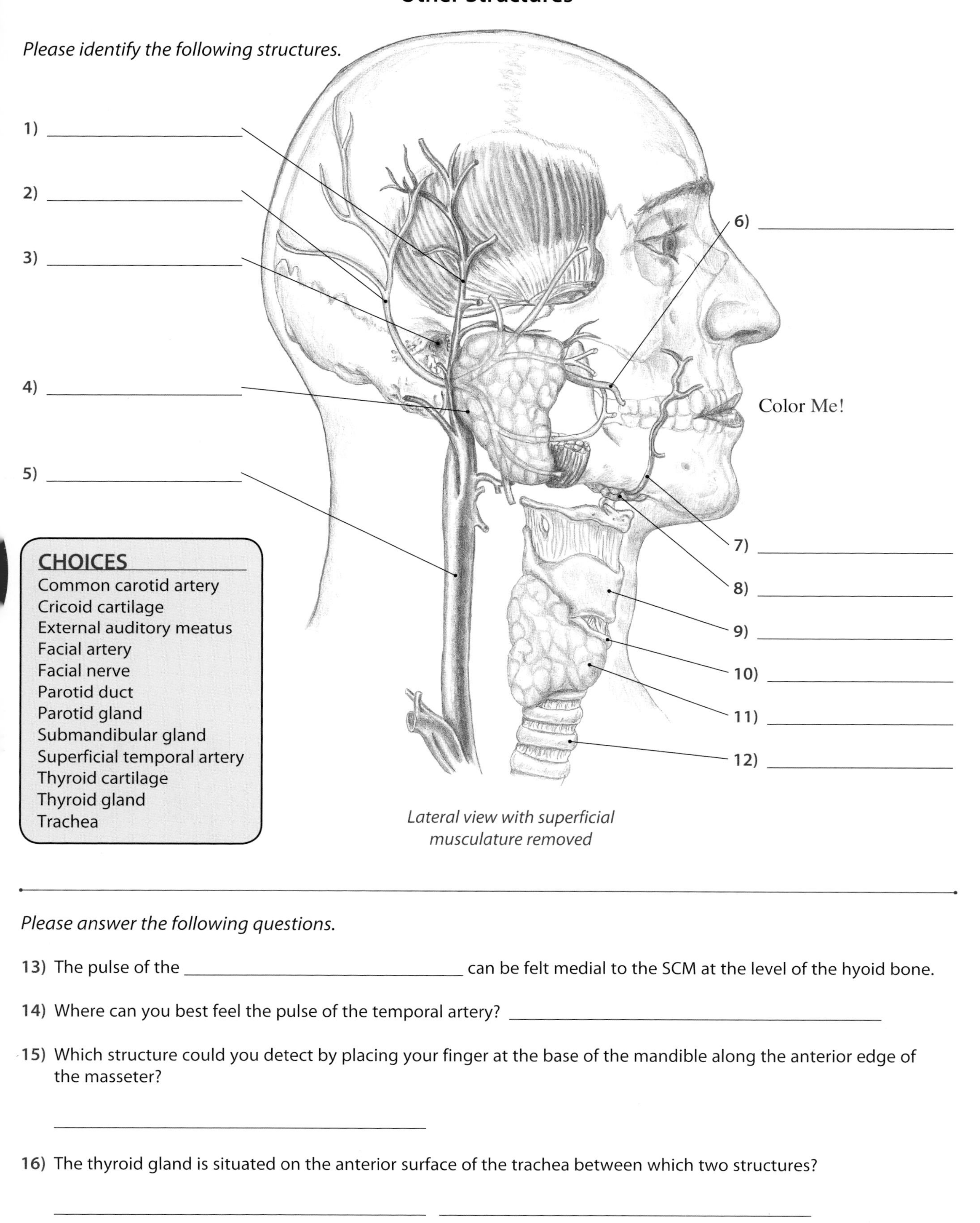

CHOICES
- Common carotid artery
- Cricoid cartilage
- External auditory meatus
- Facial artery
- Facial nerve
- Parotid duct
- Parotid gland
- Submandibular gland
- Superficial temporal artery
- Thyroid cartilage
- Thyroid gland
- Trachea

Lateral view with superficial musculature removed

Please answer the following questions.

13) The pulse of the ______________________ can be felt medial to the SCM at the level of the hyoid bone.

14) Where can you best feel the pulse of the temporal artery? ______________________

15) Which structure could you detect by placing your finger at the base of the mandible along the anterior edge of the masseter?

____

16) The thyroid gland is situated on the anterior surface of the trachea between which two structures?

______________________ ______________________

Pelvis and Thigh
Topographical Views

Please identify the following structures.

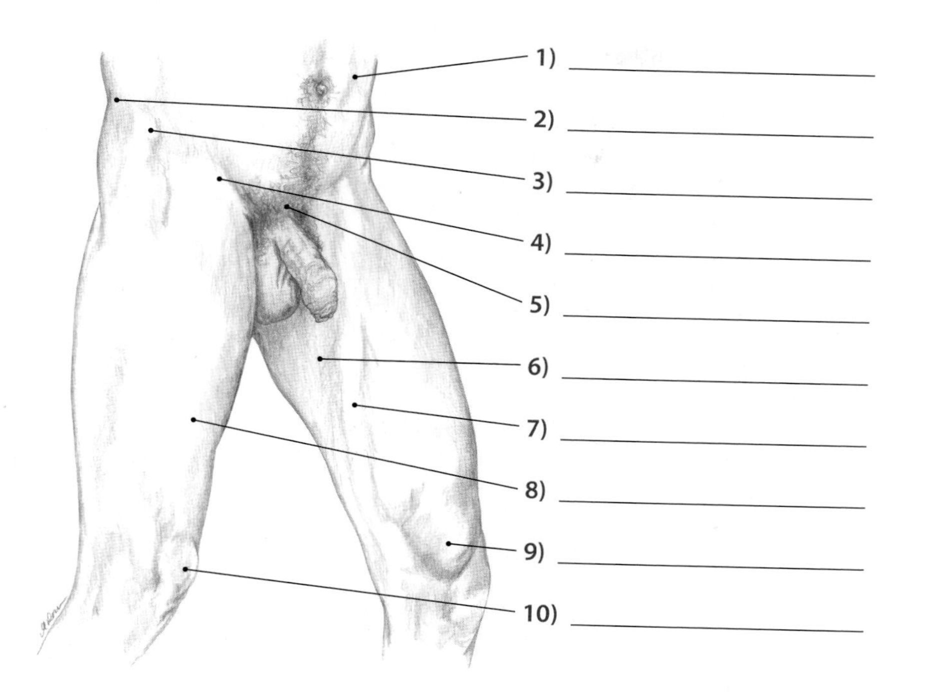

Anterior/lateral view

CHOICES
Adductors
Anterior superior iliac spine
Coccyx
Gluteal cleft
Gluteal fold
Gluteus maximus
Gluteus medius
Greater trochanter
Hamstring tendons
Hamstrings
Iliac crest
Iliotibial tract
Inguinal ligament
Patella
Popliteal fossa
Posterior superior iliac spine
Pubic crest
Rectus abdominis
Rectus femoris
Sacrum
Sartorius
Vastus lateralis
Vastus medialis

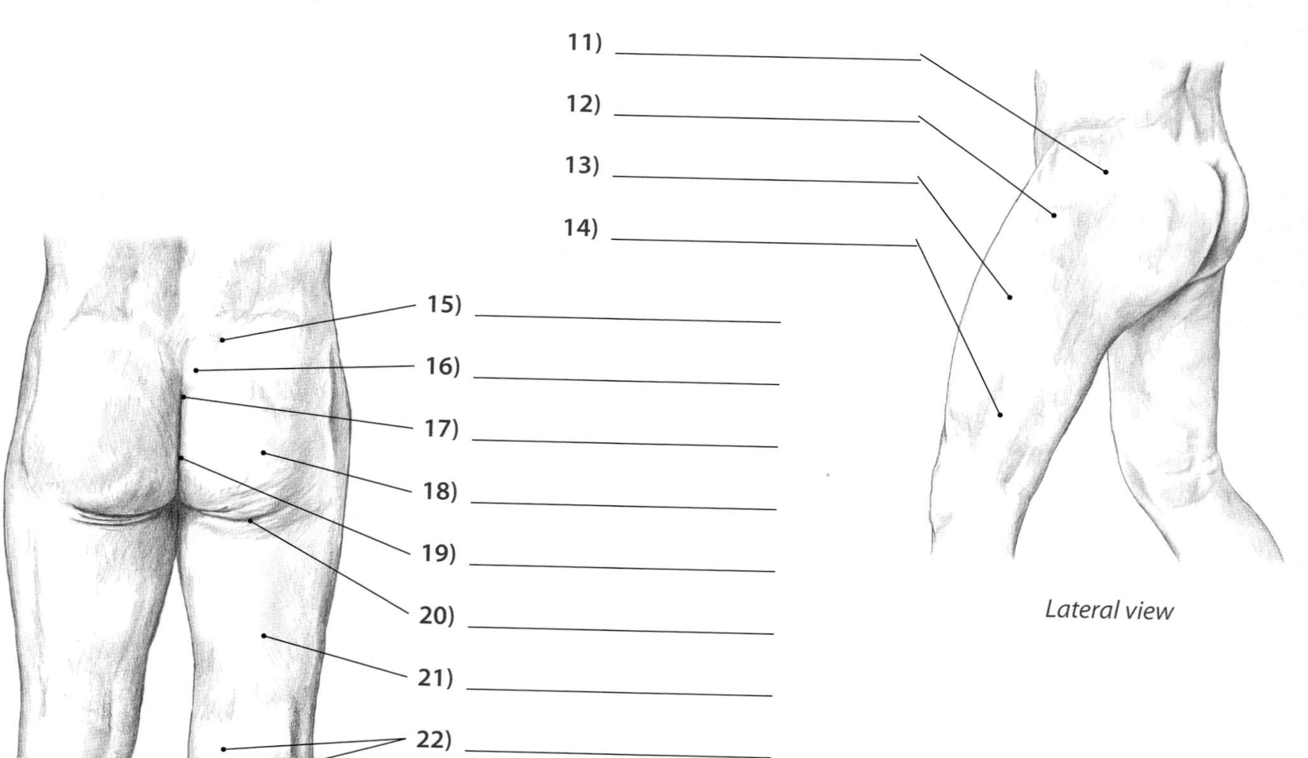

Lateral view

Posterior view

Bones and Bony Landmarks of the Pelvis #1

Please answer the following questions.

1) Name the three bones which make up the hip (coxal) bone.

_______________ _______________ _______________

2) The _______________ and _______________ are considered part of both the pelvis and the vertebral column.

3) Describe the difference between a typical male and female pelvis.

____

4) The _______________ can be palpated by following the superior pelvis from the ASIS to the PSIS on the side of the torso.

5) Which pair of bony landmarks can be visually identified by two small dimples at the base of the lower back?

____

6) The _______________ are often called the "sits bones."

7) Which large bony landmark can be located by sliding your fingerpads inferiorly four to six inches along the lateral side of the thigh? _______________

8) The _______________ is located on the medial surface of the ilium and serves as an attachment site for the iliacus muscle.

Let's Palpate!

Remember—there are no right or wrong answers here

Locate and explore the **anterior superior iliac spines (ASISs)** on three individuals. Then write three words that describe what you feel. (See p. 288 in *Trail Guide.*)

Person #1 _______________	Person #2 _______________	Person #3 _______________
_______________	_______________	_______________
_______________	_______________	_______________
_______________	_______________	_______________

Pelvis and Thigh

Bones and Bony Landmarks of the Pelvis #2

*Please identify the following structures. Numbers in **black** indicate bones, numbers in **red** are bony landmarks.*

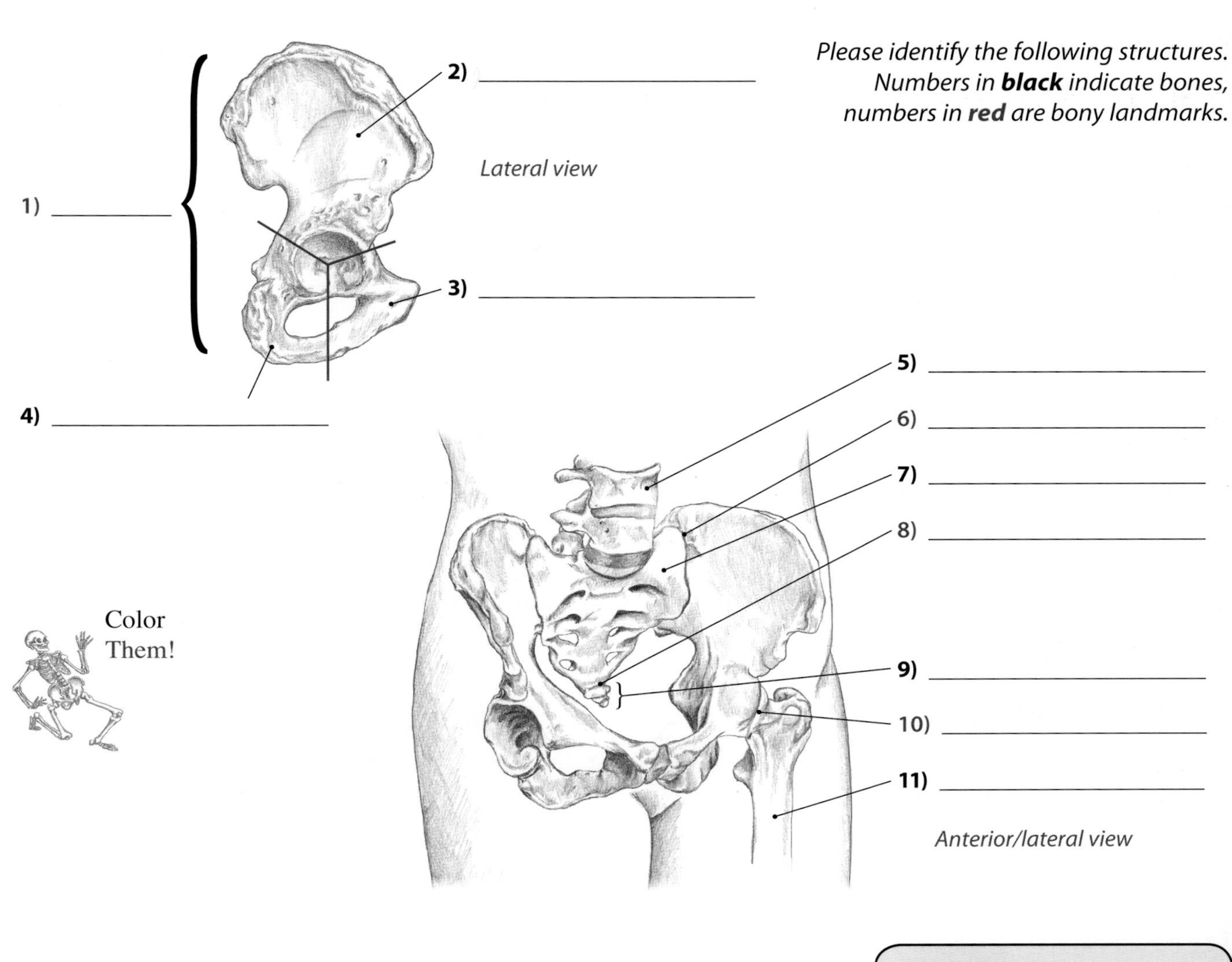

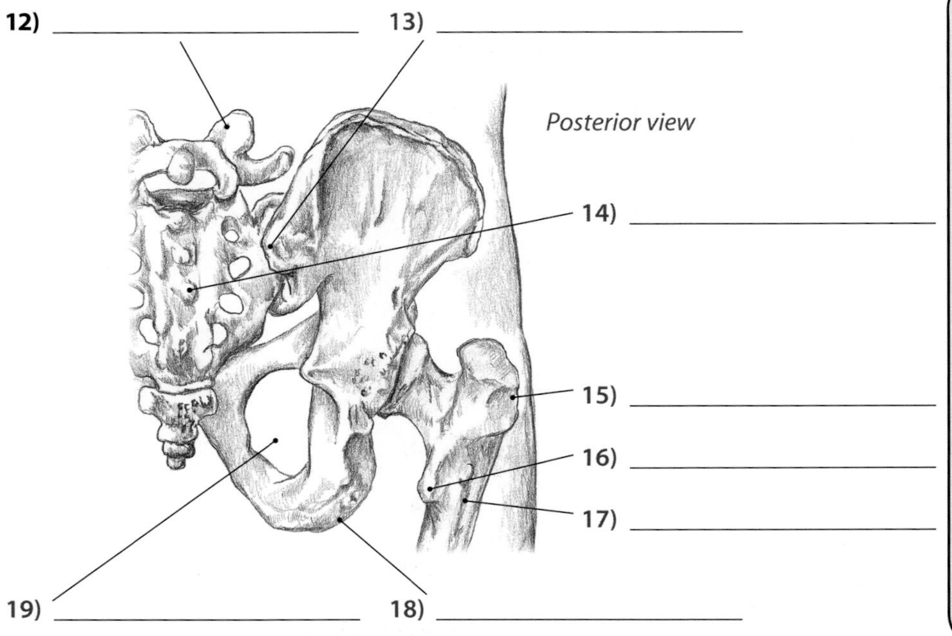

CHOICES

- Coccyx
- Coxal (hip) joint
- Femur
- Fifth lumbar vertebra
- Gluteal tuberosity
- Greater trochanter
- Hip
- Ilium
- Ischial tuberosity
- Ischium
- Lesser trochanter
- Lumbar vertebra
- Median sacral crest
- Obturator foramen
- Posterior superior iliac spine (PSIS)
- Pubis
- Sacrococcygeal joint
- Sacroiliac joint
- Sacrum

Please identify the following structures.

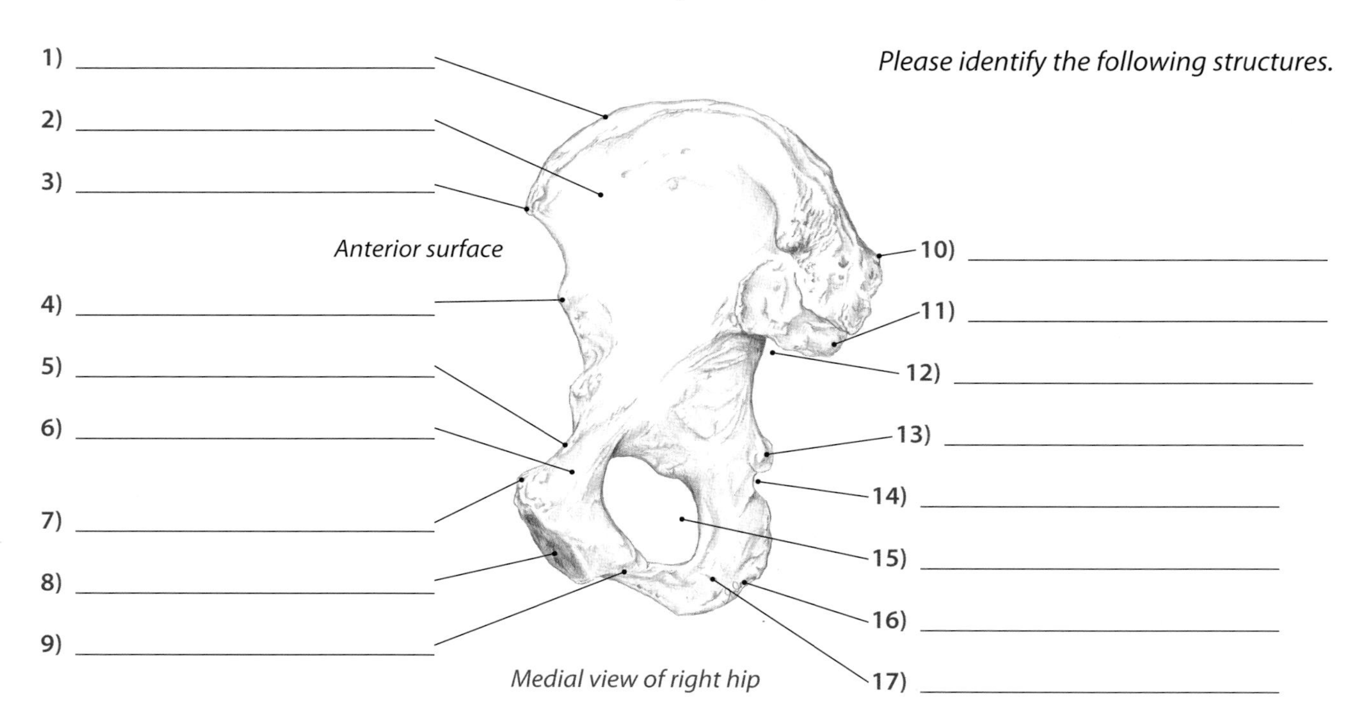

Medial view of right hip

CHOICES

Acetabulum	Inferior gluteal line	Posterior gluteal line
Anterior gluteal line	Inferior ramus of pubis (2)	Posterior inferior iliac spine (2)
Anterior inferior iliac spine (2)	Ischial spine (2)	Posterior superior iliac spine (2)
Anterior superior iliac spine (2)	Ischial tuberosity (2)	Pubic tubercle (2)
Greater sciatic notch (2)	Lesser sciatic notch (2)	Ramus of the ischium
Iliac crest (2)	Obturator foramen (2)	Superior ramus of pubis (2)
Iliac fossa	Pectineal line	Symphyseal surface
Iliac tubercle		

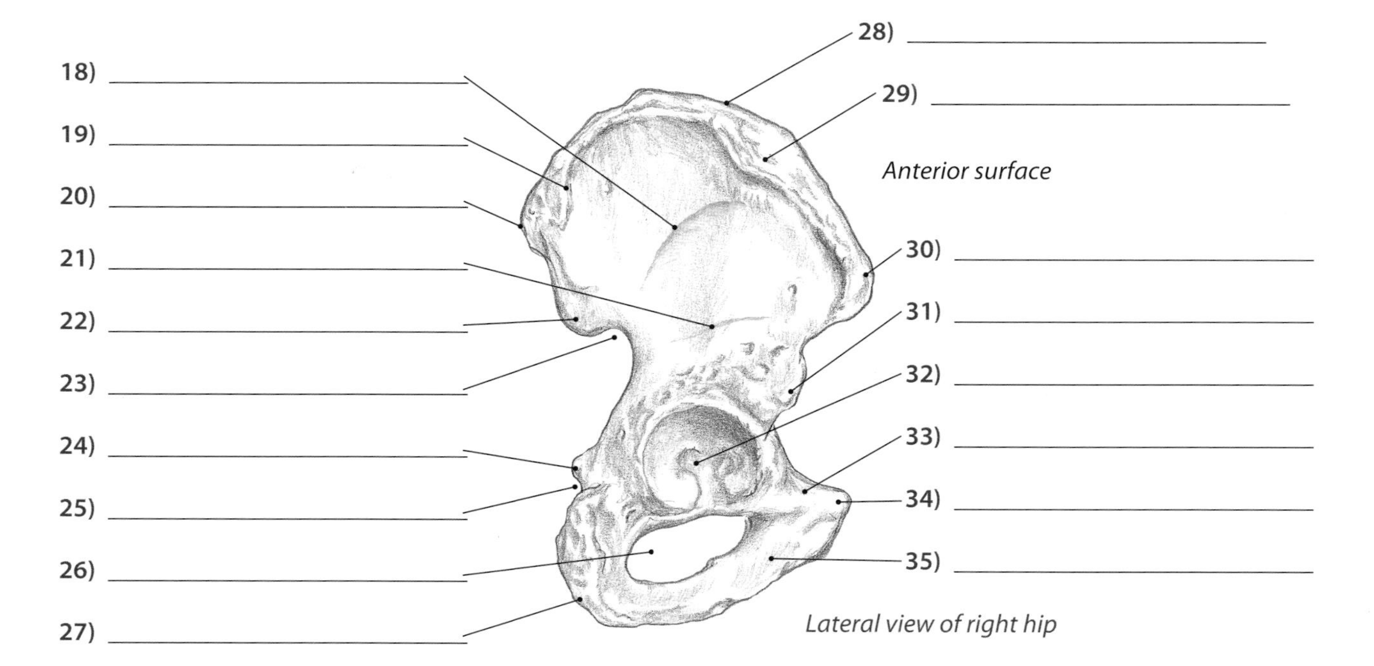

Lateral view of right hip

Pelvis and Thigh

Pelvis and Sacrum

Please identify the following structures.

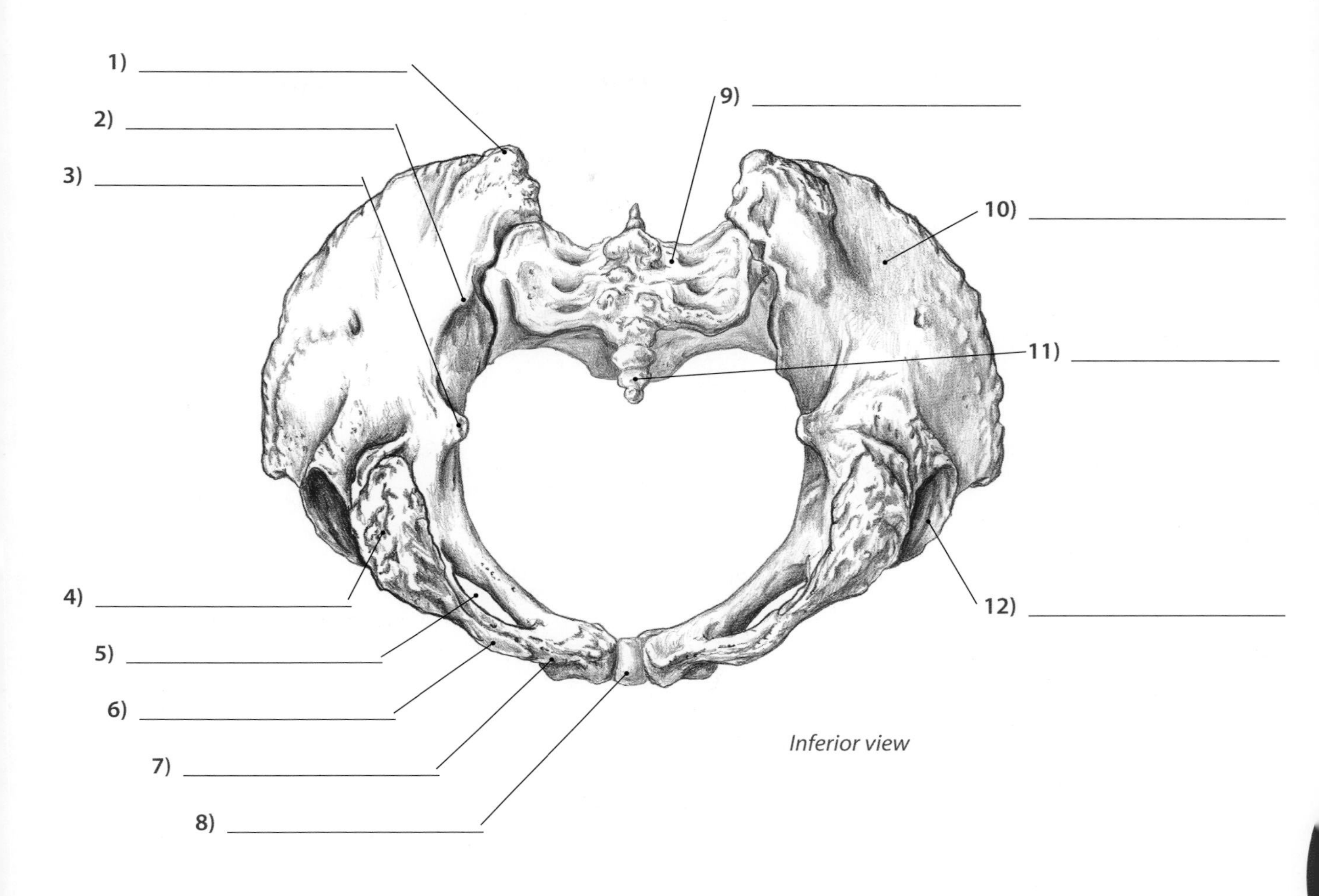

Inferior view

CHOICES

- Acetabulum
- Coccyx
- Gluteal surface of ilium
- Inferior ramus of the pubis
- Ischial spine
- Ischial tuberosity
- Obturator foramen
- Posterior inferior iliac spine
- Posterior superior iliac spine
- Pubic symphysis
- Ramus of ischium
- Sacrum

Pelvis and Thigh

Femur

Please identify the following structures.

CHOICES

- Adductor tubercle (2)
- Fovea of head
- Gluteal tuberosity
- Greater trochanter (2)
- Head (2)
- Intercondylar fossa
- Intertrochanteric crest
- Intertrochanteric line
- Lateral condyle (2)
- Lateral epicondyle (2)
- Lateral lip of linea aspera
- Lesser trochanter (2)
- Medial condyle (2)
- Medial epicondyle (2)
- Medial lip of linea aspera
- Neck (2)
- Patellar surface
- Pectineal line
- Shaft
- Trochanteric fossa

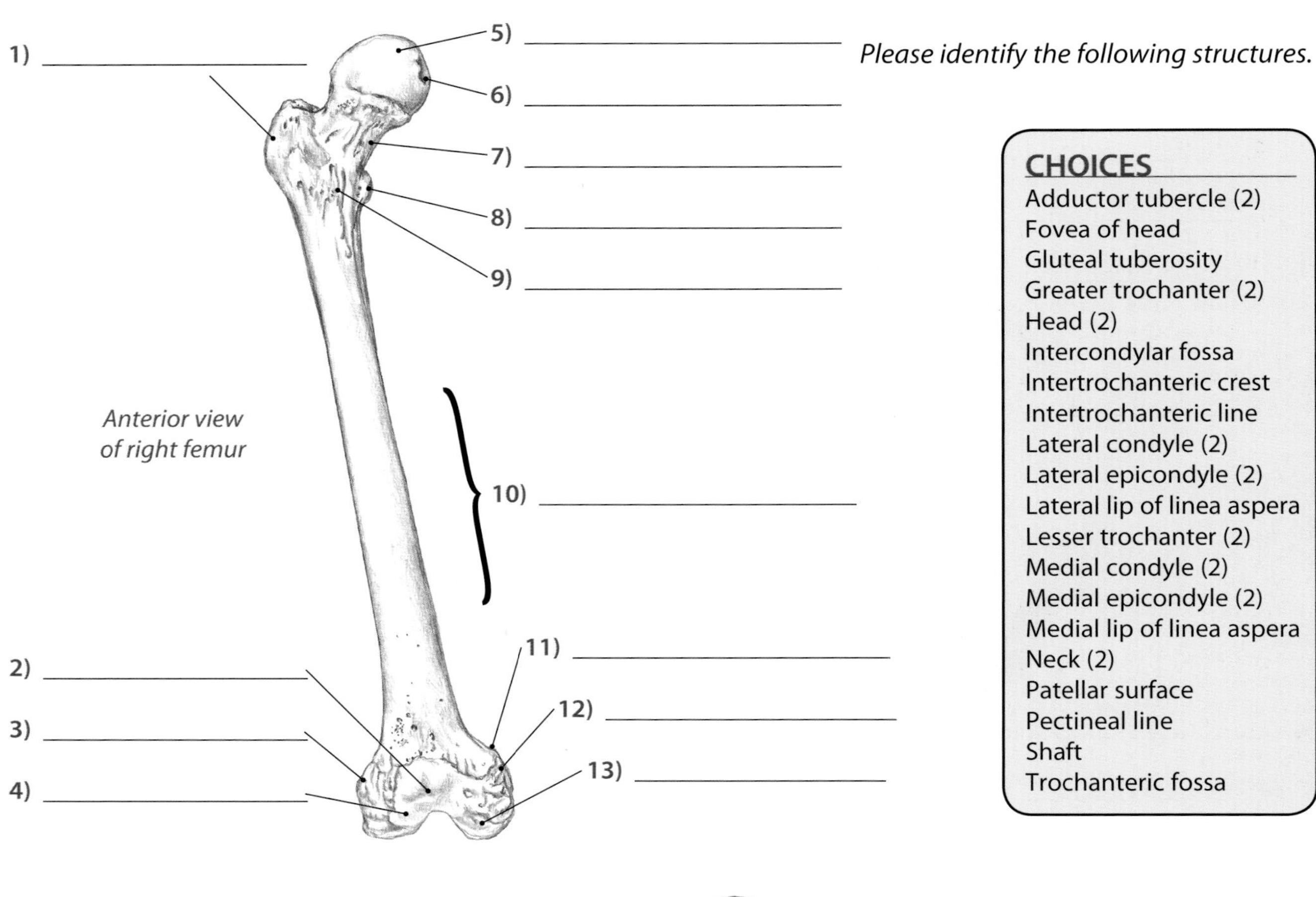

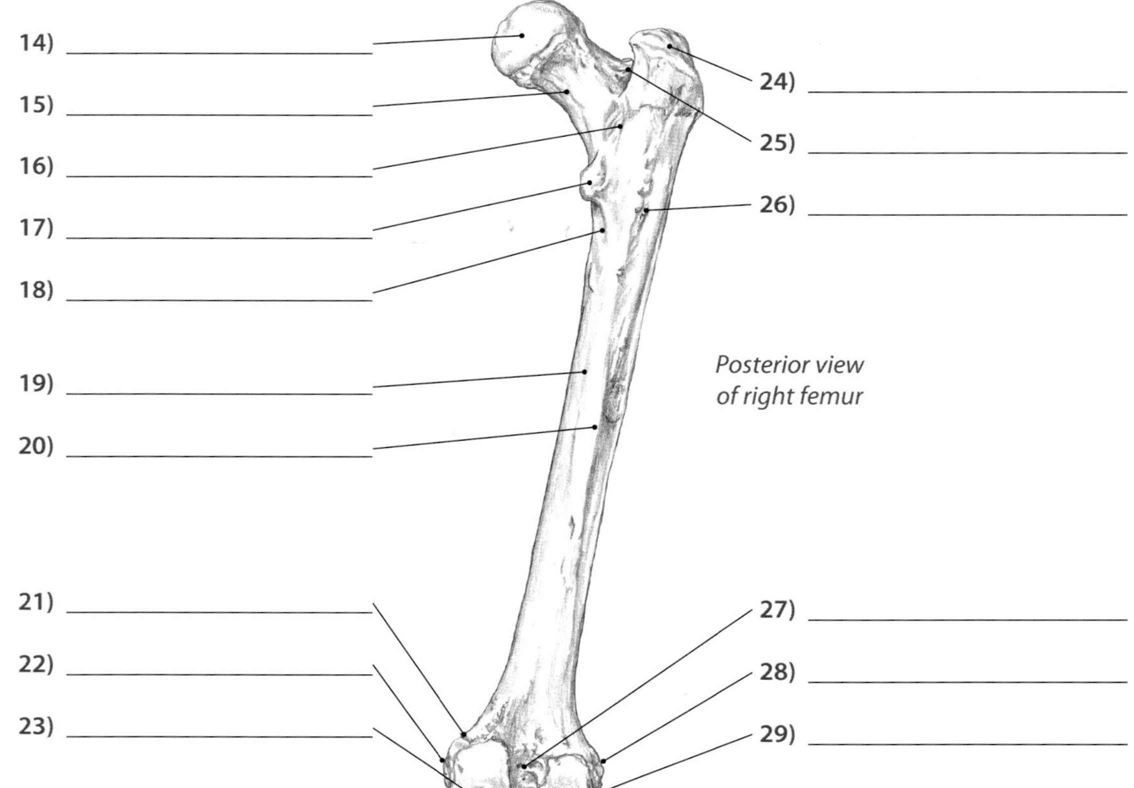

Pelvis and Thigh

Bones and Bony Landmarks of the Pelvis #3

Please answer the following questions.

1) The ______________________ is comprised of 4-5 fused vertebrae and the ______________________ is comprised of 3-4 fused bones.

2) The ridge running down the center of the sacrum is the ______________________.

3) The coccyx is located nearest to which topographical feature? ______________________

4) The ______________________ joint can be found just inferior and medial to the PSIS.

5) With your partner prone, what passive positional adjustment and motion will help you to feel movement in the sacroiliac joint?

____

6) Which bony landmark can be found just distal to the greater trochanter and directly lateral to the ischial tuberosity?

____

7) What are a couple ways to increase comfort for both you and your partner while palpating in the pubic region?

____

8) The ______________________ are the bony prominences located on the superior part of the pubic crest.

9) The superior ramus of the pubis forms a ridge that serves as an attachment site for the ______________________.

10) The rami of the pubis form a bridge between the ______________________ and the ______________________.

11) What is the recommended position of your partner while palpating the pubic rami?

____

12) The ______________________ is the horizontal line between the buttock and thigh.

Let's Palpate!

Remember—there are no right or wrong answers here

Locate and explore the **pubic crest and tubercles** on three individuals. Then write three words that describe what you feel. (See p. 293 in *Trail Guide.*)

Person #1 ____________	Person #2 ____________	Person #3 ____________
____________	____________	____________
____________	____________	____________
____________	____________	____________

Muscles of the Pelvis and Thigh #1

Please identify the following structures.

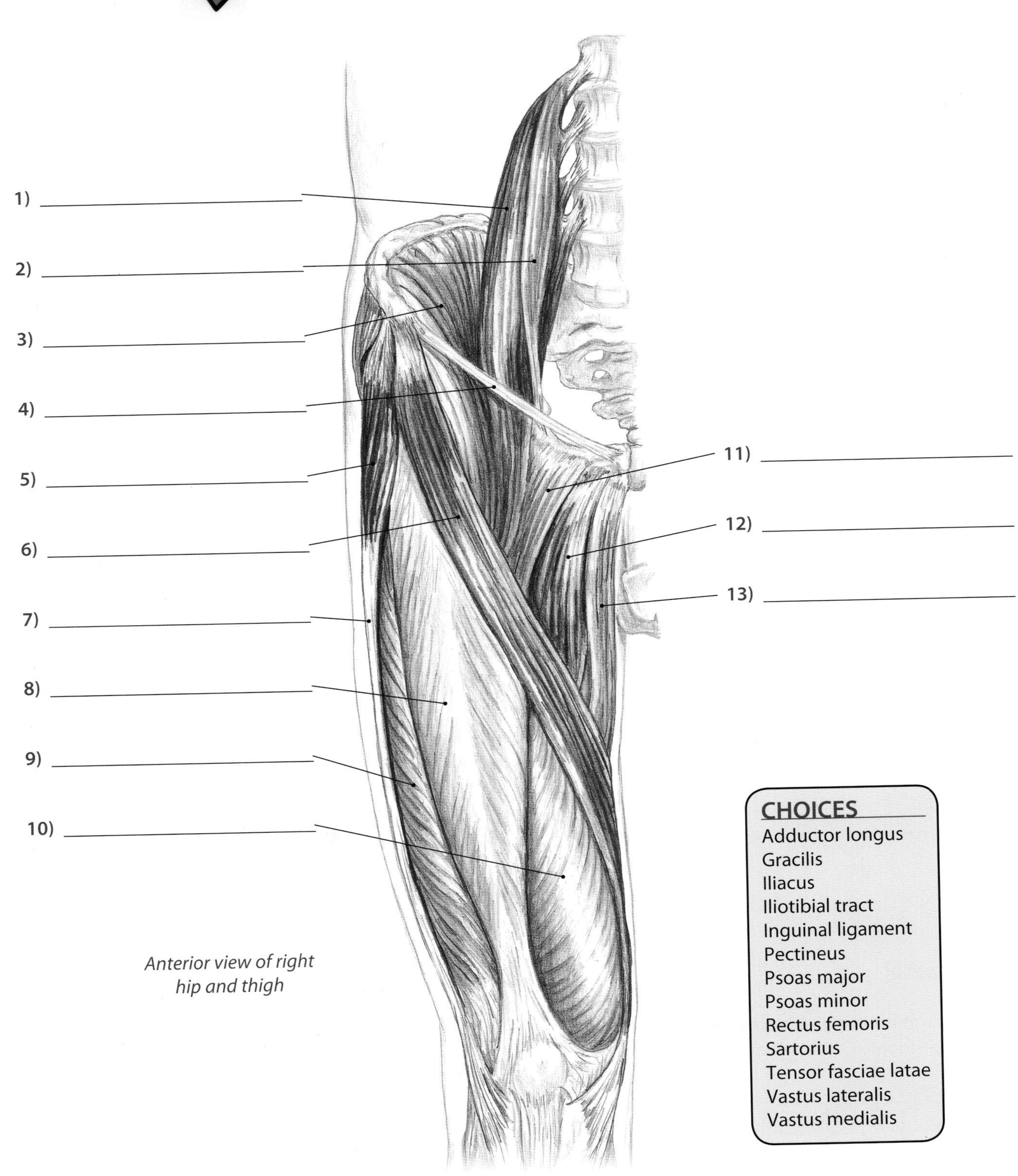

Anterior view of right hip and thigh

CHOICES
- Adductor longus
- Gracilis
- Iliacus
- Iliotibial tract
- Inguinal ligament
- Pectineus
- Psoas major
- Psoas minor
- Rectus femoris
- Sartorius
- Tensor fasciae latae
- Vastus lateralis
- Vastus medialis

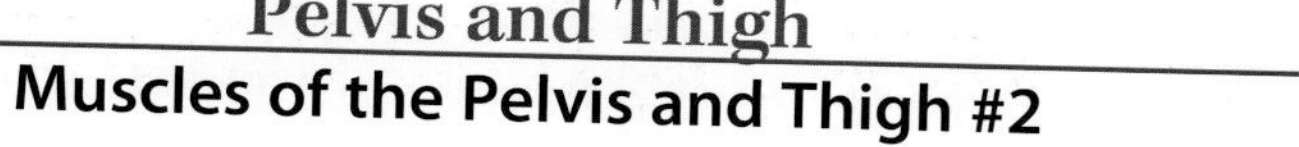

Pelvis and Thigh

Muscles of the Pelvis and Thigh #2

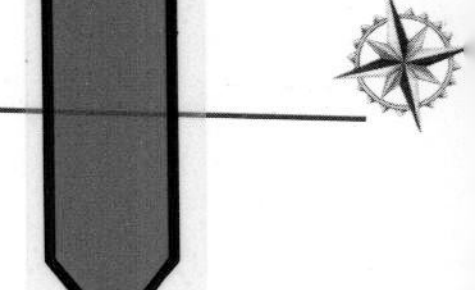

Please identify the following structures.

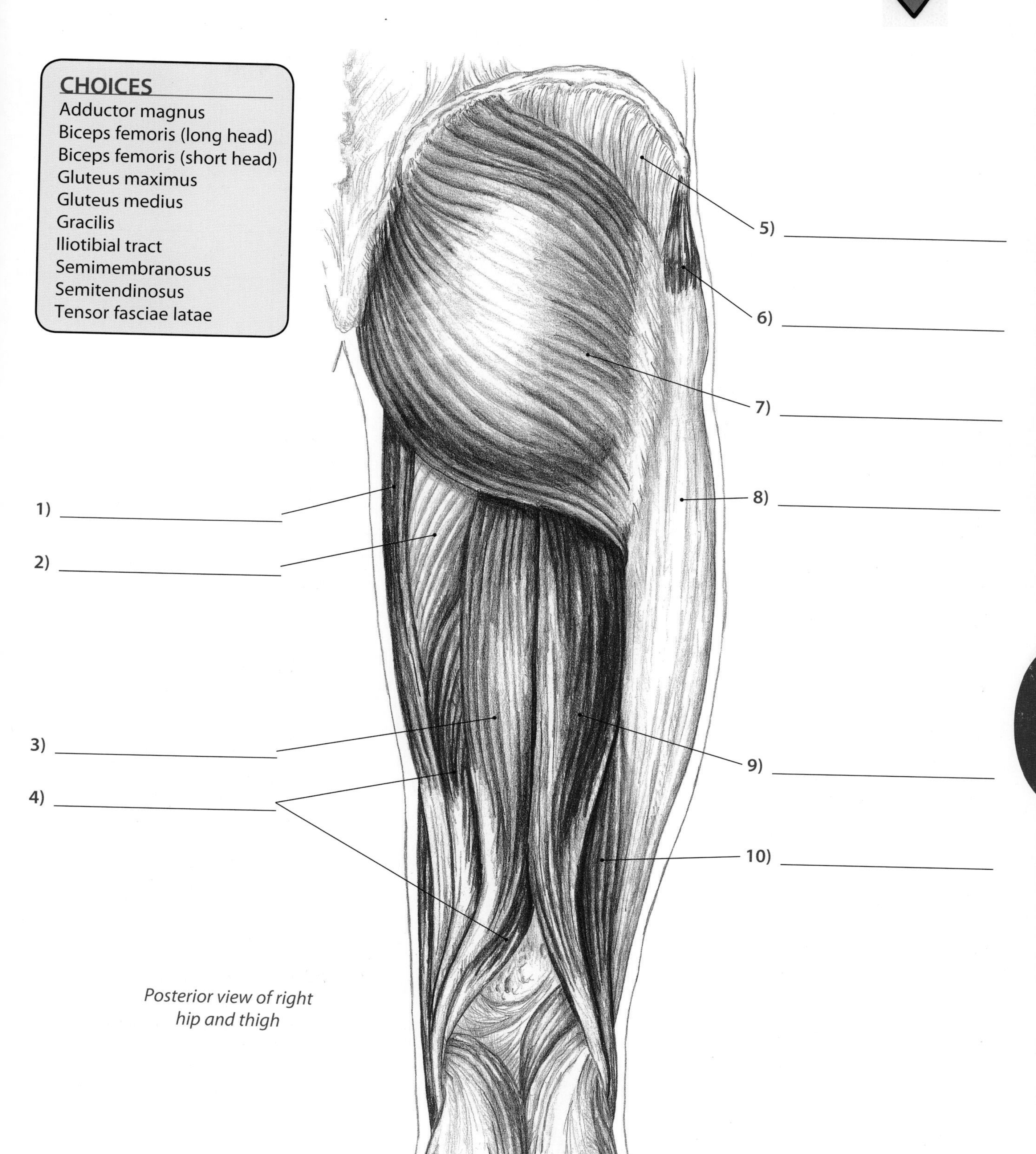

Posterior view of right hip and thigh

Muscles of the Pelvis and Thigh #3

Please identify the following structures.

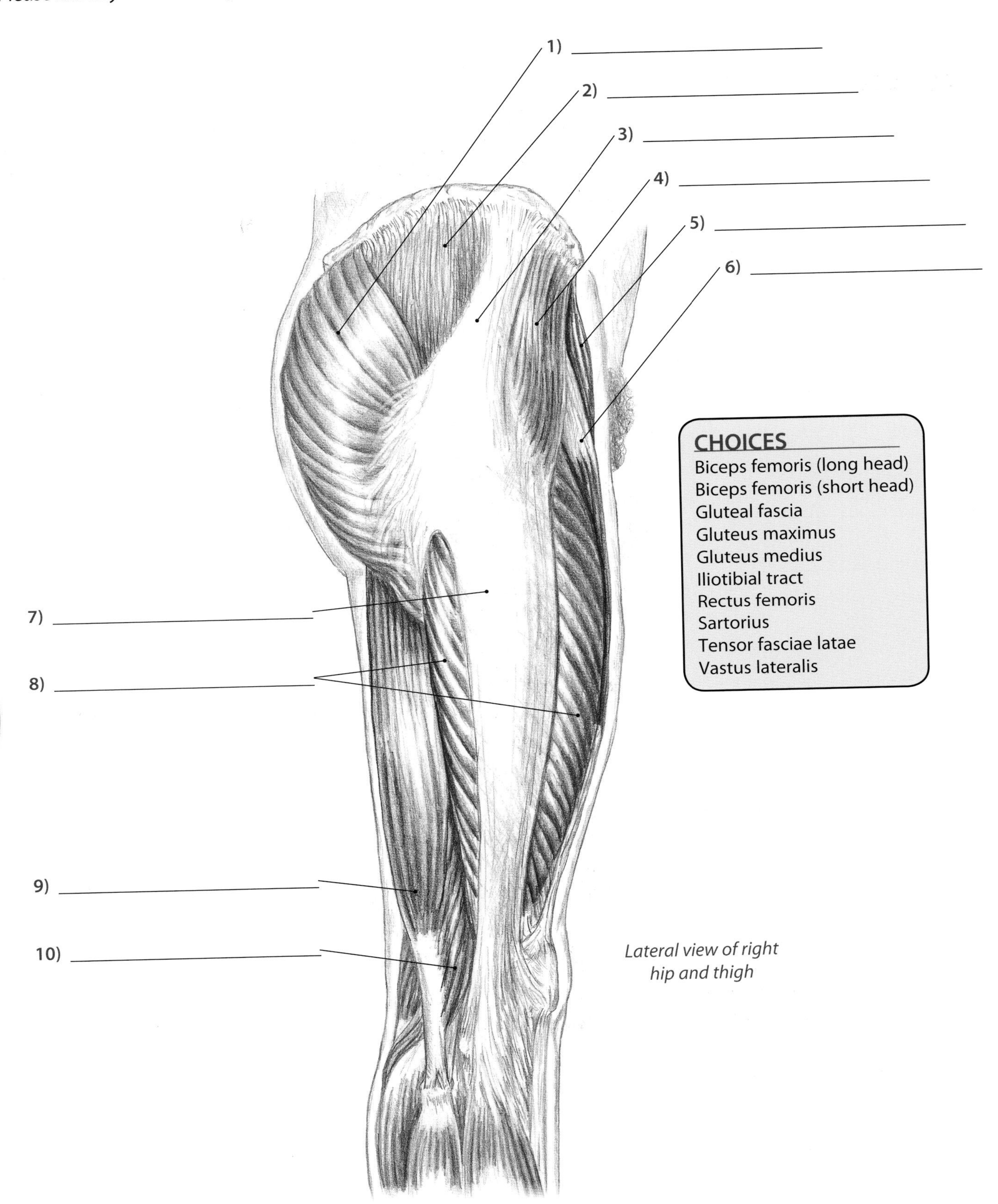

Lateral view of right hip and thigh

Pelvis and Thigh

Muscles of the Pelvis and Thigh #4

Please identify the following structures.

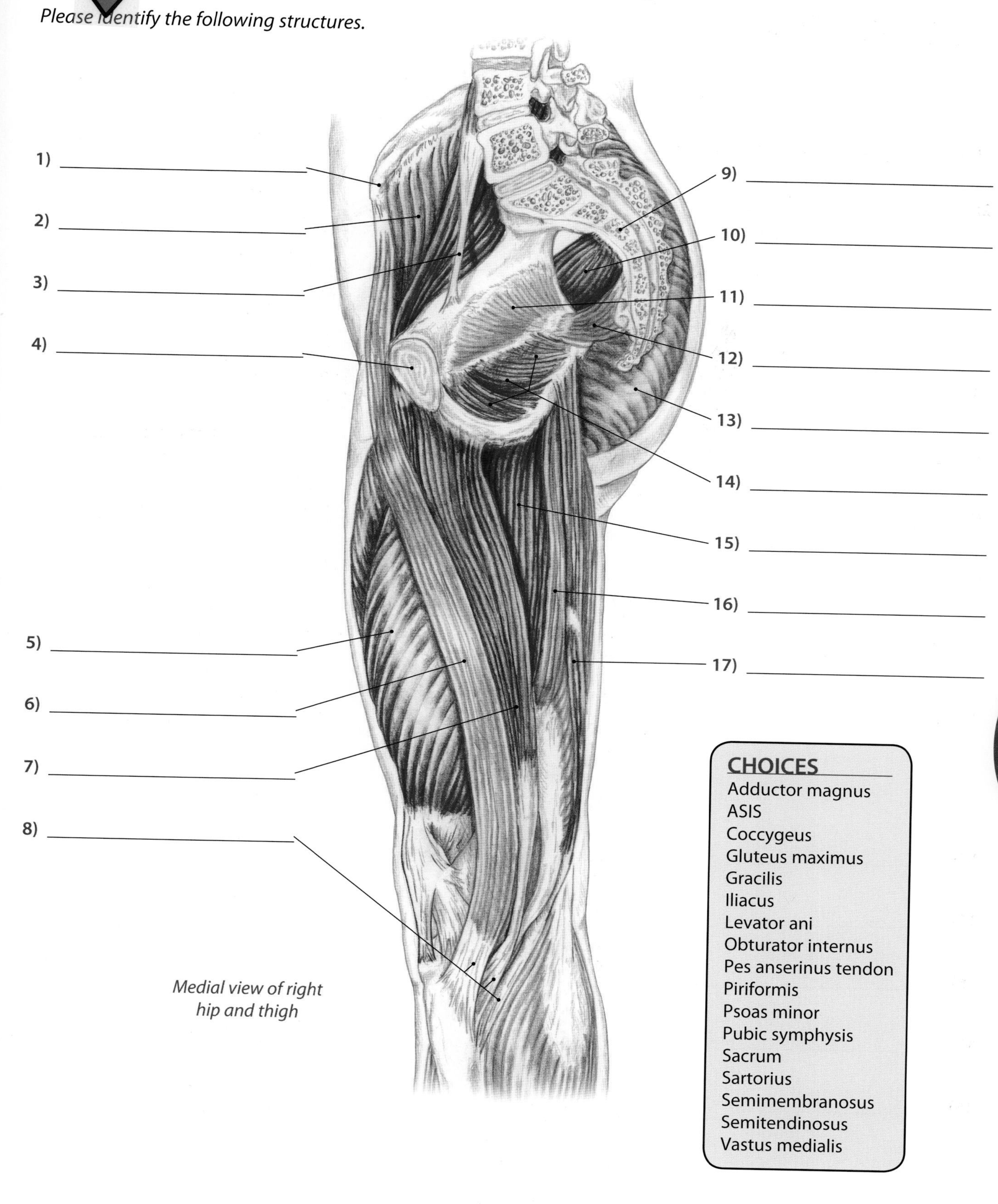

Medial view of right hip and thigh

CHOICES

Adductor magnus
ASIS
Coccygeus
Gluteus maximus
Gracilis
Iliacus
Levator ani
Obturator internus
Pes anserinus tendon
Piriformis
Psoas minor
Pubic symphysis
Sacrum
Sartorius
Semimembranosus
Semitendinosus
Vastus medialis

Muscles of the Pelvis and Thigh #5

Please identify the following structures.

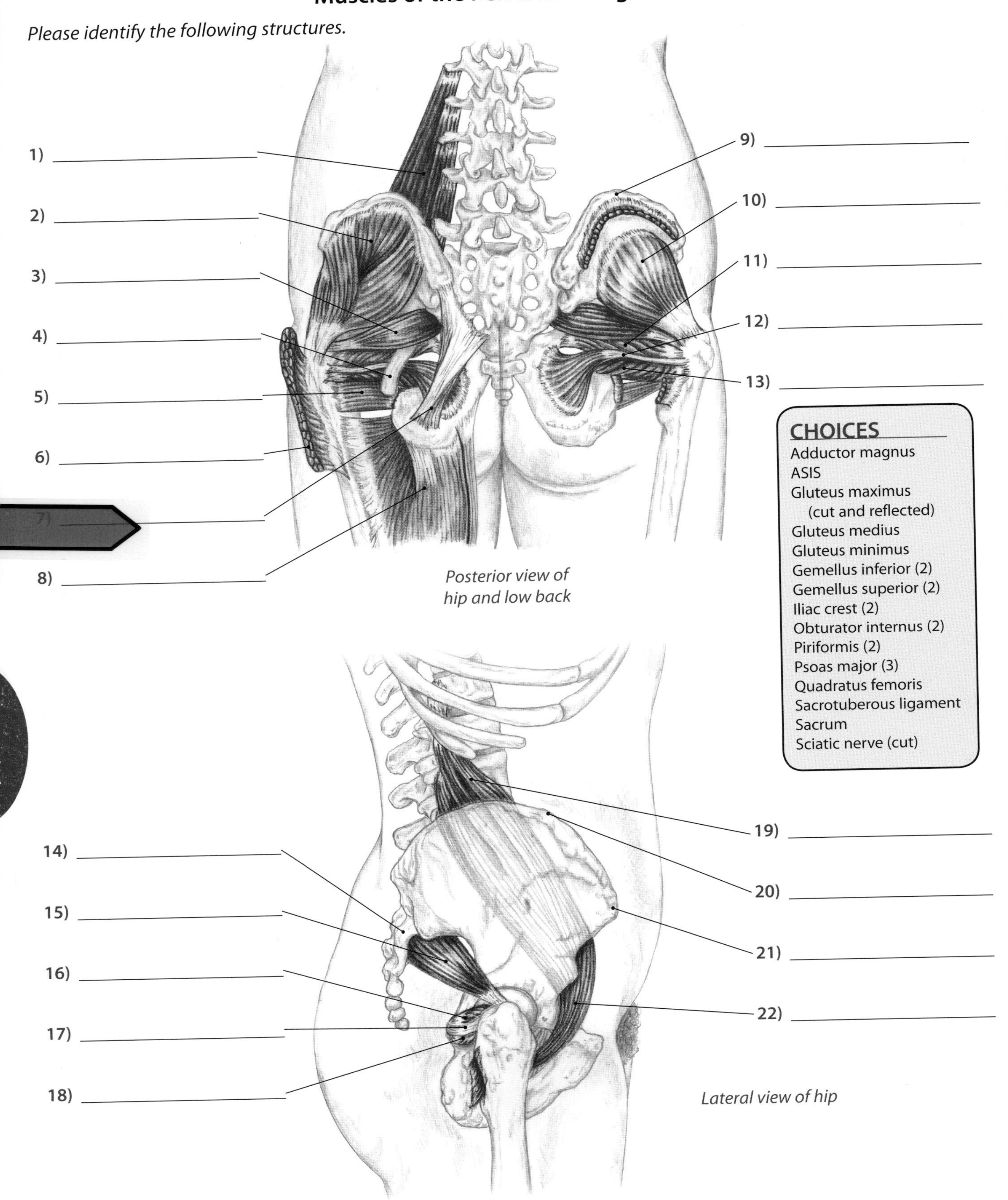

Posterior view of hip and low back

Lateral view of hip

p. 300-301

Pelvis and Thigh

Perineum and Pelvic Floor

Please identify the following structures.

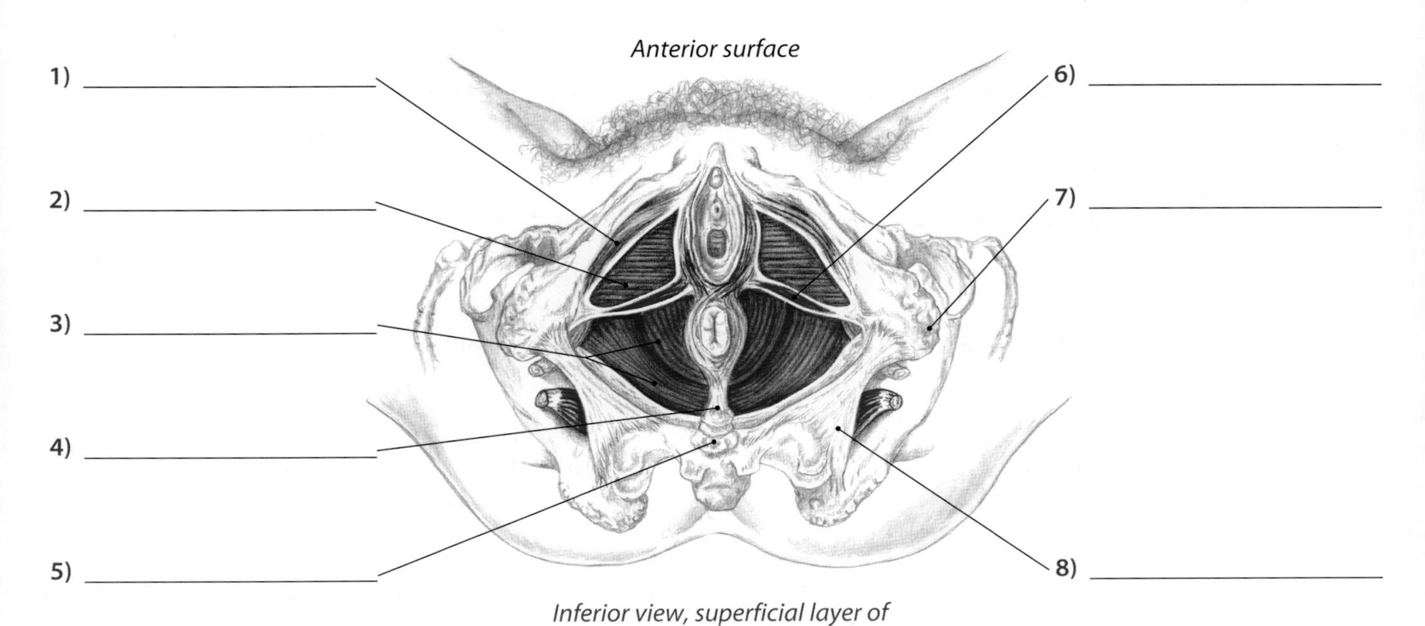

Inferior view, superficial layer of female pelvic floor (hips abducted)

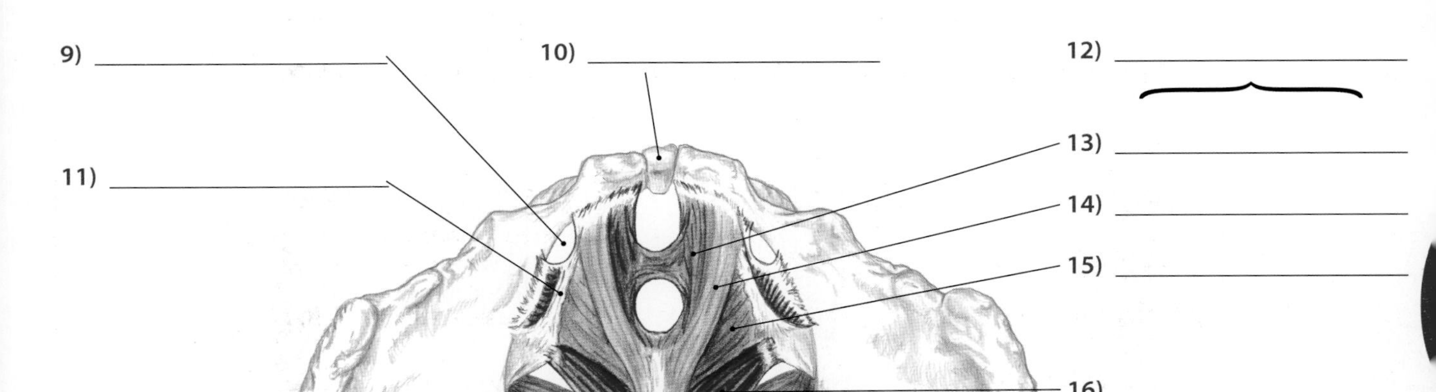

Superior view, superficial layer of female pelvic floor

CHOICES

Anococcygeal ligament	Ischiocavernosus	Puborectalis
Coccygeus	Levator ani (2)	Sacrotuberous ligament
Coccyx (2)	Obturator canal	Sacrum
Deep transverse perineal	Piriformis	Superficial transverse perineal
Iliococcygeus	Pubic symphysis	Tendinous arch of levator ani
Ischial tuberosity	Pubococcygeus	

Color the Muscles #1

p. 296

Using different colors, please fill in and label the muscles and other structures listed below.

Adductor longus
Gracilis
Iliacus
Iliotibial tract
Inguinal ligament
Pectineus
Psoas major
Psoas minor
Rectus femoris
Sartorius
Tensor fasciae latae
Vastus lateralis
Vastus medialis

Anterior view of right hip and thigh

Pelvis and Thigh

Color the Muscles #2

Using different colors, please fill in and label the muscles and other structures listed below.

- Adductor magnus
- Biceps femoris (long head)
- Biceps femoris (short head)
- Gluteus maximus
- Gluteus medius
- Gracilis
- Iliotibial tract
- Semimembranosus
- Semitendinosus
- Tensor fasciae latae

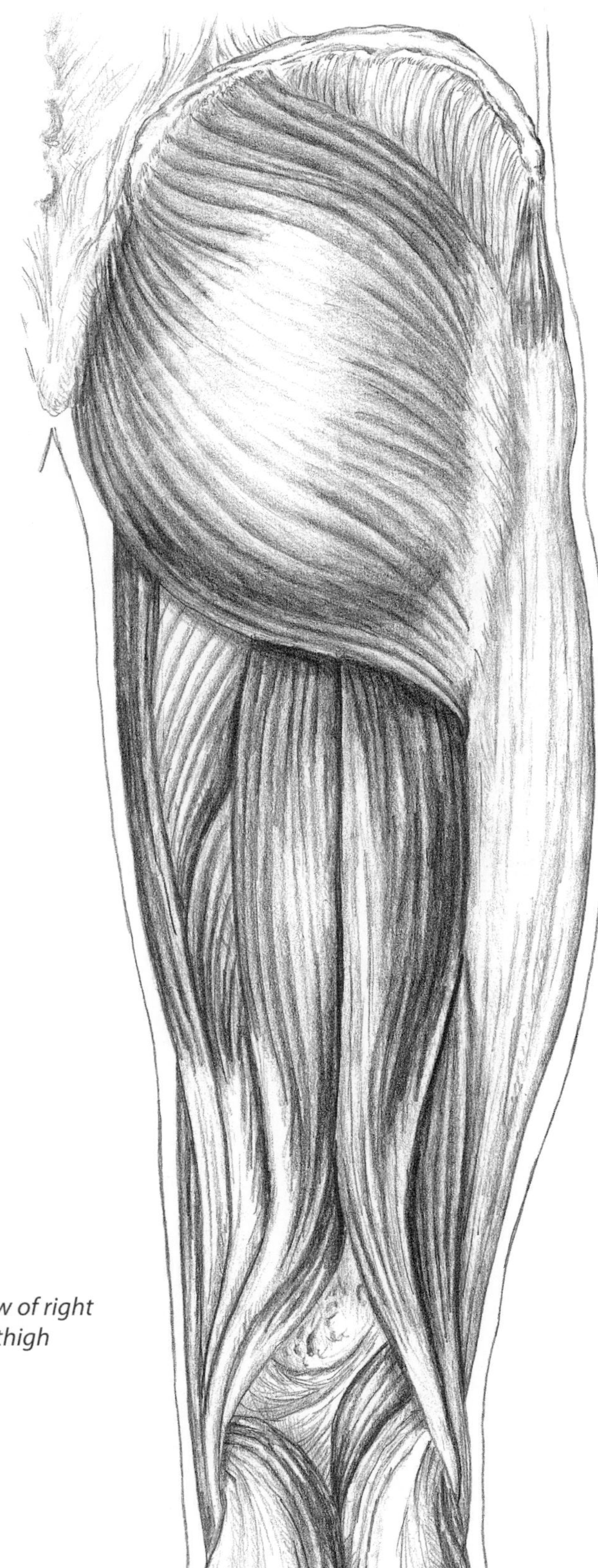

Posterior view of right hip and thigh

p. 297

Color the Muscles #3

Using different colors, please fill in and label the muscles and other structures listed below.

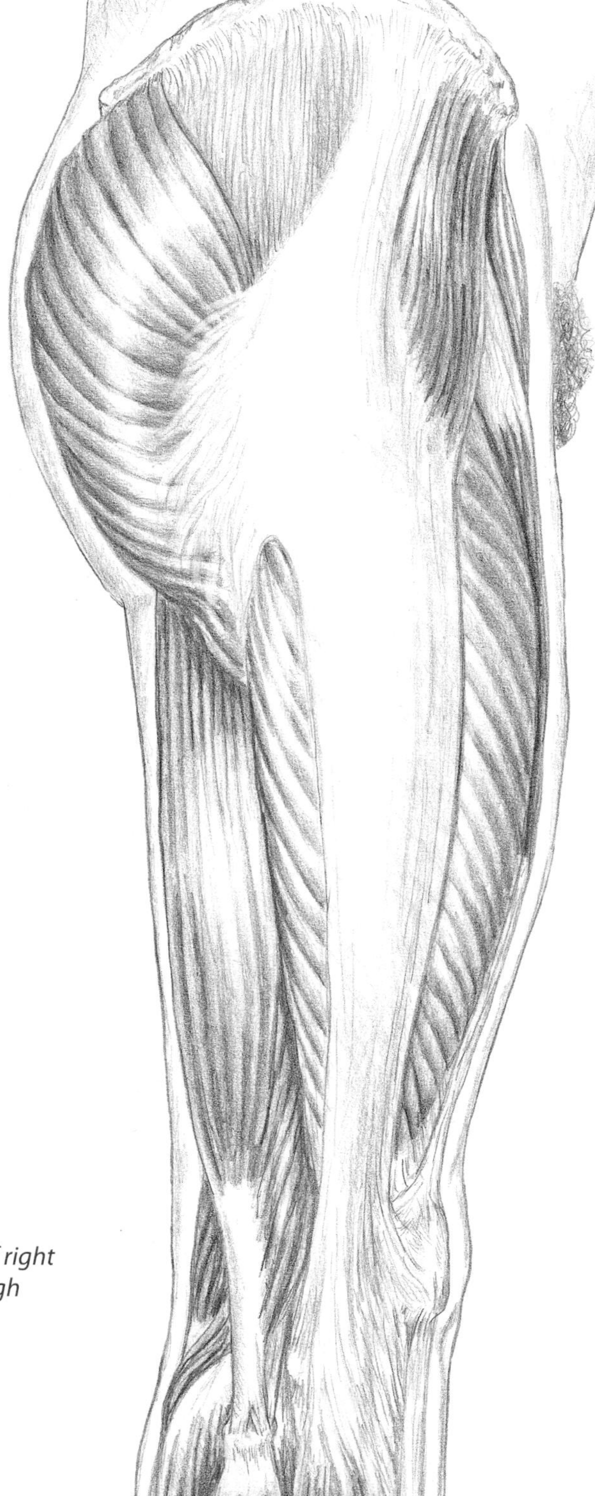

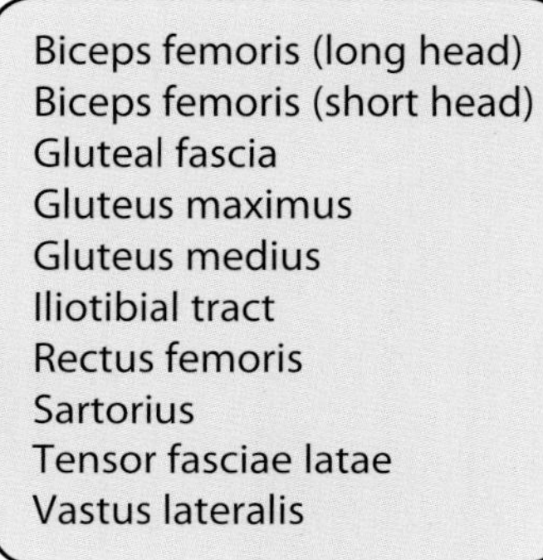

Biceps femoris (long head)
Biceps femoris (short head)
Gluteal fascia
Gluteus maximus
Gluteus medius
Iliotibial tract
Rectus femoris
Sartorius
Tensor fasciae latae
Vastus lateralis

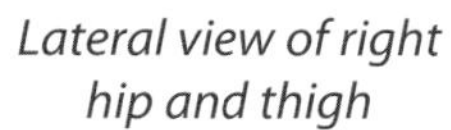

Lateral view of right hip and thigh

Pelvis and Thigh

Color the Muscles #4

Using different colors, please fill in and label the muscles and other structures listed below.

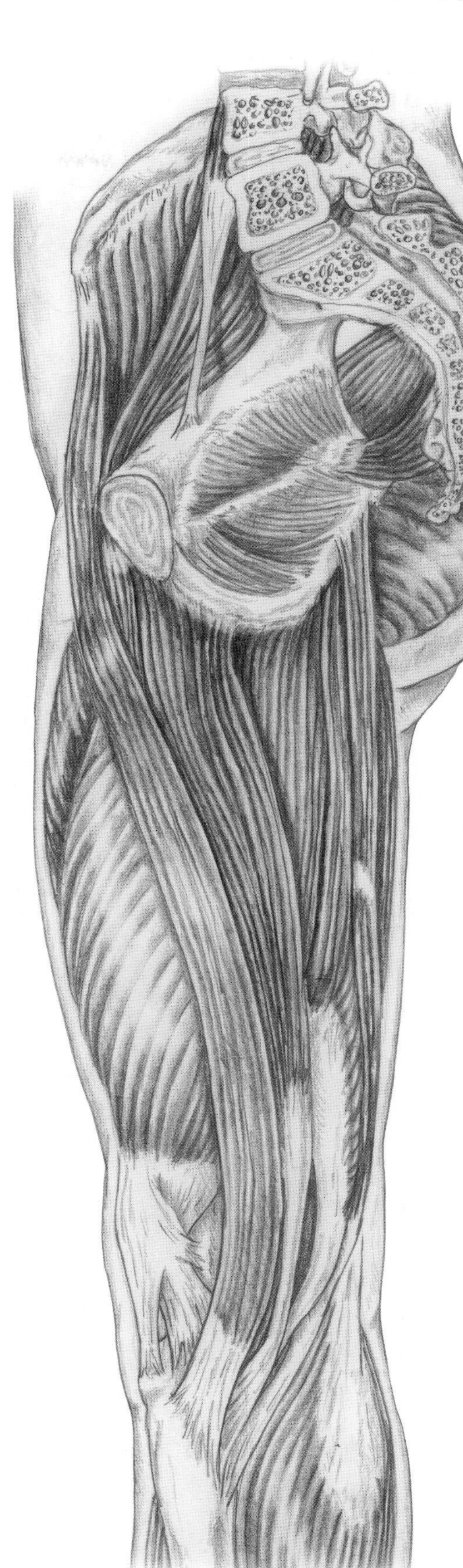

Adductor longus
Adductor magnus
Gracilis
Pes anserinus tendon
Sartorius
Semimembranosus
Semitendinosus
Tendon of semitendinosus
Vastus medialis

Medial view of right hip and thigh

Muscles and Movements #1

Please list the action demonstrated, two synergists and one antagonist. The first letter of each muscle has been provided.

1) This action happens at which joint?

____

2) Action

____

3) Synergists

B ______________________

A ______________________

4) Antagonist

R ______________________

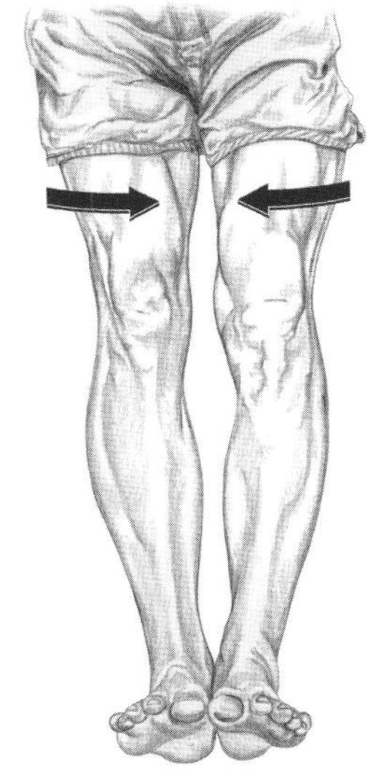

5) Action

____

6) Synergists

G ______________________

P ______________________

7) Antagonist

T ______________________

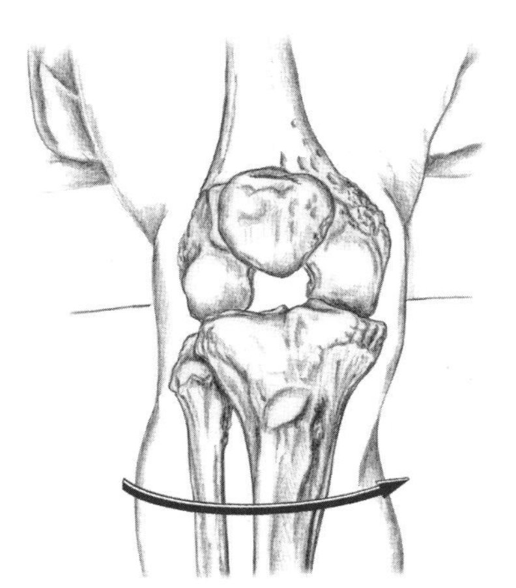

8) Action

____

9) Synergists

S ______________________

P ______________________

10) Antagonist

B ______________________

Pelvis and Thigh

Muscles and Movements #2

Please list the action demonstrated, two synergists and one antagonist.
The first letter of each muscle has been provided.

1) This action happens at which joint?

____

2) Action

____

3) Synergists

V ______________________________

V ______________________________

4) Antagonist

G ______________________________

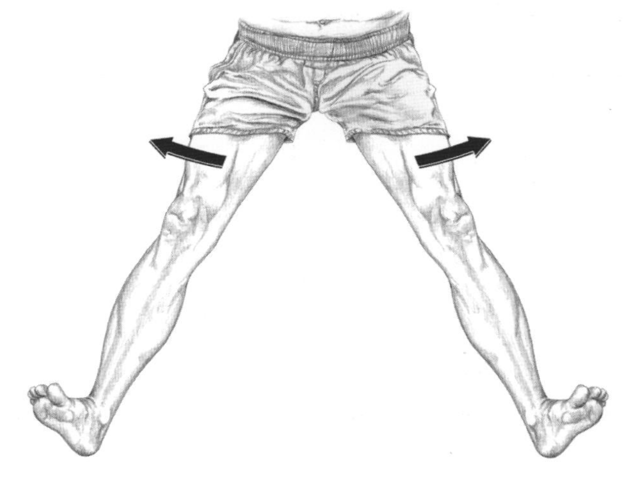

5) Action

____

6) Synergists

G ______________________________

G ______________________________

7) Antagonist

P ______________________________

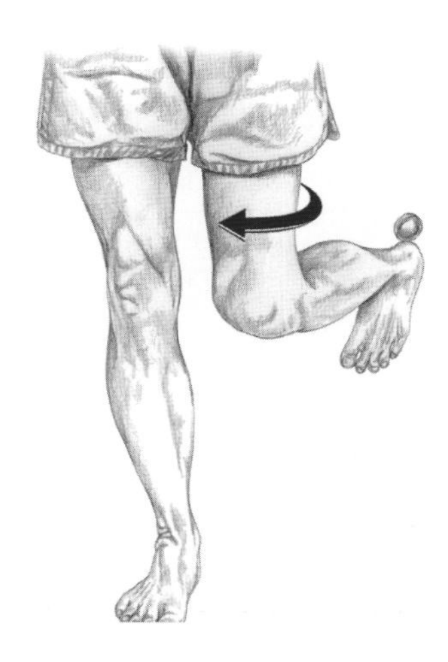

8) Action

____

9) Synergists

G ______________________________

P ______________________________

10) Antagonist

B ______________________________

Muscles and Movements #3

p. 302-305

Please list the action demonstrated, synergist(s) and antagonist(s).
The first letter of each muscle has been provided.

1) This action happens at which joint?

____

2) Action

____

3) Synergists

B ______________________

G ______________________

4) Antagonist

R ______________________

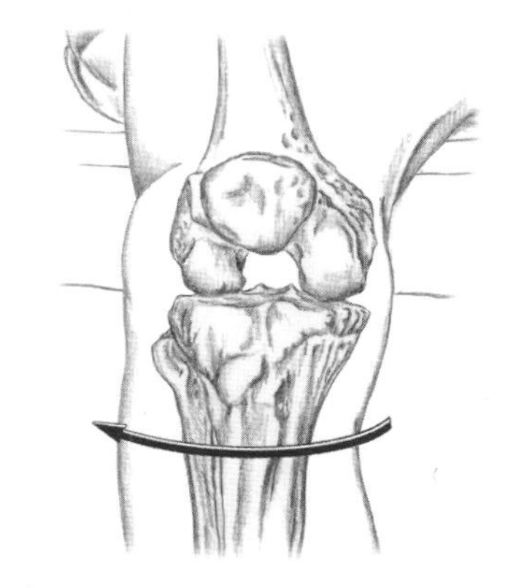

5) Action

____

6) Synergist

B ______________________

7) Antagonists

S ______________________

S ______________________

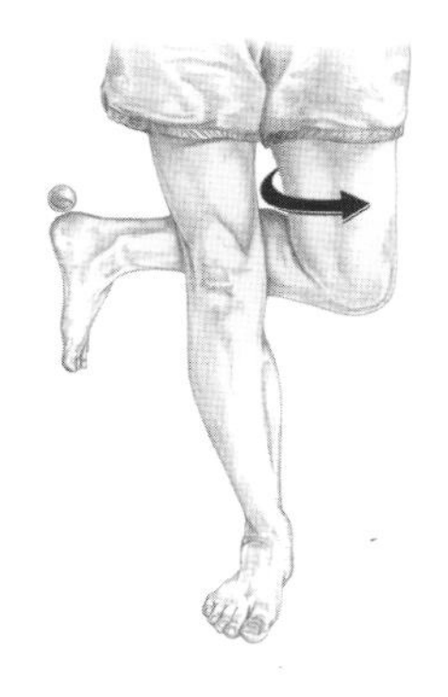

8) Action

____

9) Synergists

P ______________________

I ______________________

10) Antagonist

A ______________________

11) Action

____

12) Synergists

T ______________________

S ______________________

13) Antagonist

G ______________________

What's the Muscle? #1

Please identify the following muscles.

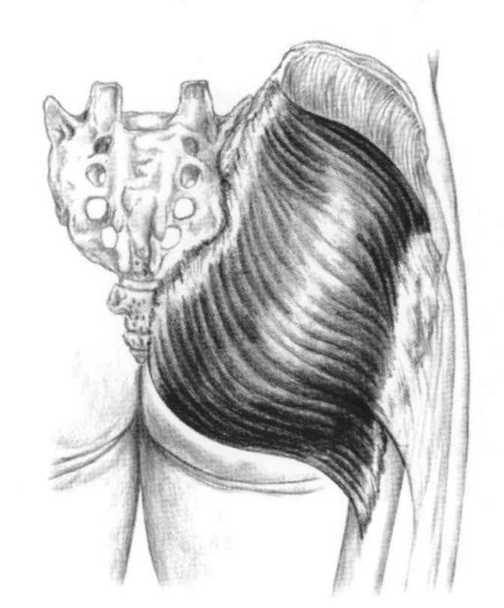

1) ______________________________

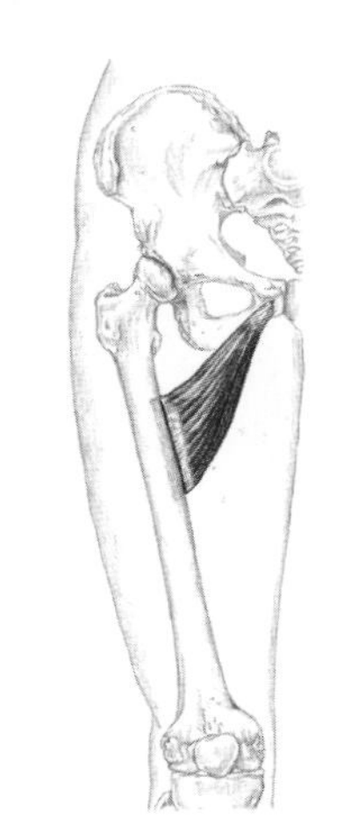

2) ______________________________

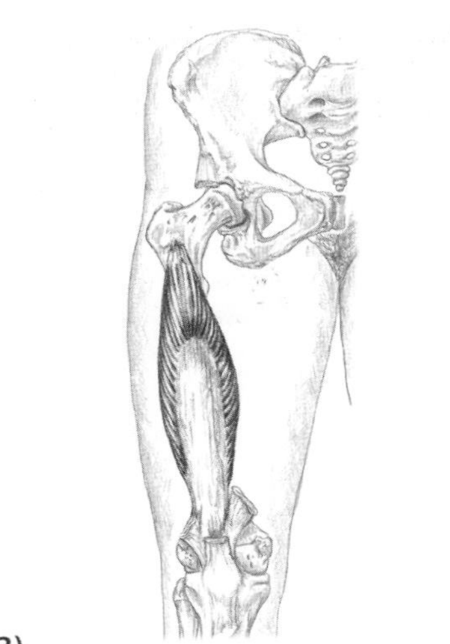

3) ______________________________

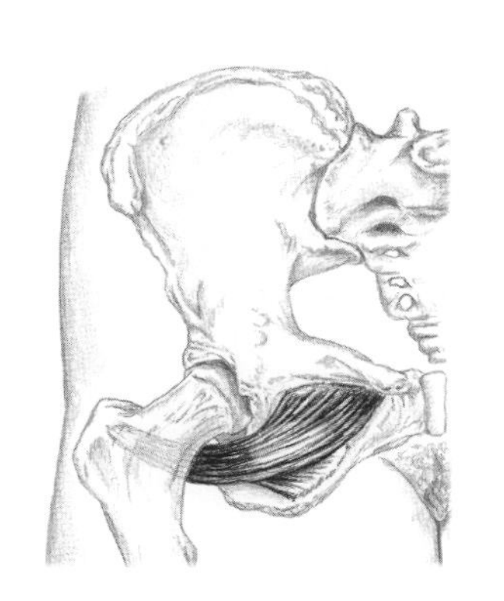

4) ______________________________

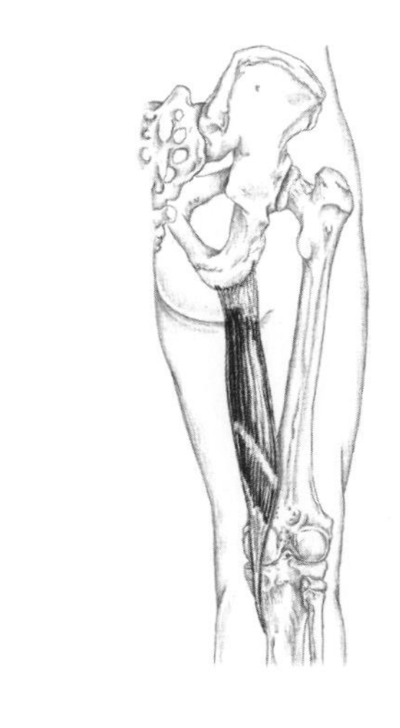

5) ______________________________

6) ______________________________

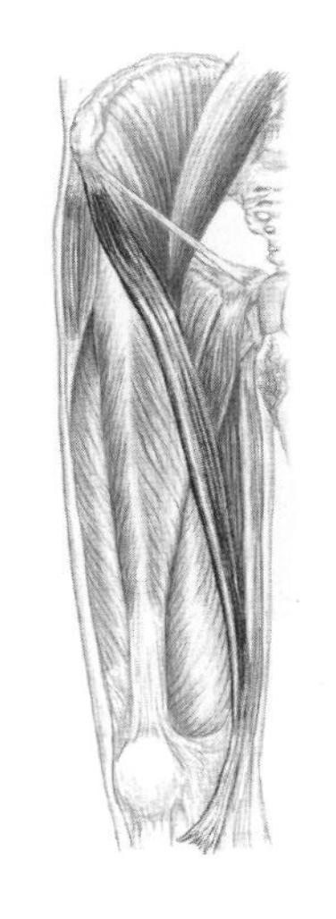

7) ______________________________

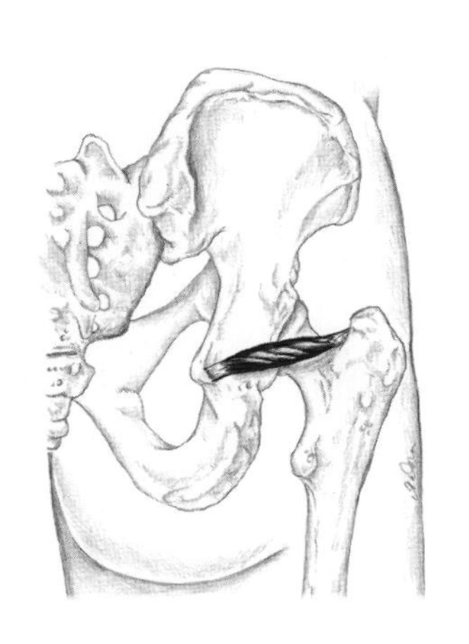

8) ______________________________

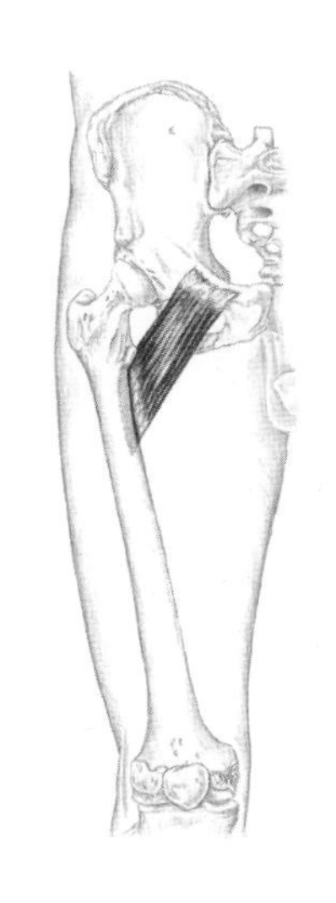

9) ______________________________

What's the Muscle? #2

Please identify the following muscles.

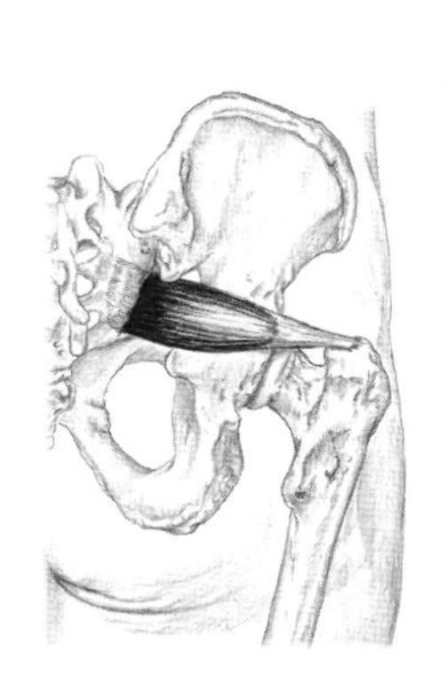

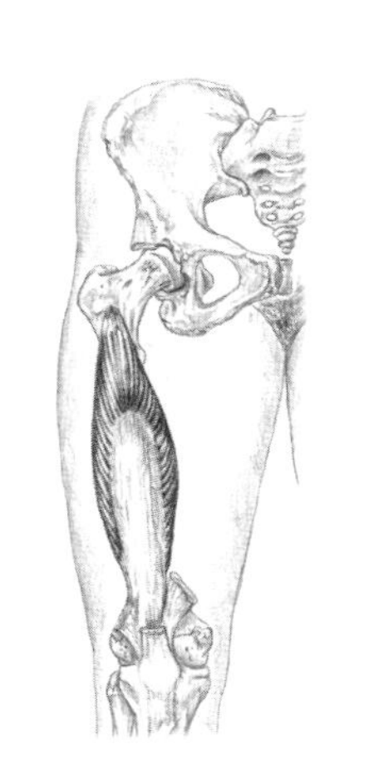

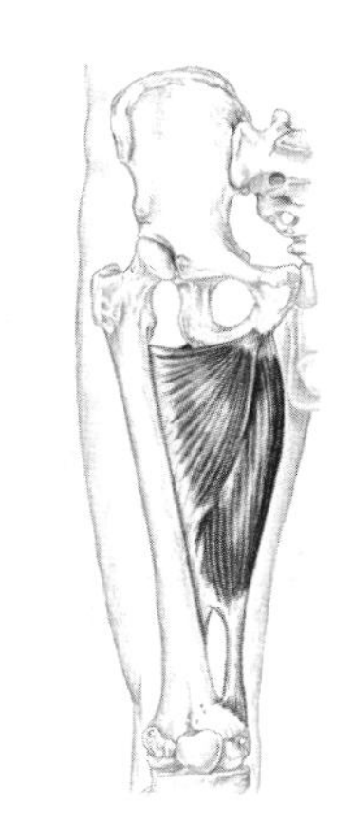

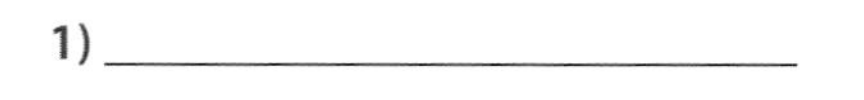

1) ____________________

2) ____________________

3) ____________________

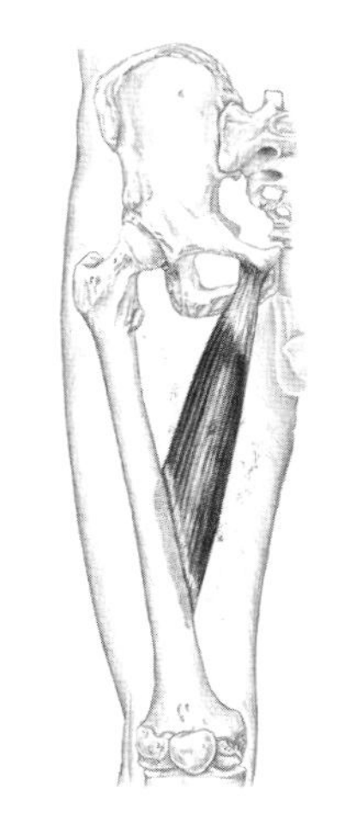

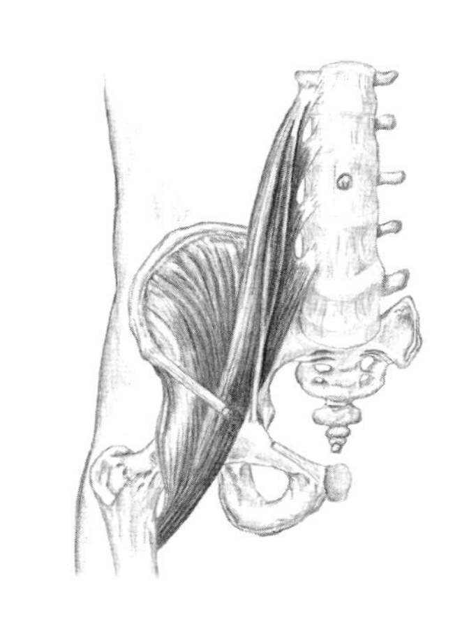

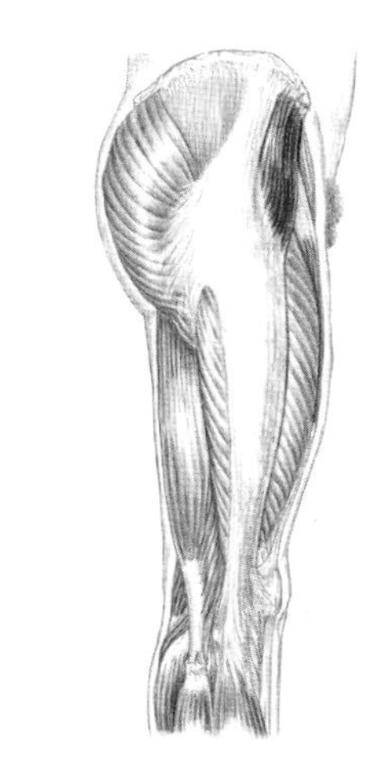

4) ____________________

5) ____________________

6) ____________________

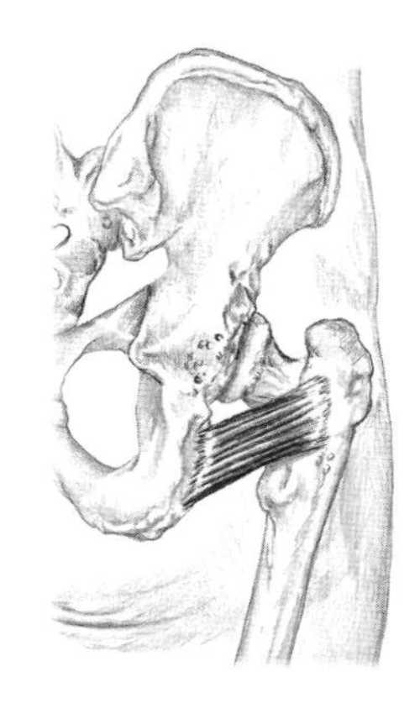

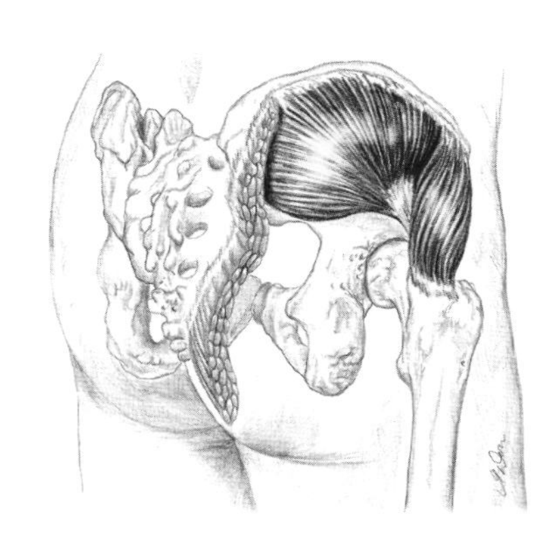

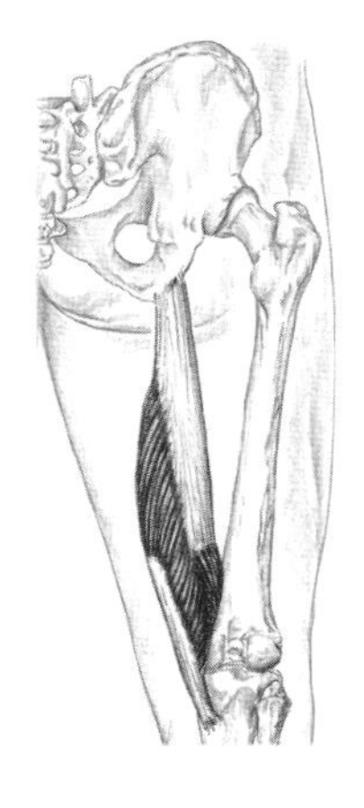

7) ____________________

8) ____________________

9) ____________________

What's the Muscle? #3

Please identify the following muscles.

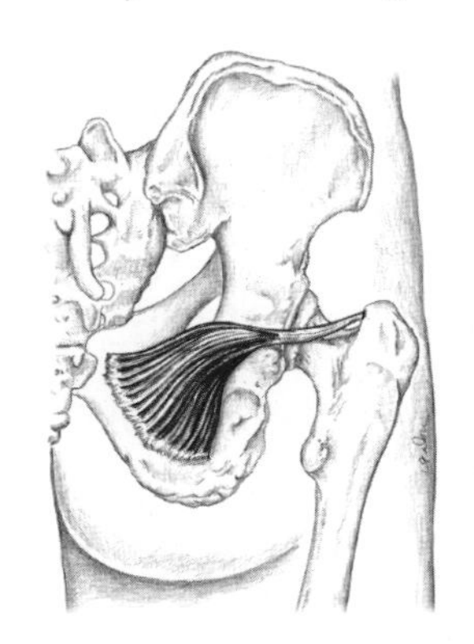

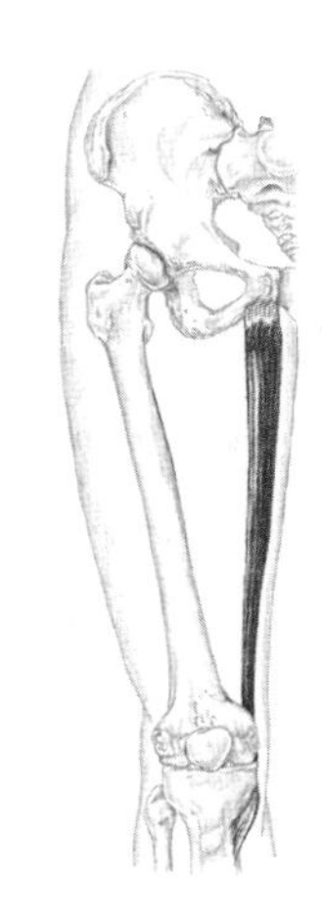

1) ______________________________ 2) ______________________________ 3) ______________________________

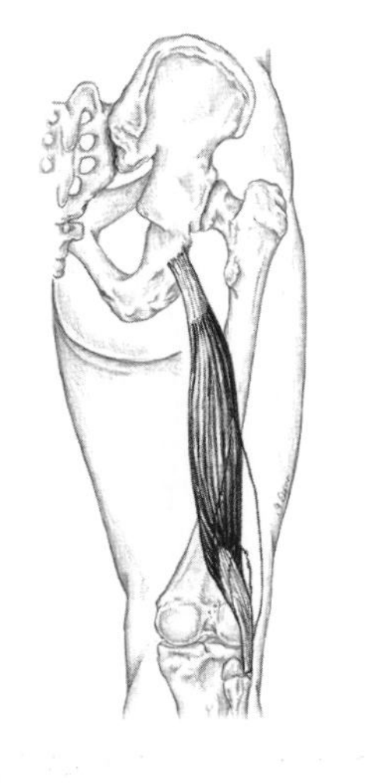

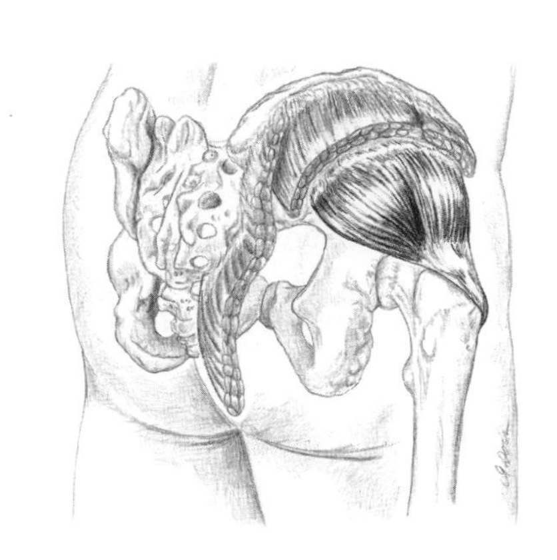

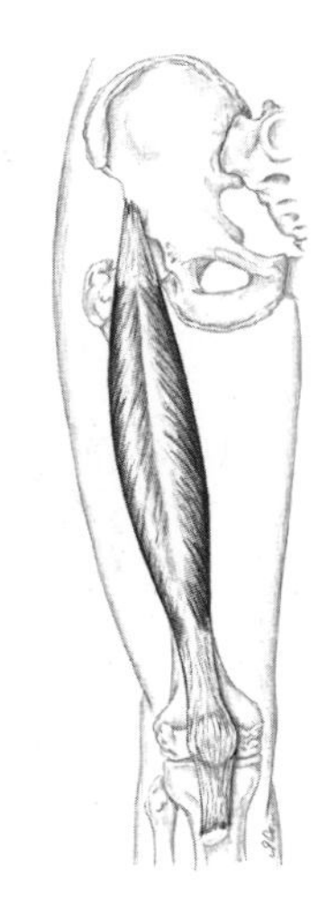

4) ______________________________ 5) ______________________________ 6) ______________________________

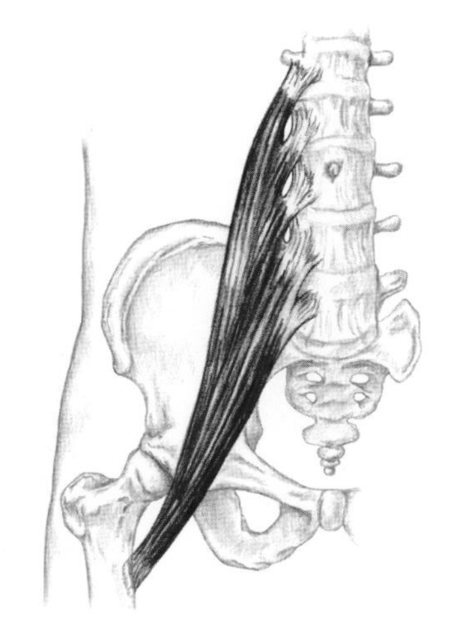

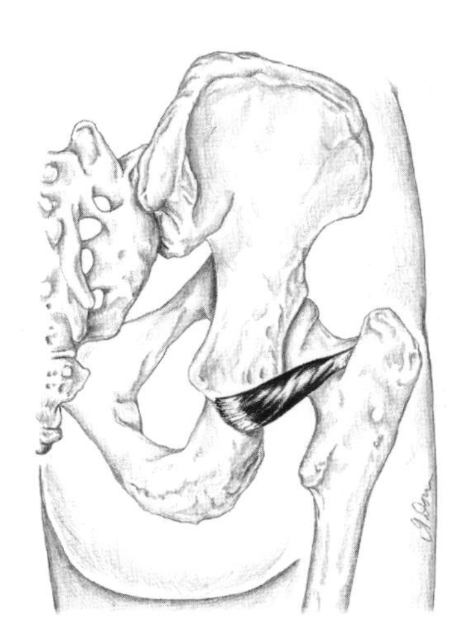

7) ______________________________ 8) ______________________________

Pelvis and Thigh, Muscle Group #1

Quadriceps and Hamstrings

p. 296-314

Please answer the following questions.

1) The muscles of the pelvis and thigh primarily create movement at the ______________________ and ______________________ joints.

2) ______________________ is the only quadriceps muscle that crosses two joints, the hip and knee.

3) Deep to the iliotibial tract, the ______________________ is the sole muscle of the lateral thigh.

4) To follow the path of the rectus femoris, it is helpful to draw an imaginary line from the ______________________ to the ______________________.

5) While your partner extends his knee, palpate just medial and proximal to the patella for the bulbous shape of the ______________________.

6) All three hamstrings share a common origin at the ______________________.

7) The hamstrings are located on the posterior thigh between the ______________________ and ______________________ muscles.

8) In which direction does biceps femoris rotate the hip? ______________________

9) The ______________________ is the more superficial of the medial hamstrings.

Let's Palpate!

Remember—there are no right or wrong answers here

Locate and explore the **hamstrings** on three individuals. Then write three words that describe what you feel. (See p. 313-314 in *Trail Guide*.)

Person #1 __________	Person #2 __________	Person #3 __________
______________	______________	______________
______________	______________	______________
______________	______________	______________

Pelvis and Thigh, Muscle Group #1

Quadriceps and Hamstrings

Matching

Match the origin and insertion to the correct muscle.

Origins

1) Anterior and lateral shaft of the femur
2) Anterior inferior iliac spine (AIIS)
3) Ischial tuberosity (2)
4) Ischial tuberosity, lateral lip of linea aspera
5) Lateral lip of linea aspera, gluteal tuberosity and greater trochanter
6) Medial lip of linea aspera

Insertions

7) Head of the fibula
8) Posterior aspect of medial condyle of tibia
9) Proximal, medial shaft of the tibia at pes anserinus tendon
10) Tibial tuberosity (via the patella and patellar ligament) (4)

Muscle	O	I
Biceps femoris	______	______
Rectus femoris	______	______
Semimembranosus	______	______
Semitendinosus	______	______
Vastus intermedius	______	______
Vastus lateralis	______	______
Vastus medialis	______	______

Shorten or Lengthen?

11) Passive flexion of the knee would ____________________ the vastus lateralis.
12) Passive tilting of the pelvis anteriorly would ____________________ the biceps femoris.
13) Passive medial rotation of the hip would ____________________ the semitendinosus.
14) Passive extension of the knee would ____________________ the vastus intermedius.
15) Passive lateral rotation of the flexed knee would ____________________ the biceps femoris.
16) Passive flexion of the hip would ____________________ the semimembranosus,

    but ____________________ the rectus femoris.

Pelvis and Thigh, Muscle Group #2

p. 315-323

Gluteals and Adductors

Please answer the following questions.

1) Of the three gluteal muscles, the ____________________ is the most posterior and superficial.

2) Which gluteal muscle has the ability to flex and extend the hip (but not simultaneously)?

    ____

3) Locating the coccyx, the posterior two inches of the iliac crest and gluteal tuberosity will help you to outline which muscle?

    ____

4) To palpate gluteus minimus, you will need to sink your fingers deep to which muscle? ____________________

5) To locate both gluteus medius and minimus in a side lying position, you could ask your partner to perform which movement? ____________________

6) The adductor tendons form a connective tissue drape along the base of the pelvis extending from which two bony landmarks? ____________________ ____________________

7) Located just anterior to the hamstrings, ____________________ is the most posterior of the adductor muscles.

8) Gracilis is the only adductor to cross which joint? ____________________

9) What are the two actions common to all the muscles of the adductor group?

    ____________________ ____________________

10) You will find the prominent tendon(s) of the gracilis and adductor longus extending off of, or nearby, which bony landmark? ____________________

11) Which muscle can be located just anterior to the prominent adductor tendon? ____________________

12) Which muscle can be located between the ischial tuberosity and the adductor tubercle?

    ____

Let's Palpate!

Remember—there are no right or wrong answers here

Locate and explore the **adductor group** on three individuals. Then write three words that describe what you feel. (See p. 319-323 in *Trail Guide*.)

Person #1 ________	Person #2 ________	Person #3 ________
________	________	________
________	________	________
________	________	________

p. 315-323

Pelvis and Thigh, Muscle Group #2
Gluteals and Adductors

Matching
Match the origin and insertion to the correct muscle.

Origins
1) Coccyx, edge of sacrum, posterior iliac crest, sacrotuberous and sacroiliac ligaments
2) Gluteal surface of the ilium between the anterior and inferior gluteal lines
3) Gluteal surface of the ilium, between posterior and anterior gluteal lines, just below iliac crest
4) Inferior ramus of pubis (2)
5) Inferior ramus of pubis, ramus of ischium and ischial tuberosity
6) Pubic tubercle
7) Superior ramus of pubis

Insertions
8) Anterior aspect of greater trochanter
9) Iliotibial tract (upper fibers) and gluteal tuberosity (lower fibers)
10) Lateral aspect of greater trochanter
11) Medial lip of linea aspera
12) Medial lip of linea aspera and adductor tubercle
13) Pectineal line and medial lip of linea aspera
14) Pectineal line of femur
15) Proximal, medial shaft of tibia at pes anserinus tendon

Muscle	O	I
Adductor brevis	______	______
Adductor longus	______	______
Adductor magnus	______	______
Gluteus maximus	______	______
Gluteus medius	______	______
Gluteus minimus	______	______
Gracilis	______	______
Pectineus	______	______

Shorten or Lengthen?

16) Passive abduction of the hip would ____________________ the adductor brevis and longus.

17) Passive lateral rotation of the hip would ____________________ the gluteus maximus.

18) Passive extension of the hip would ____________________ the posterior fibers of the adductor magnus.

19) Passive adduction of the hip would ____________________ the gluteus medius.

20) Passive lateral rotation of the hip would ____________________ the gluteus minimus.

21) Passive extension and lateral rotation of the hip would ____________________ the gracilis.

22) Passive medial rotation of the hip would ____________________ the adductors.

23) Passive flexion of the hip would ____________________ the gluteus maximus.

Pelvis and Thigh, Muscle Group #3

TFL, Sartorius, Lateral Rotators and Iliopsoas

p. 324-335

Please answer the following questions.

1) Which muscle is most accessible between the upper fibers of the rectus femoris and gluteus medius?

____

2) Which cablelike band of fascia can be isolated just anterior to the biceps femoris tendon?

____

3) In order to feel the tensor fasciae latae contract, position your partner in a supine position and ask him to perform what action? _______________________

4) Which muscle stretches from the anterior superior iliac spine (ASIS) to the medial knee?

____

5) The proximal fibers of the sartorius are just lateral to which artery? _______________________

6) Which three tendons blend together to become the pes anserinus tendon?

_______________ _______________ _______________

7) Which muscle lies superficial to the sciatic nerve and can compress the nerve if overcontracted?

____

8) To locate the piriformis, form a "T" with which three bony landmarks?

_______________ _______________ _______________

9) Which rectangular muscle can be isolated by placing your fingerpads between the distal, posterior aspect of the greater trochanter and the ischial tuberosity? _______________________

10) Which muscle spans from the anterior surface of the lumbar vertebrae to the lesser trochanter?

____

11) To access the psoas major, place your fingerpads between the _______________ and _______________ before slowly compressing toward the muscle.

12) What are some ways to ensure your partner's comfort during palpation of the psoas major?

_______________ _______________

_______________ _______________

_______________ _______________

13) What action could you ask your partner to perform to confirm that you have located the psoas major?

____

Pelvis and Thigh, Muscle Group #3

TFL, Sartorius, Lateral Rotators and Iliopsoas

Matching

Match the origin and insertion to the correct muscle.

Origins

1) Anterior superior iliac spine (ASIS)
2) Anterior surface of sacrum
3) Body and transverse process of first lumbar vertebra
4) Bodies and transverse processes of lumbar vertebrae
5) Iliac crest, posterior to the ASIS
6) Iliac fossa
7) Ischial spine
8) Ischial tuberosity
9) Lateral border of ischial tuberosity
10) Obturator membrane and inferior surface of obturator foramen
11) Rami of pubis and ischium, obturator membrane

Insertions

12) Superior aspect of greater trochanter
13) Iliotibial tract
14) Lesser trochanter (2)
15) Medial surface of greater trochanter (3)
16) Intertrochanteric crest, between the greater and lesser trochanters
17) Proximal, medial shaft of the tibia at pes anserinus tendon
18) Superior ramus of pubis
19) Trochanteric fossa of femur

Muscle	O	I
Gemellus inferior	______	______
Gemellus superior	______	______
Iliacus	______	______
Obturator externus	______	______
Obturator internus	______	______
Piriformis	______	______
Psoas major	______	______
Psoas minor	______	______
Quadratus femoris	______	______
Sartorius	______	______
Tensor fasciae latae	______	______

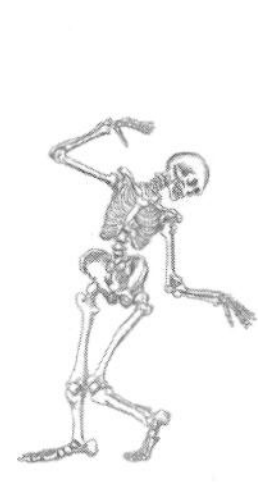

Let's Palpate!

Remember—there are no right or wrong answers here

Locate and explore the **tensor fasciae latae and iliotibial tract** on three individuals. Then write three words that describe what you feel. (See p. 324-325 in *Trail Guide.*)

Person #1 ______________	Person #2 ______________	Person #3 ______________
______________________	______________________	______________________
______________________	______________________	______________________
______________________	______________________	______________________

TFL, Sartorius, Lateral Rotators and Iliopsoas

p. 324-335

Shorten or Lengthen?

1) Passive medial rotation of the flexed knee would ____________________ the sartorius.

2) Passive adduction of the hip would ____________________ the tensor fasciae latae.

3) Passive extension of the hip would ____________________ the iliopsoas muscles.

4) Passive lateral rotation of the hip would ____________________ the piriformis.

5) Passive extension of the hip would ____________________ the psoas major.

6) Passive flexion of the hip would ____________________ the sartorius.

7) Passive lateral rotation of the hip would ____________________ the iliacus.

8) Passive medial rotation of the hip would ____________________ the tensor fasciae latae.

9) Passive medial rotation of the hip would ____________________ the quadratus femoris.

10) Passive abduction of the hip would ____________________ the sartorius.

Let's Palpate!

Remember—there are no right or wrong answers here

Locate and explore the **psoas major** on three individuals. Then write three words that describe what you feel. (See p. 332-334 in *Trail Guide*.)

Person #1 ______________	Person #2 ______________	Person #3 ______________
____________________	____________________	____________________
____________________	____________________	____________________
____________________	____________________	____________________

Pelvis and Thigh

Other Structures #1

Please identify the following structures.

CHOICES
Adductor longus (2)
Femoral artery
Femoral nerve
Femoral vein
Great saphenous vein
Inguinal ligament (2)
Inguinal lymph nodes
Sartorius (2)

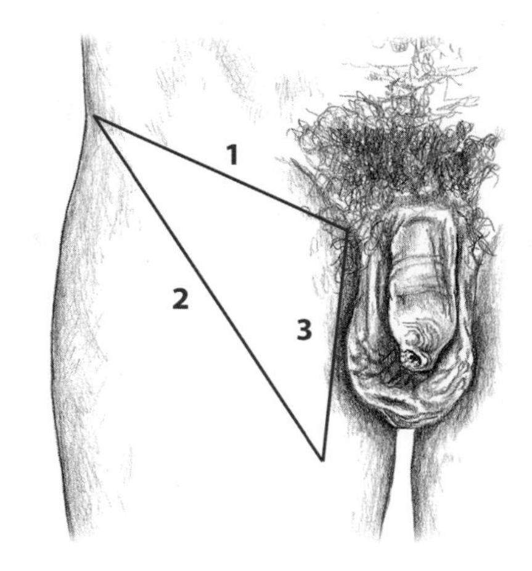

1) ______________

2) ______________

3) ______________

The three borders of the femoral triangle

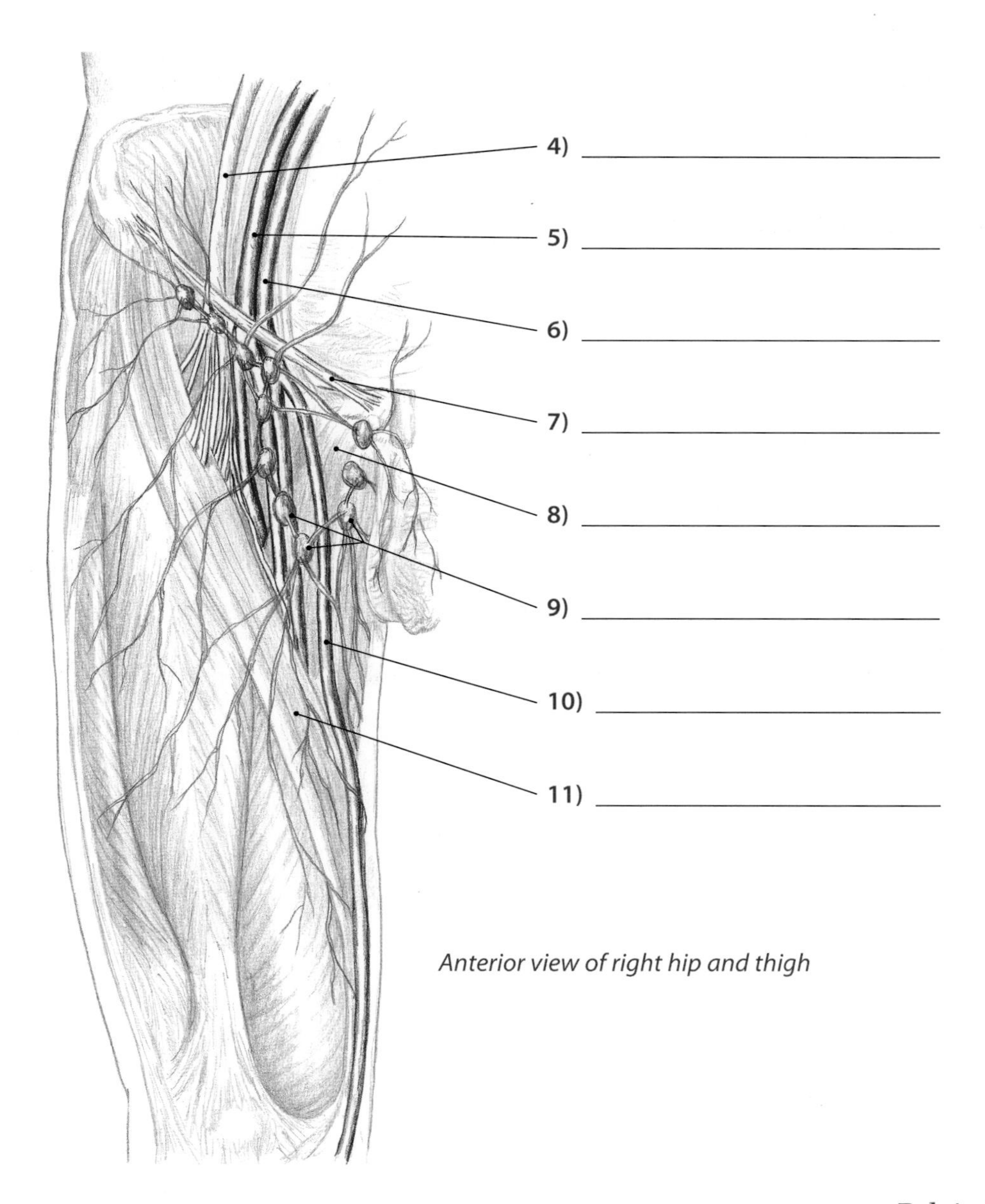

4) ______________

5) ______________

6) ______________

7) ______________

8) ______________

9) ______________

10) ______________

11) ______________

Anterior view of right hip and thigh

Pelvis and Thigh

Joints and Ligaments #1

Please identify the following structures.

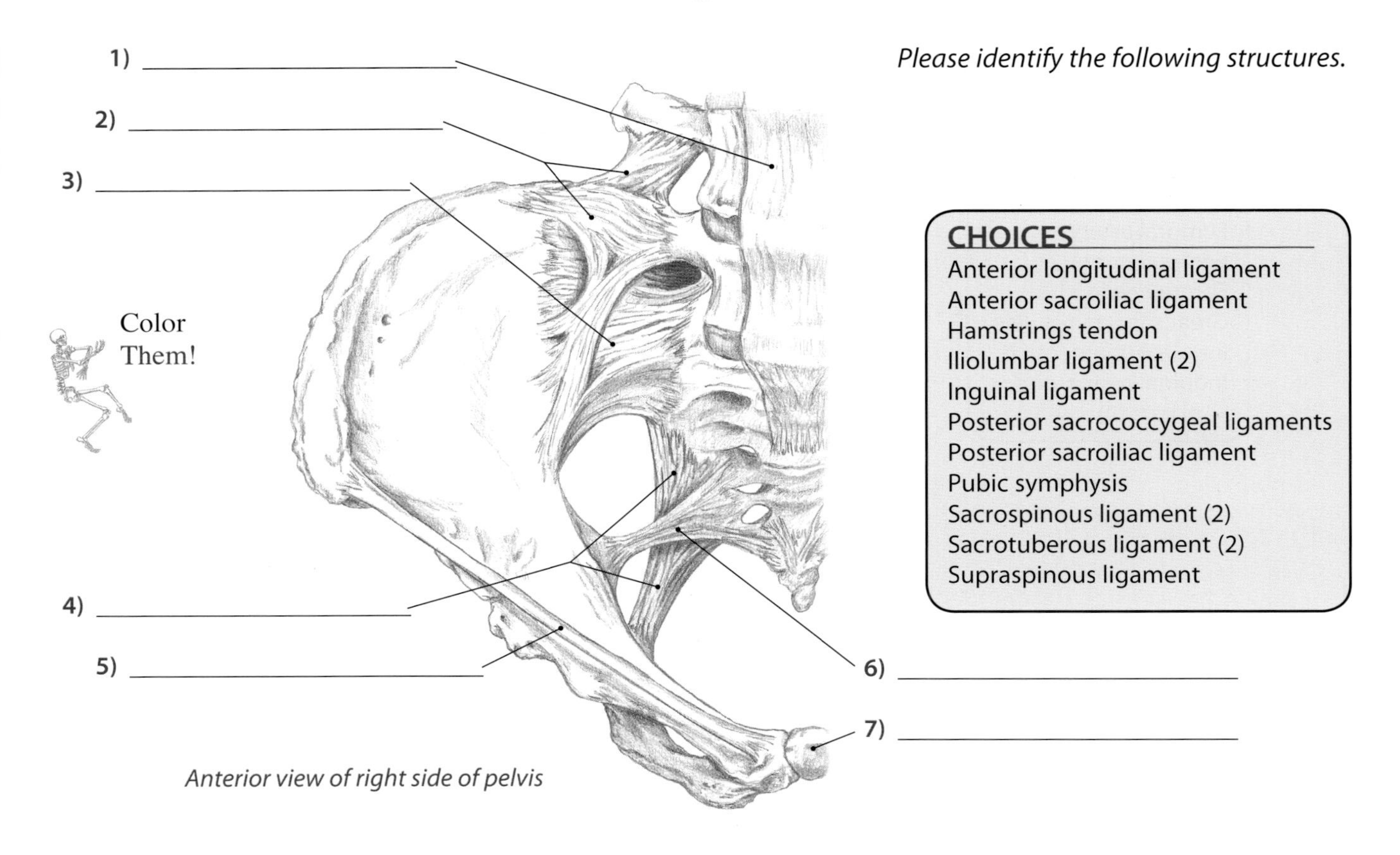

CHOICES

Anterior longitudinal ligament
Anterior sacroiliac ligament
Hamstrings tendon
Iliolumbar ligament (2)
Inguinal ligament
Posterior sacrococcygeal ligaments
Posterior sacroiliac ligament
Pubic symphysis
Sacrospinous ligament (2)
Sacrotuberous ligament (2)
Supraspinous ligament

Anterior view of right side of pelvis

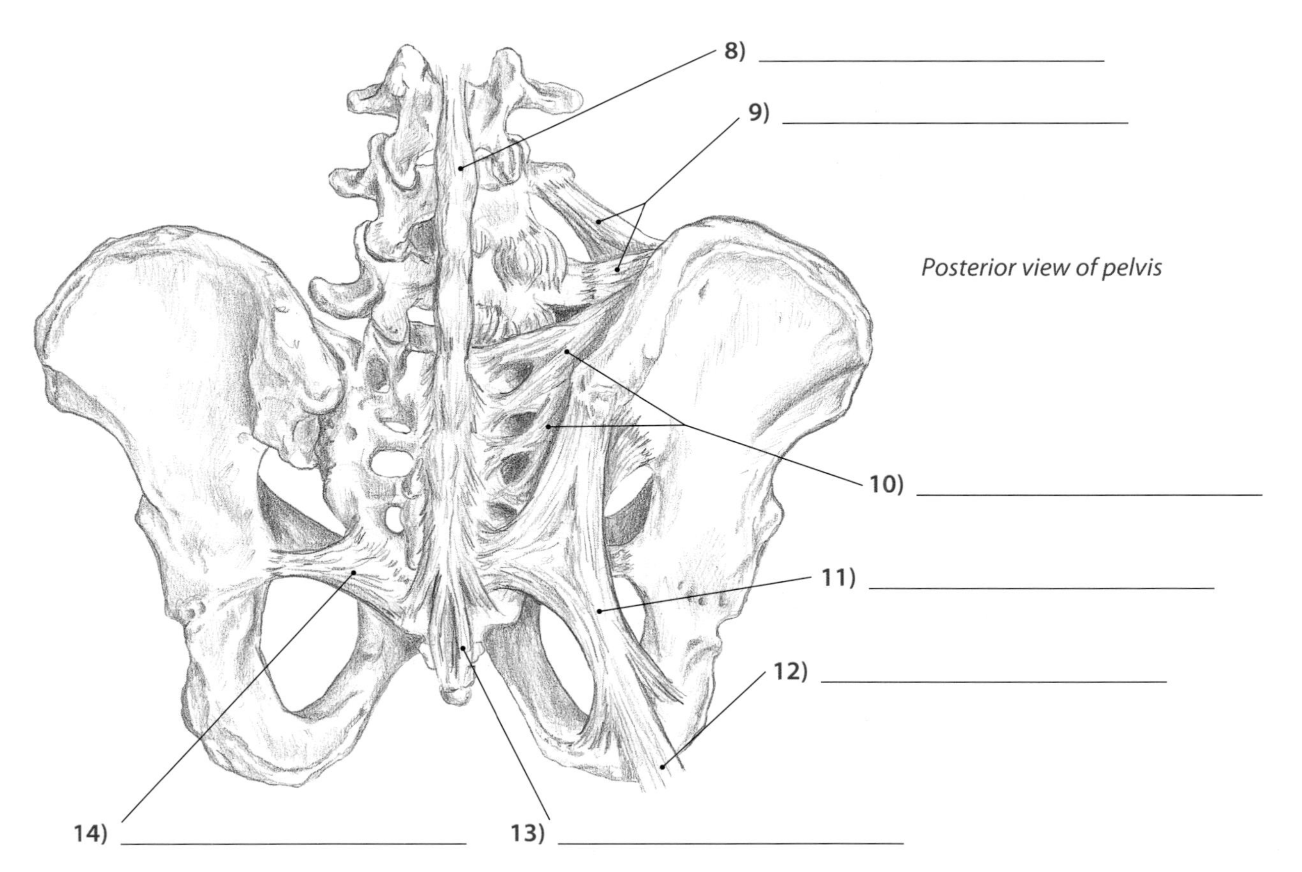

Posterior view of pelvis

Pelvis and Thigh

Joints and Ligaments #2

Please identify the following structures.

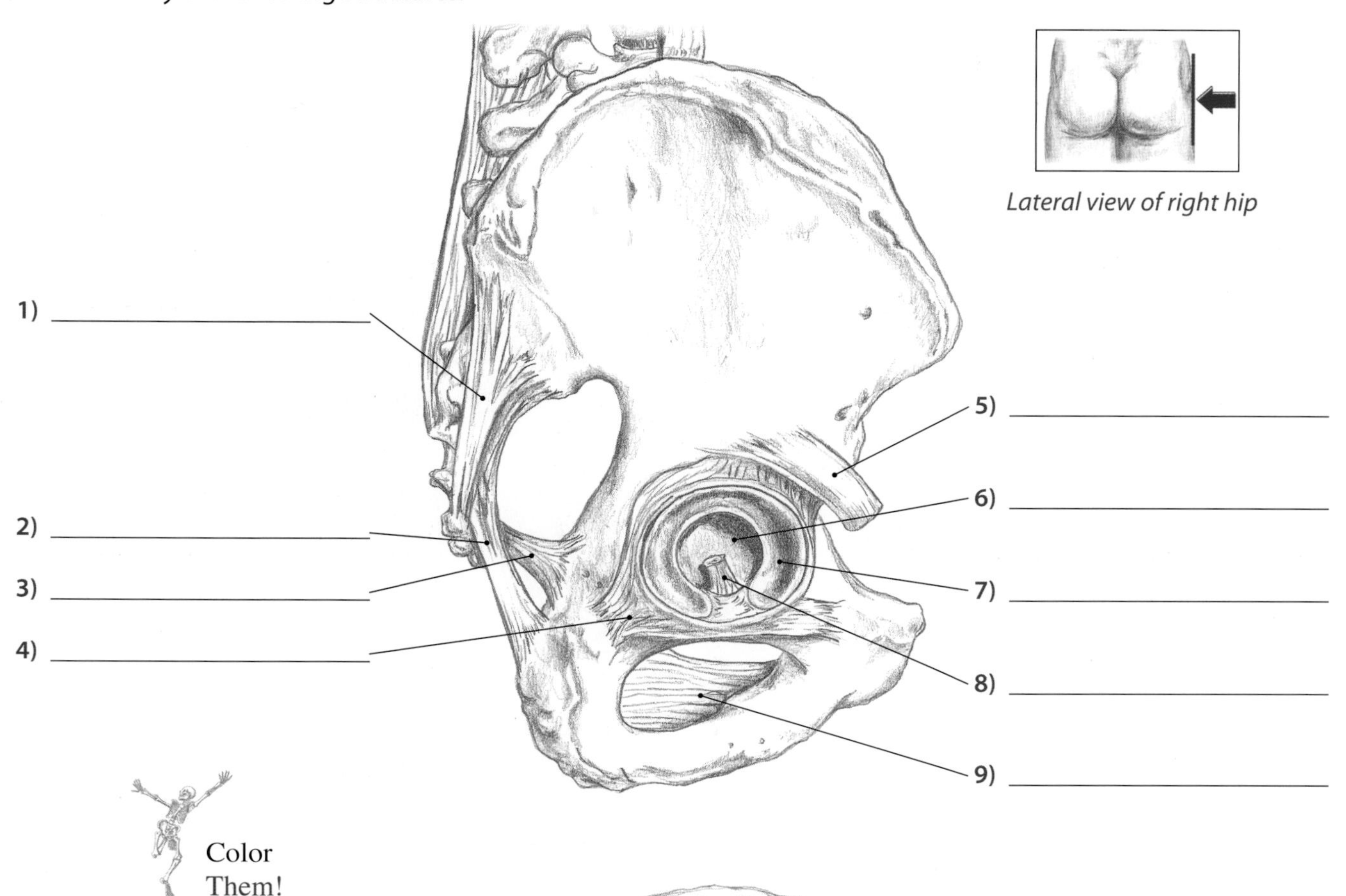

Lateral view of right hip

CHOICES

- Acetabulum
- Anterior sacroiliac ligament
- Articular capsule of coxal joint
- Lunate surface of acetabulum
- Obturator membrane (2)
- Posterior sacroiliac ligament
- Pubic symphysis
- Round ligament (ligamentum capitis femoris—cut)
- Sacrospinous ligament (2)
- Sacrotuberous ligament (2)
- Tendon of rectus femoris (cut)

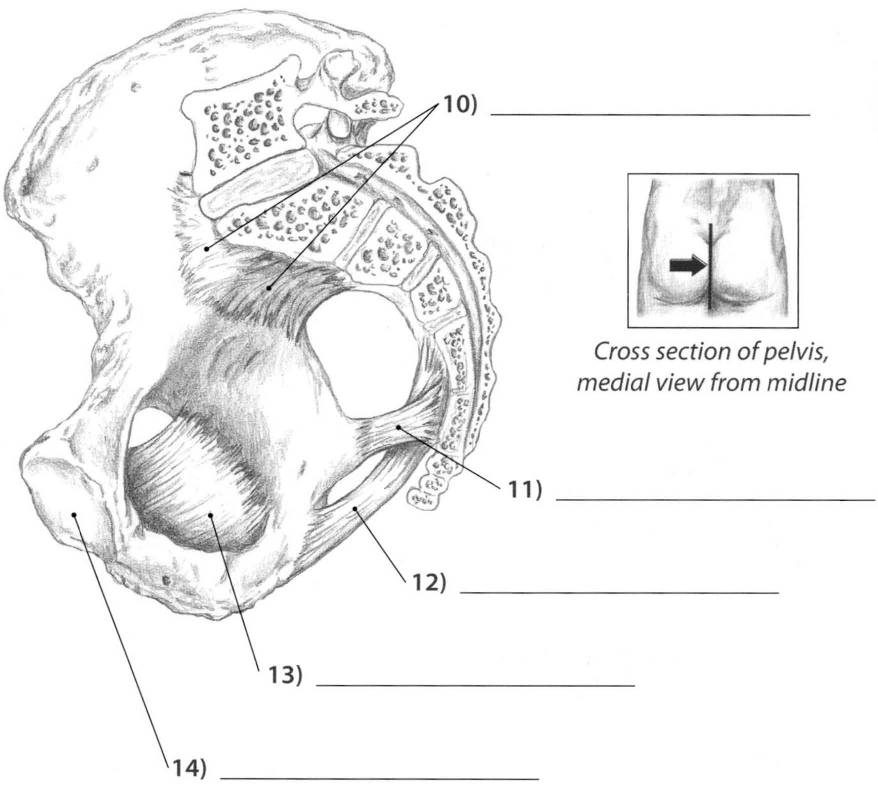

Cross section of pelvis, medial view from midline

Coxal Joint

p. 338

Please identify the following structures.

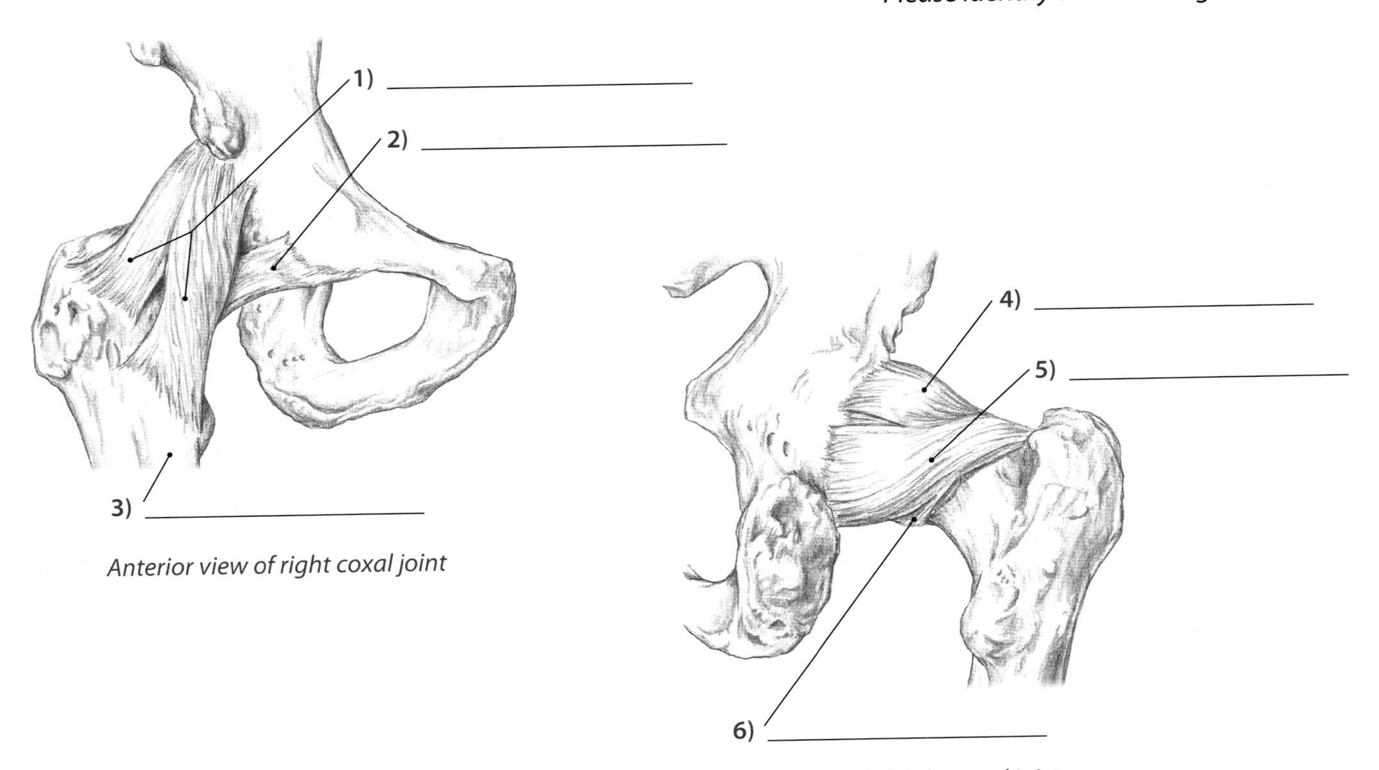

Anterior view of right coxal joint

Posterior view of right coxal joint

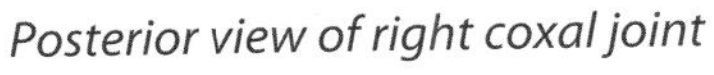

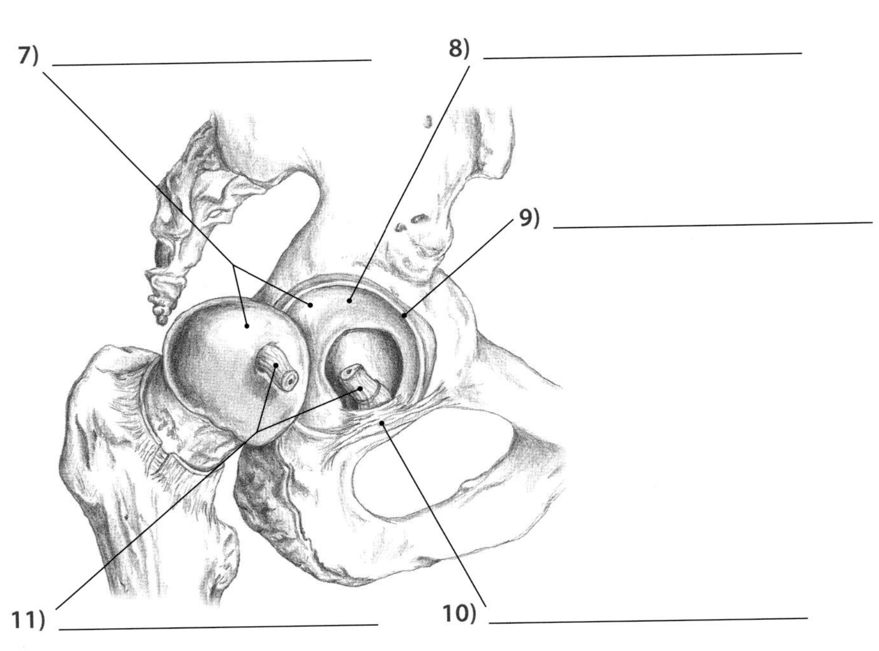

Lateral view of right coxal joint, femur reflected

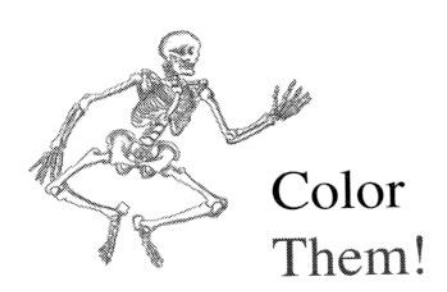

CHOICES

Acetabular labrum
Articular cartilage
Femur
Iliofemoral ligament (2)
Ischiofemoral ligament
Lunate surface of acetabulum
Pubofemoral ligament
Round ligament
(ligamentum capitis femoris—cut)
Transverse acetabular ligament
Zona orbicularis

Pelvis and Thigh

Other Structures #2

Please answer the following questions.

1) The inguinal ligament stretches from the ______________________ to the ______________________.

2) Which three vessels pass through the femoral triangle?

______________________________ ______________________________

____

3) Where should you position your fingers to feel the pulse of the femoral artery?

____

4) What structure spans from the ischial tuberosity to the edge of the sacrum? ______________________

5) The ______________________ ligaments help to reinforce the union of the sacrum and the ilium.

6) The transverse processes of the fourth and fifth lumbar vertebrae and the posterior iliac crest are helpful landmarks in finding which ligament? ______________________________

7) Which structure spans from the lower lumbar vertebrae, between the ischial tuberosity and greater trochanter and down the posterior thigh? ______________________________

8) Which structure reduces friction between the greater trochanter and the gluteus maximus?

____

Let's Palpate!

Remember—there are no right or wrong answers here

Locate and explore the **sacrotuberous ligament** on three individuals. Then write three words that describe what you feel. (See p. 340 in *Trail Guide*.)

Person #1 ____________	Person #2 ____________	Person #3 ____________
____________________	____________________	____________________
____________________	____________________	____________________
____________________	____________________	____________________

Notes

Leg and Foot
Topographical Views

Please identify the following structures.

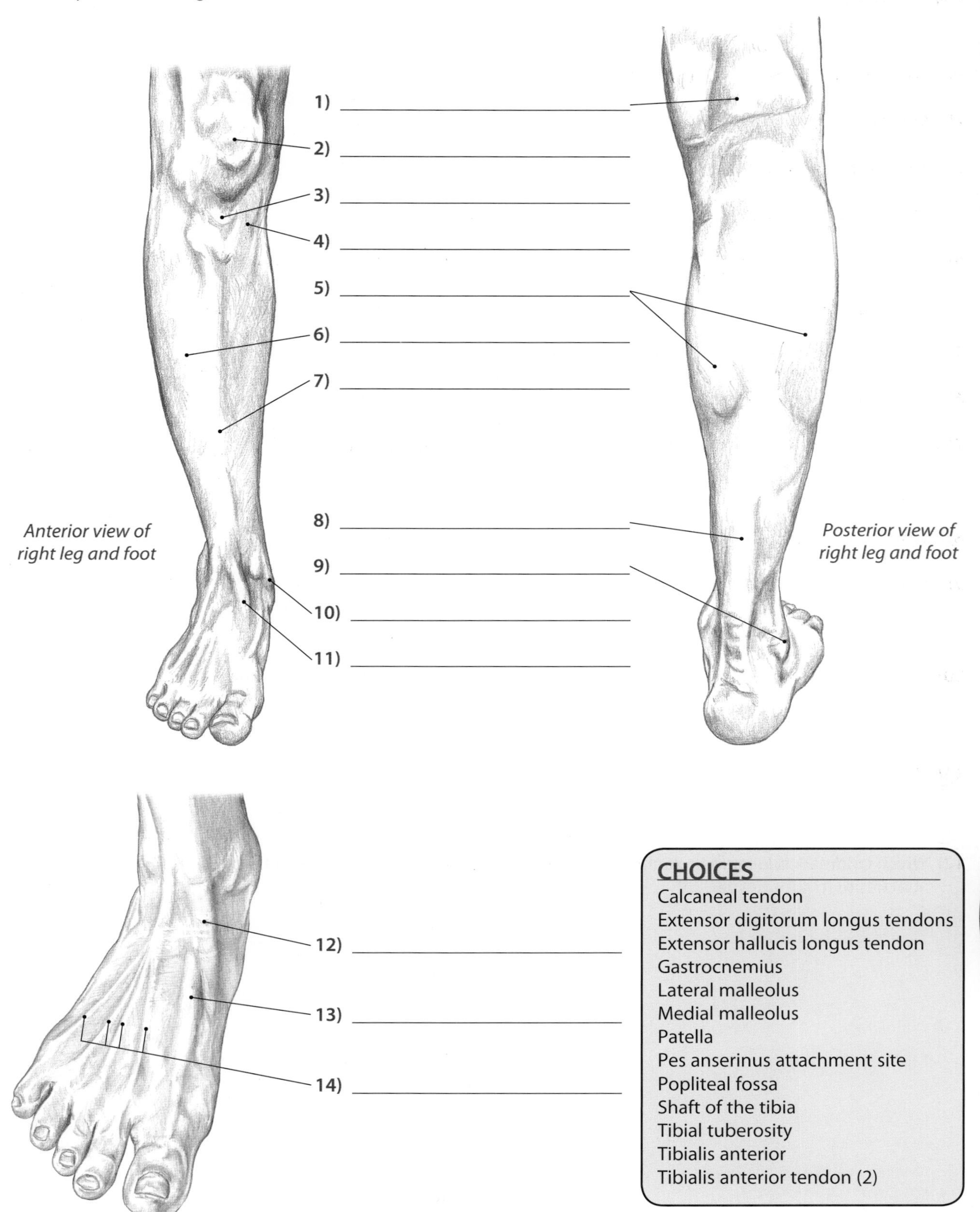

Anterior view of right leg and foot

Posterior view of right leg and foot

Dorsal view of right foot

CHOICES

Calcaneal tendon
Extensor digitorum longus tendons
Extensor hallucis longus tendon
Gastrocnemius
Lateral malleolus
Medial malleolus
Patella
Pes anserinus attachment site
Popliteal fossa
Shaft of the tibia
Tibial tuberosity
Tibialis anterior
Tibialis anterior tendon (2)

p. 346-353

Bones and Bony Landmarks of the Knee and Leg

Please answer the following questions.

1) The anatomical name for the knee is the ______________________________ joint.

2) Medial and lateral rotation of the knee can occur when the knee is in a ______________________________ position.

3) The bone running superficially down the anterior surface of the leg is the ______________________________ while the bone buried deep to the surrounding muscle tissue on the leg is the ______________________________.

4) The patella seems to disappear when the knee is flexed. It sinks between which two landmarks?

______________________________ ______________________________

5) The bony landmark located distal to the patella is the ______________________________.

6) The connective tissue structure connecting the patella to the tibial tuberosity is the ______________________________.

7) The head of the fibula is the attachment site for which two muscles and ligament?

____________________ ____________________ ____________________

8) Which portion of the tibial plateaus can be accessed? ______________________________

9) Which three tendons form the pes anserinus tendon?

____________________ ____________________ ____________________

10) With the knee fully extended and the patella shifted medially or laterally, what structures can be found deep to the patella? ______________________________

11) To locate the lateral epicondyle of the femur, you would need to palpate deep to what structure?

____

12) Which landmark is located proximal to the medial epicondyle of the femur, and the tendon of what muscle attaches to it?

______________________________ ______________________________

Leg and Foot

Bones of the Knee, Leg and Foot

Please identify the following ***bones.*** *(Questions 1-8)*

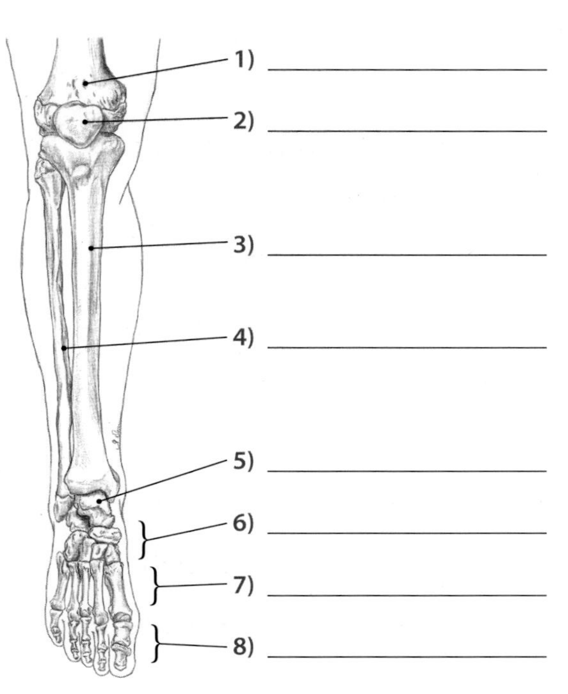

Anterior view of right knee, leg and foot, foot plantar flexed

CHOICES
- Femur
- Fibula
- Fossa of lateral malleolus
- Head of the fibula
- Lateral condyle (2)
- Lateral malleolus (2)
- Medial and lateral intercondylar tubercles
- Medial condyle
- Medial malleolus (2)
- Metatarsals
- Patella
- Phalanges
- Soleal line
- Talus
- Tarsals
- Tibia
- Tibial tuberosity

Please identify the following ***bony landmarks.*** *(Questions 9-20)*

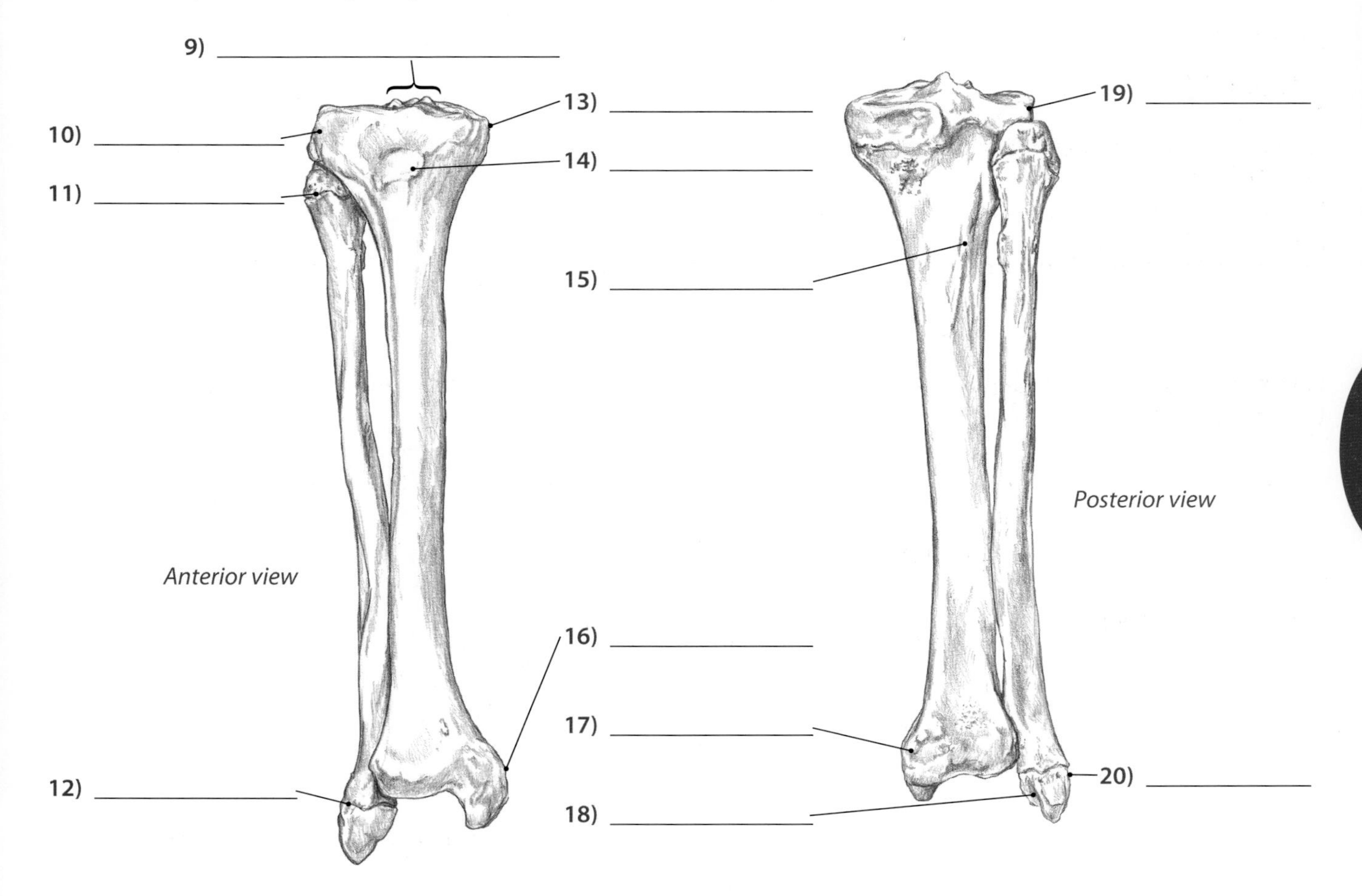

Bony Landmarks of the Knee and Leg

Please identify the following structures.

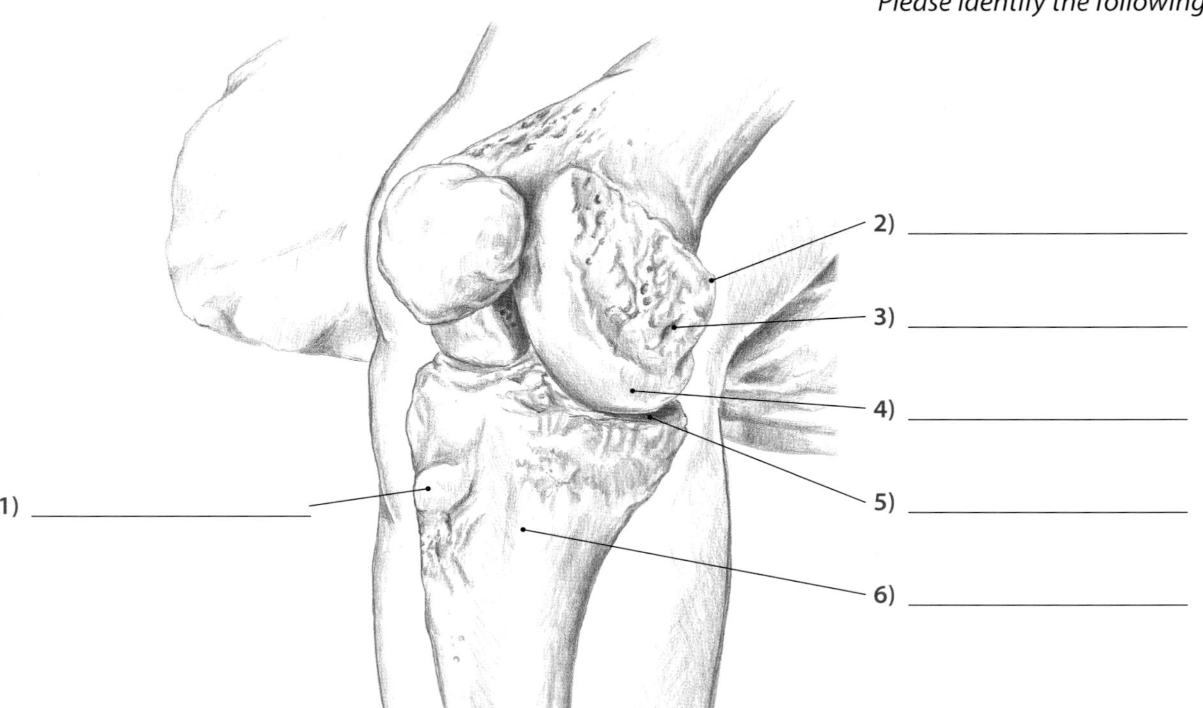

Anterior/medial view of right knee

CHOICES

Adductor tubercle
Head of the fibula
Lateral condyle
Lateral epicondyle
Medial condyle
Medial epicondyle
Pes anserinus attachment site
Tibia
Tibial plateau (2)
Tibial tubercle
Tibial tuberosity (2)

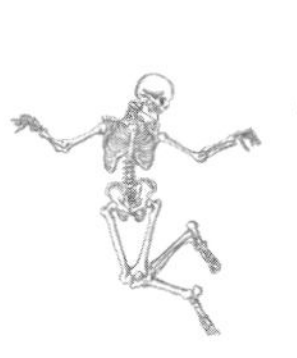

Color Them!

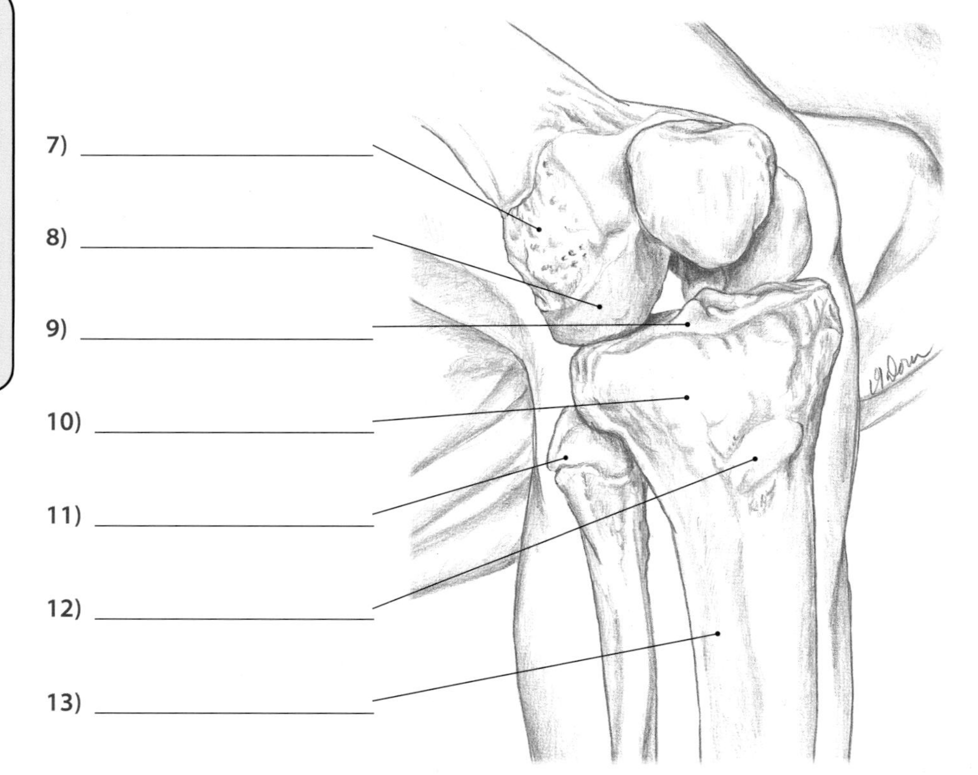

Anterior/lateral view of right knee

Leg and Foot

Bones of the Foot

Please identify the following structures.

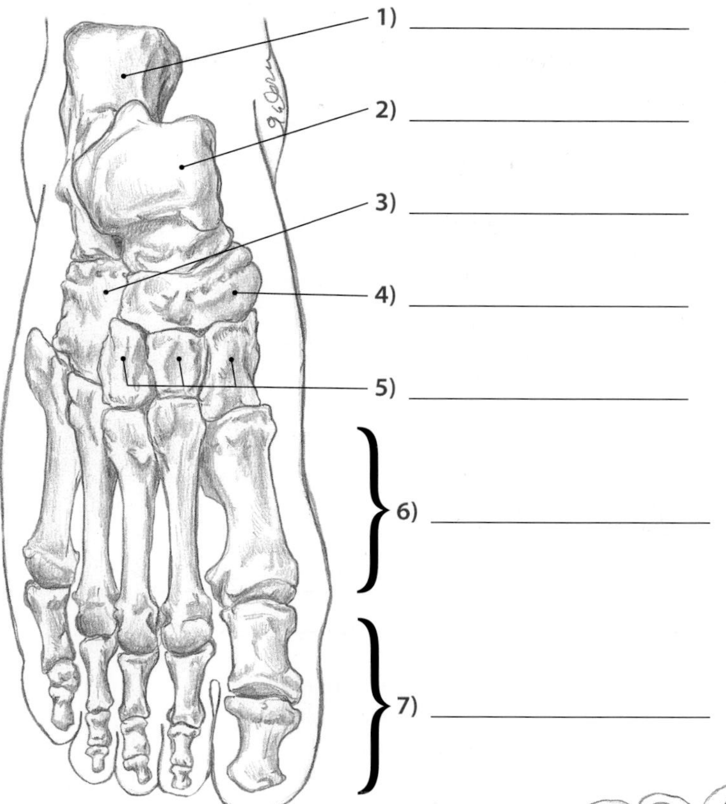

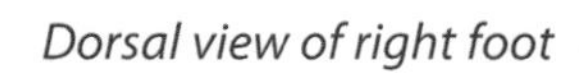
Dorsal view of right foot

CHOICES

- Calcaneus (2)
- Cuboid (2)
- Cuneiforms (2)
- Metatarsals (2)
- Navicular (2)
- Phalanges (2)
- Sesamoid bones
- Talus (2)

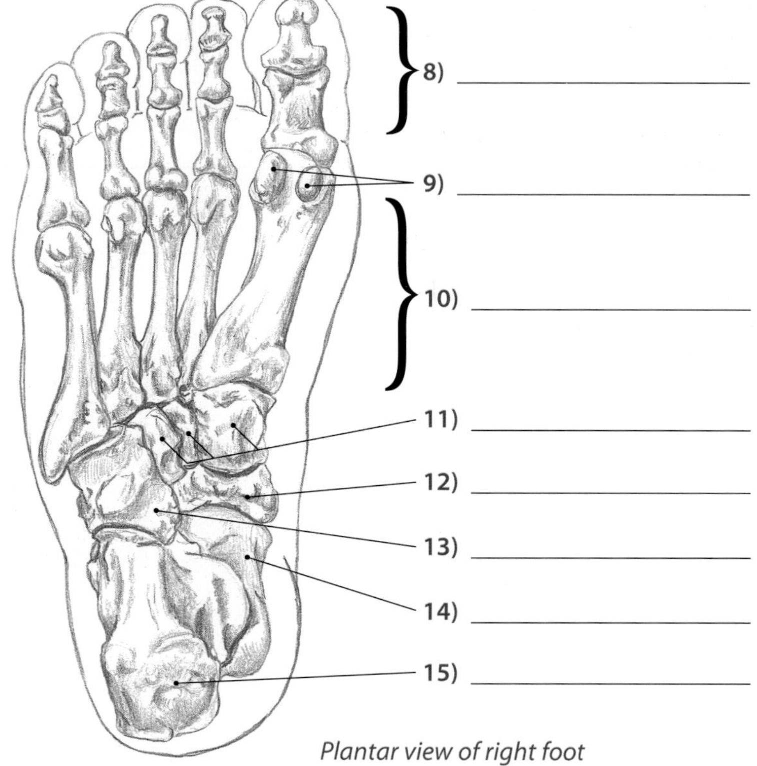

Plantar view of right foot

Bones and Bony Landmarks of the Foot #1

*Please identify the following structures. Numbers in **black** indicate bones, numbers in **red** are bony landmarks.*

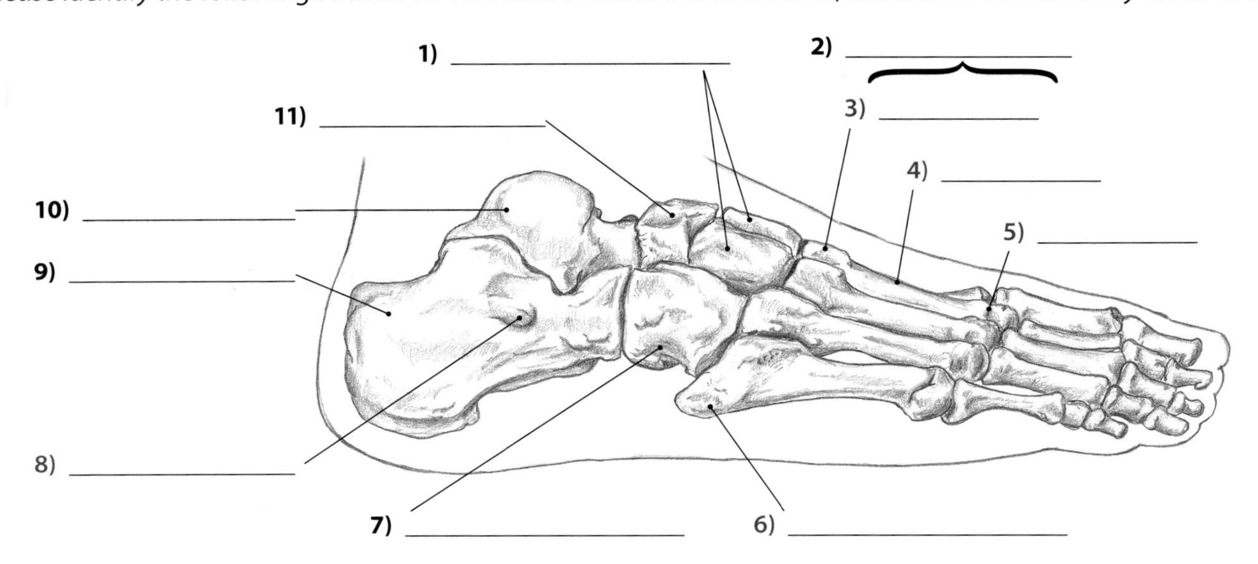

Lateral view of right foot

CHOICES	
Base (2)	Navicular
Base of first metatarsal	Navicular tubercle
Calcaneus (2)	Peroneal trochlea
Cuboid	Phalanges
Head (2)	Shaft (2)
Head of the talus	Sustentaculum tali
Lateral and middle cuneiform	Talus (2)
Medial cuneiform	Trochlea of the talus
Medial tubercle of talus	Tuberosity of calcaneus
Metatarsals	Tuberosity of fifth metatarsal

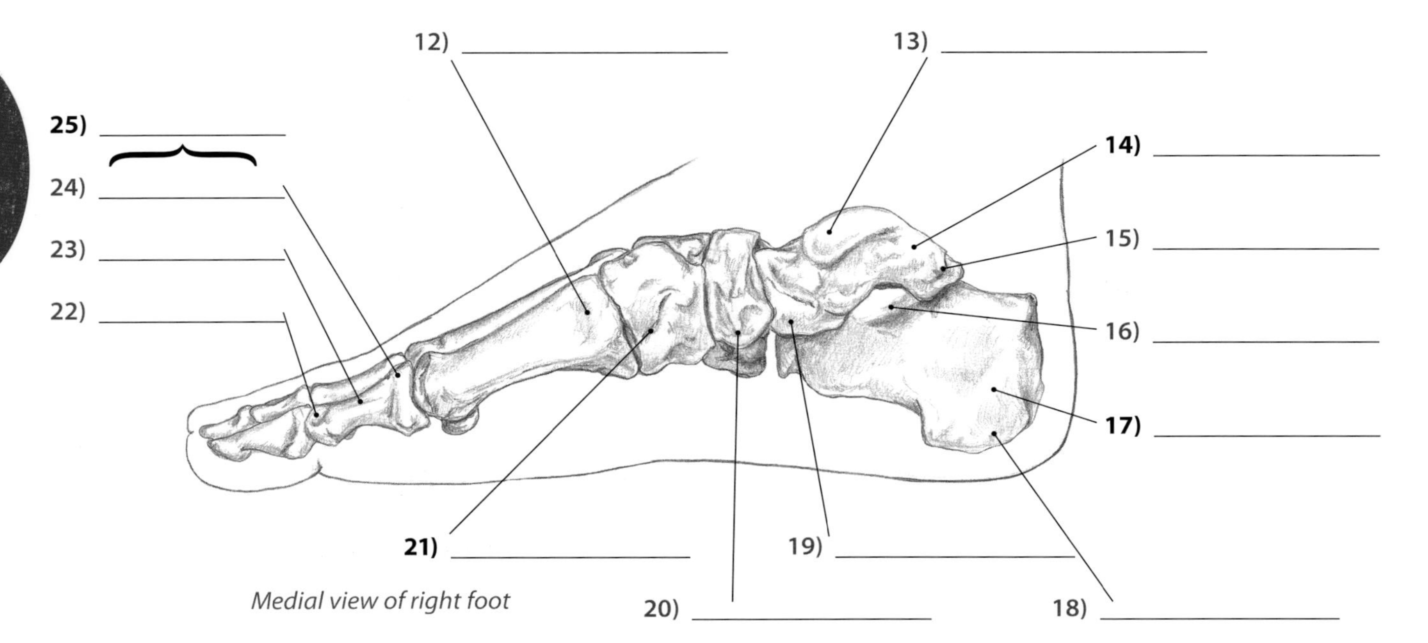

Medial view of right foot

Leg and Foot

Calcaneus and Talus

Please identify the following structures.

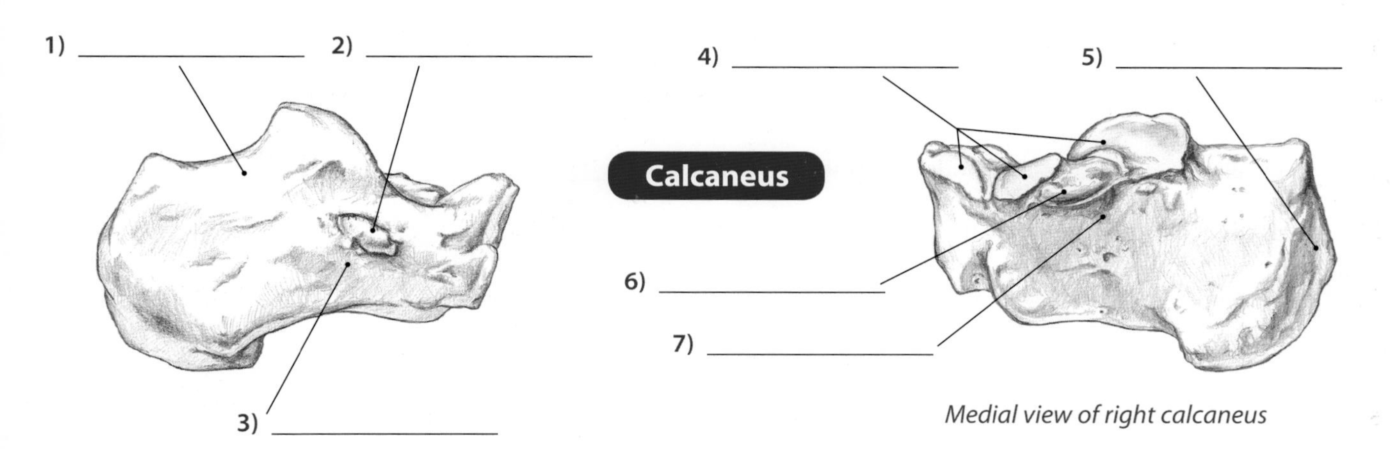

Lateral view of right calcaneus

Medial view of right calcaneus

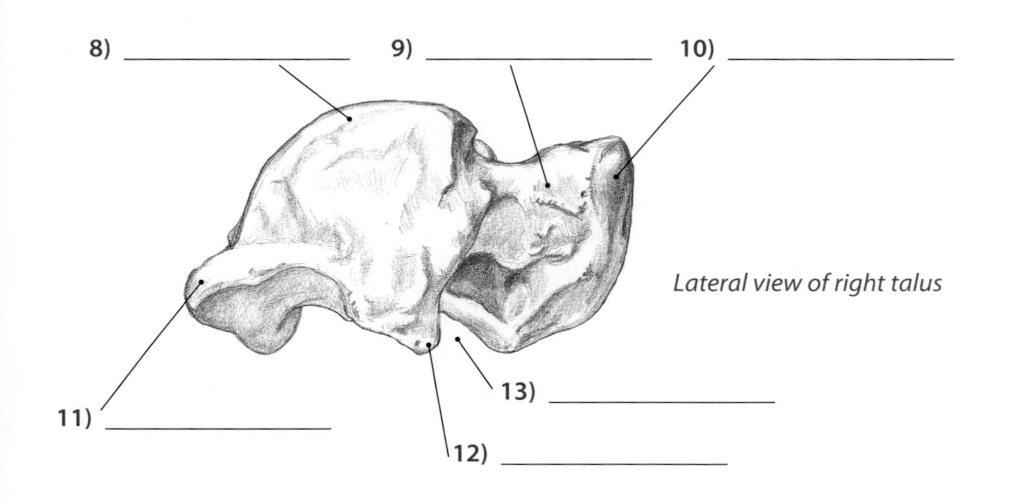

Lateral view of right talus

CHOICES

- Articular surfaces for talus
- Body
- Groove for flexor hallucis longus tendon
- Groove for peroneus longus tendon
- Head (2)
- Lateral process
- Lateral tubercle
- Medial tubercle
- Neck (2)
- Peroneal trochlea
- Sustentaculum tali
- Tarsal sinus
- Trochlea (2)
- Tuberosity

Talus

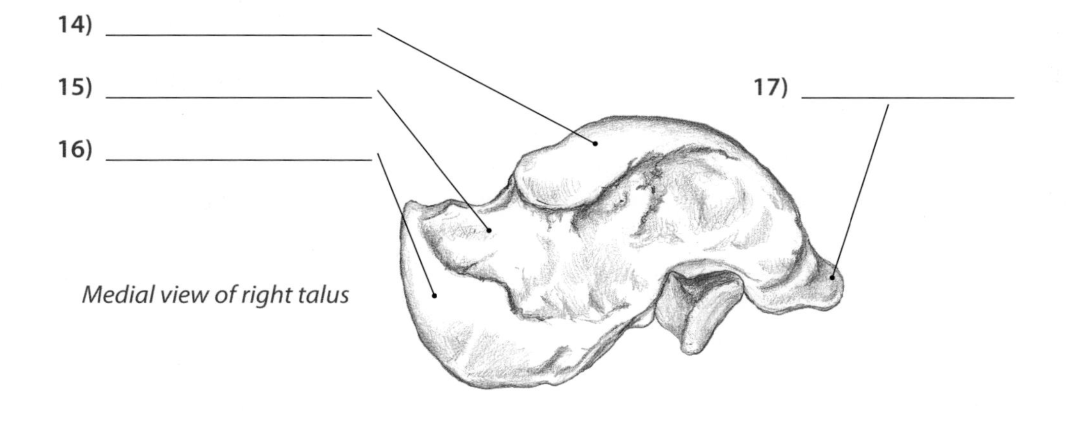

Medial view of right talus

p. 354-365

Bones and Bony Landmarks of the Foot #2

Please answer the following questions.

1) The bone at the heel of the foot is the ______________________, while the bone that articulates with the tibia and fibula is the ______________________.

2) The tarsals are most accessible along which surface of the foot? ______________________

3) As you palpate both malleoli of the ankle, which extends further distally? ______________________

4) How would you passively position the foot to shorten the surrounding tissue of the medial malleolar groove?

____

5) In which direction from the medial malleolus would you move your thumb to locate the sustentaculum tali? And approximately how far would you move?

______________________ ______________________

6) The head of the talus can be located between which two bony landmarks?

______________________ ______________________

7) How would you passively position the foot to best locate the trochlea of the talus?

____

8) The proximal end of the first metatarsal articulates with which bone? ______________________

9) Which two superficial surfaces of the first metatarsal are easily accessible?

______________________ ______________________

10) Spelled out, what do the acronyms "pip" and "dip" stand for?

______________________ ______________________

11) The tuberosity of the fifth metatarsal is the attachment site for which muscle? ______________________

12) Which tendon could you follow along the dorsal surface of the ankle to locate the medial cuneiform?

____

13) Locate the navicular and tuberosity of the fifth metatarsal. Which is situated further distally on the foot?

____

14) Between which two bony landmarks can you draw a line to locate the cuboid?

______________________ ______________________

Muscles of the Leg and Foot #1

Please identify the following structures.

CHOICES
- Calcaneal tendon (2)
- Flexor retinaculum
- Gastrocnemius
- Gastrocnemius (cut) (2)
- Peroneal tendons
- Plantaris (2)
- Popliteus
- Soleus (2)
- Superior peroneal retinaculum
- Tendons of flexors of ankle and toes

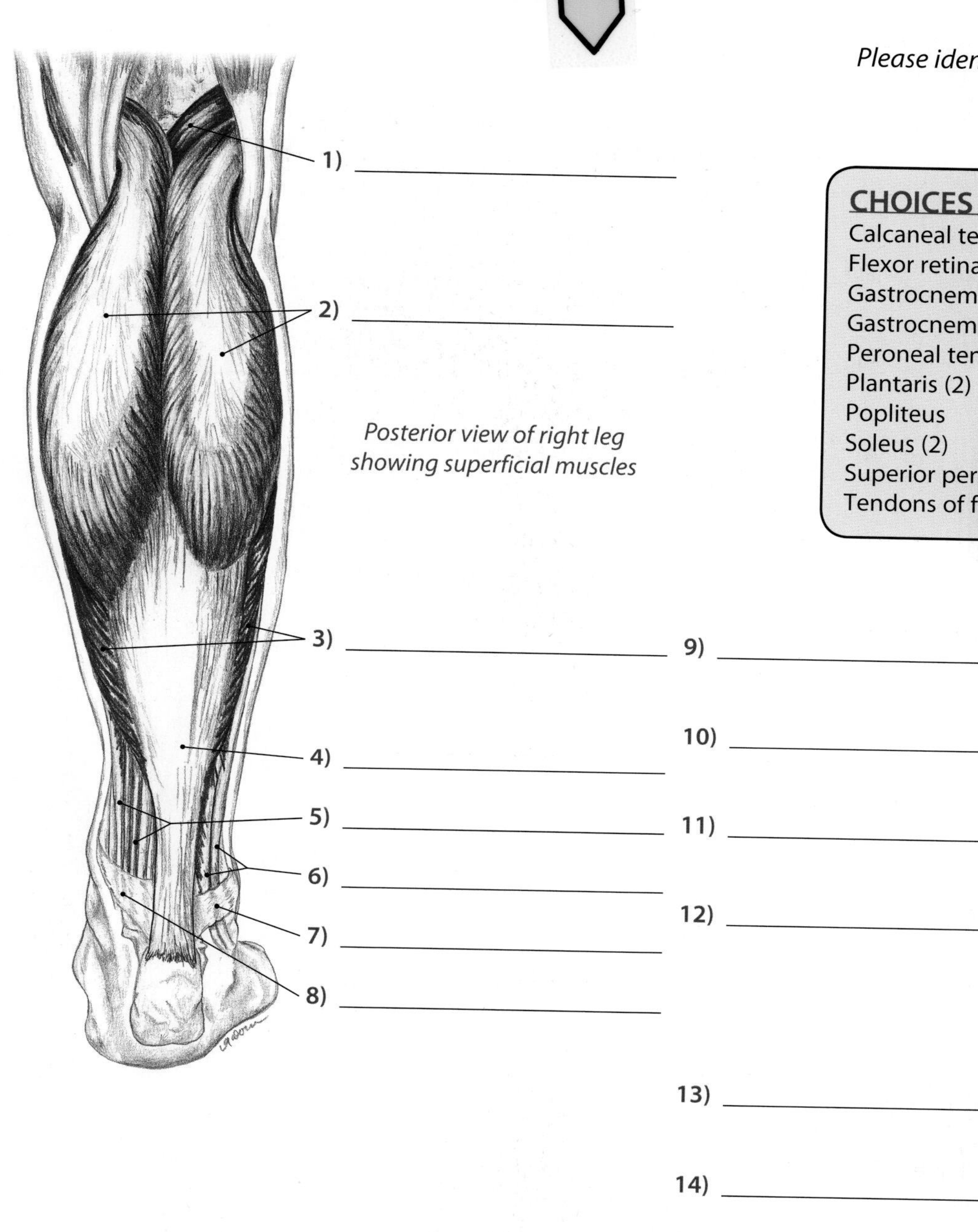

Posterior view of right leg showing superficial muscles

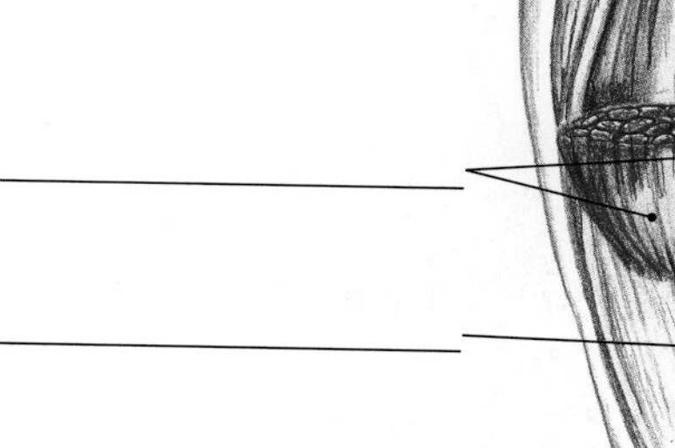

Posterior view of right leg showing deeper muscles

Muscles of the Leg and Foot #2

Please identify the following structures.

CHOICES
Extensor digitorum longus (2)
Extensor hallucis longus
Gastrocnemius (2)
Peroneus brevis (2)
Peroneus longus (2)
Soleus (2)
Tibialis anterior (2)

1) ____________________

2) ____________________

3) ____________________

4) ____________________

5) ____________________

6) ____________________

Lateral view of right leg and foot

7) ____________________

8) ____________________

9) ____________________

10) ____________________

11) ____________________

12) ____________________

13) ____________________

Anterior view of right leg and foot

Leg and Foot

Muscles of the Leg and Foot #3

Please identify the following structures.

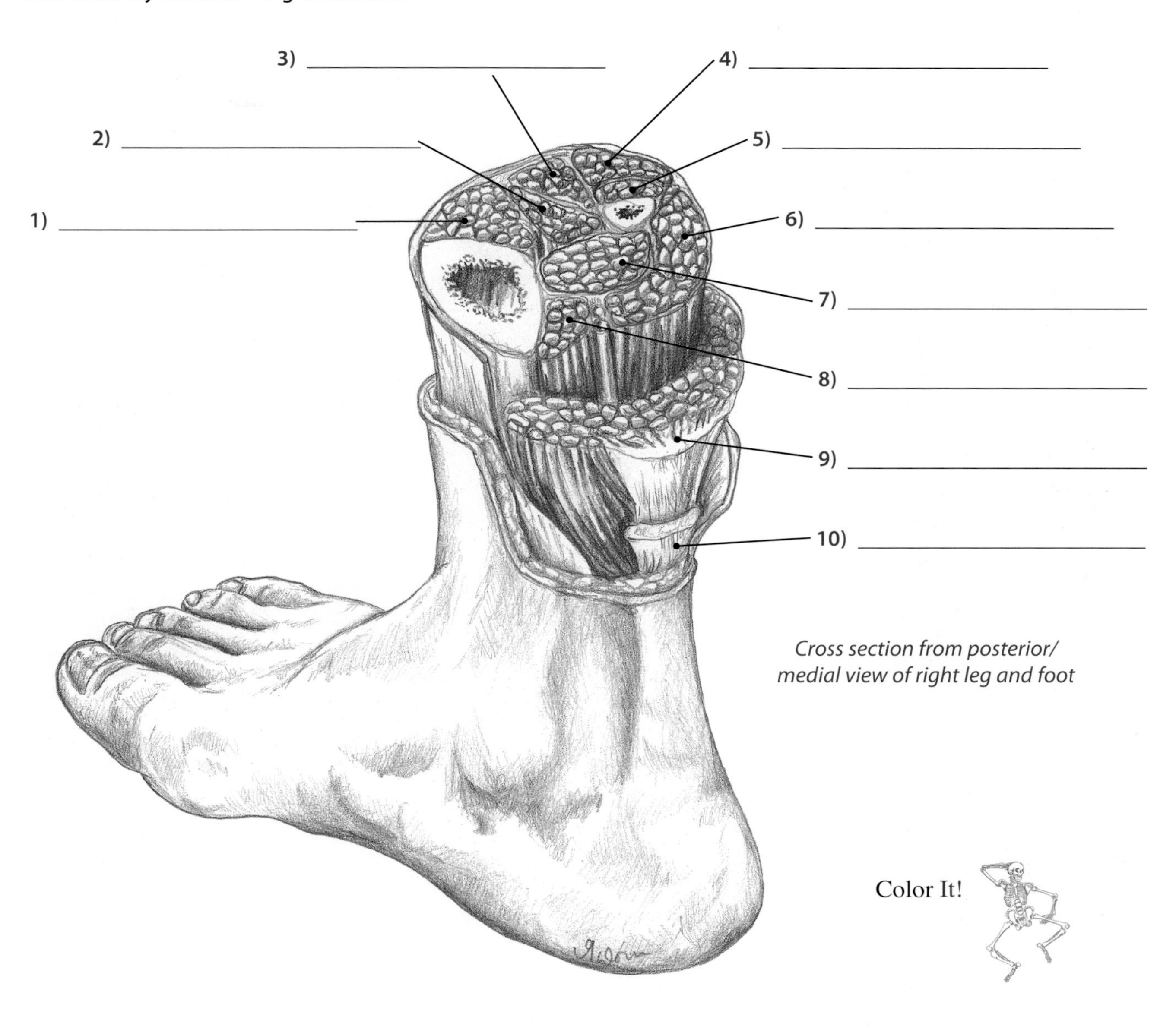

Cross section from posterior/ medial view of right leg and foot

Color It!

CHOICES

Calcaneal tendon
Extensor digitorum longus
Extensor hallucis longus
Flexor digitorum longus
Flexor hallucis longus
Peroneus brevis
Peroneus longus
Soleus
Tibialis anterior
Tibialis posterior

Color the Muscles #1

p. 366

Using different colors, please fill in and label the muscles and other structures listed below.

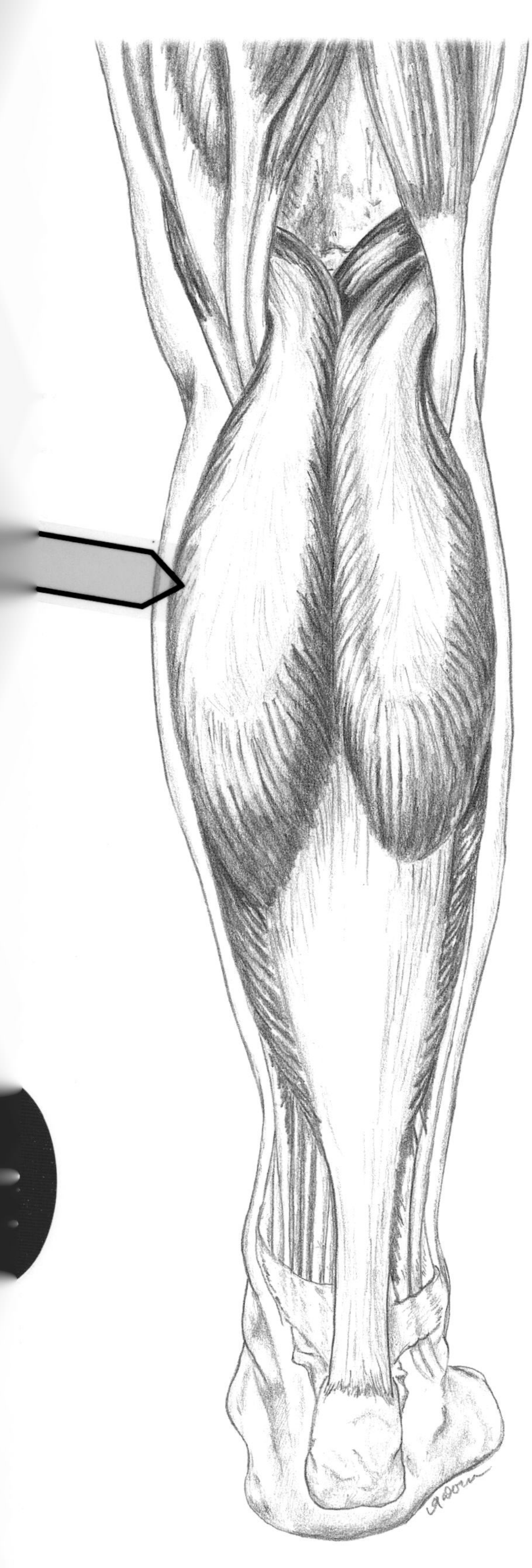

Posterior view of right leg showing superficial muscles

Calcaneal tendon
Flexor tendons
Gastrocnemius
Gastrocnemius (cut)
Peroneal tendons
Plantaris
Popliteus
Soleus

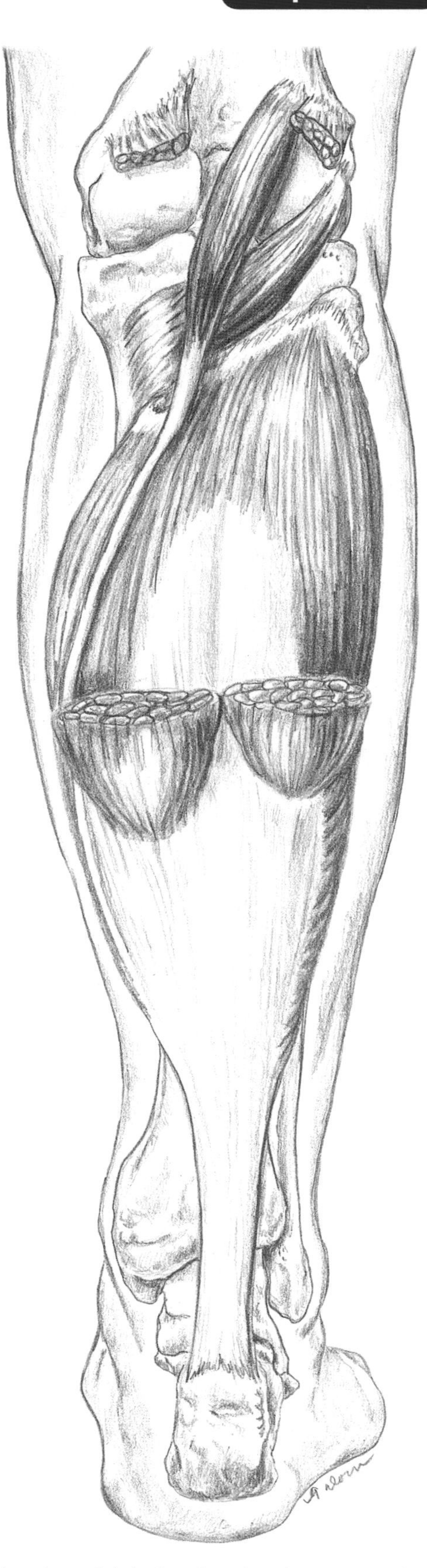

Posterior view of right leg showing deeper muscles

Leg and Foot

Color the Muscles #2

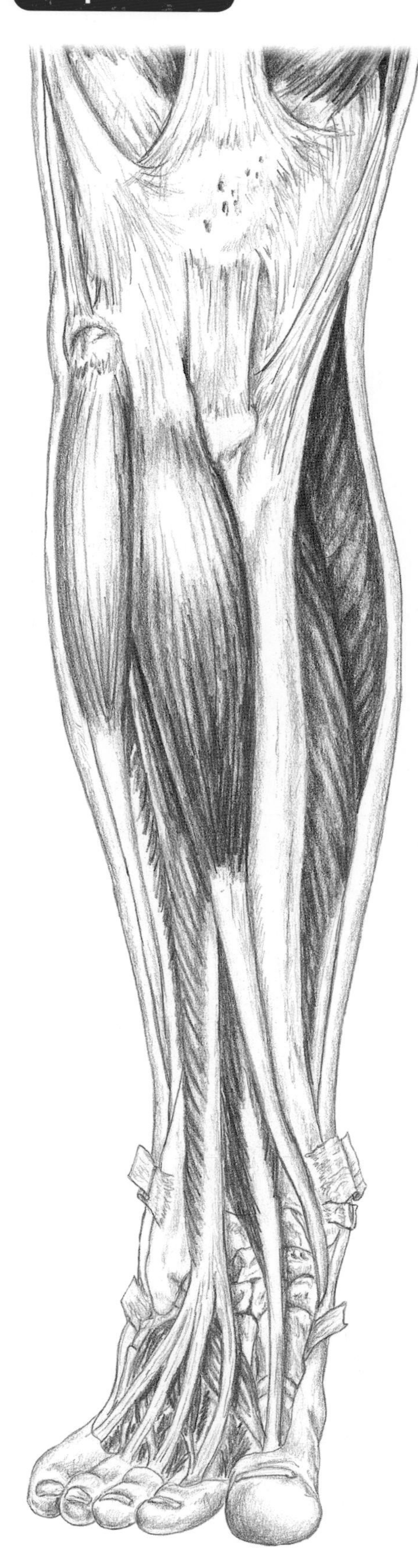

Anterior view of right leg and foot

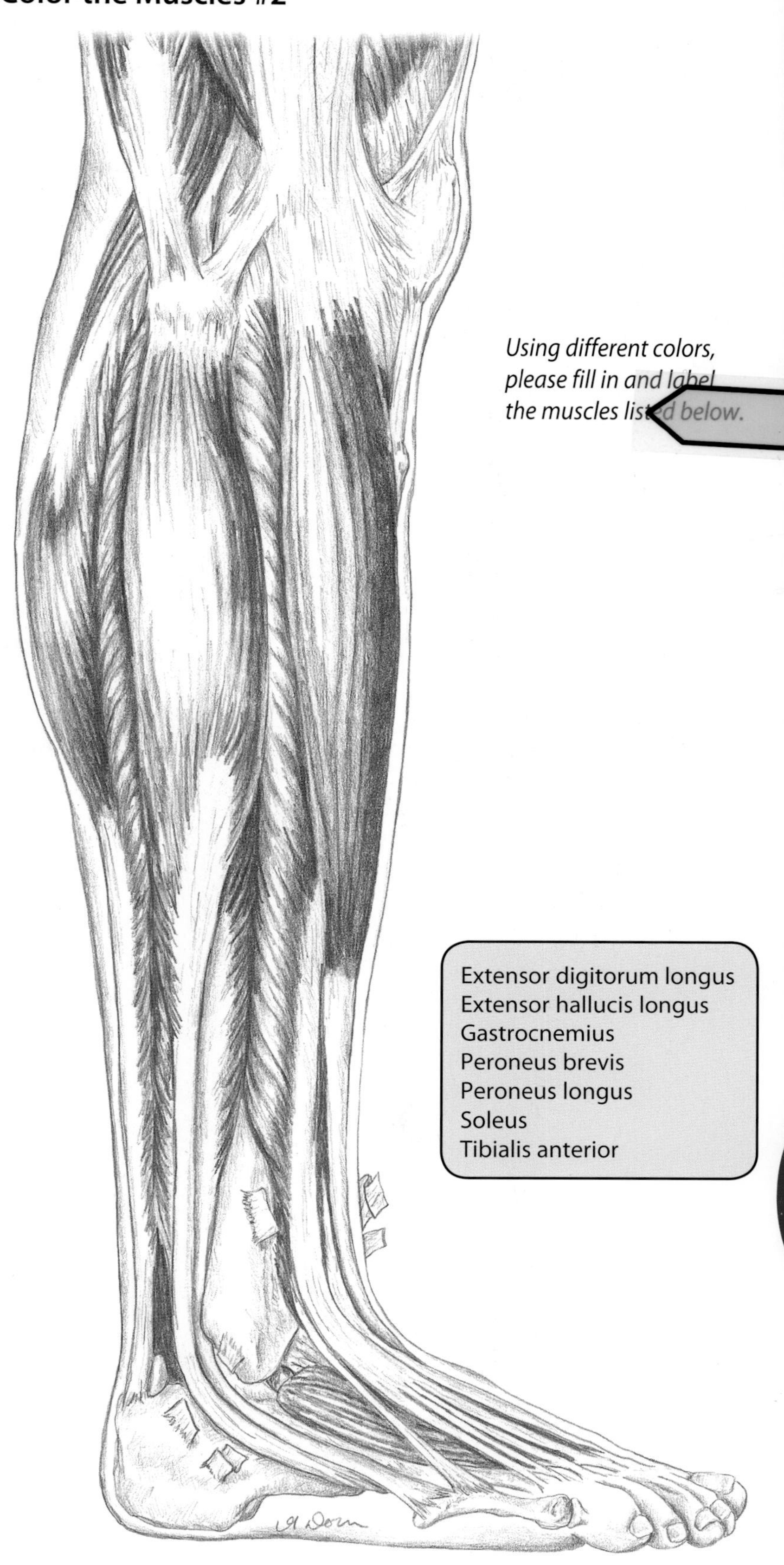

Lateral view of right leg and foot

Using different colors, please fill in and label the muscles listed below.

- Extensor digitorum longus
- Extensor hallucis longus
- Gastrocnemius
- Peroneus brevis
- Peroneus longus
- Soleus
- Tibialis anterior

Muscles and Movements #1

Please list the action demonstrated, two synergists and one antagonist. The first letter of each muscle has been provided.

1) This action happens at which joint?

____

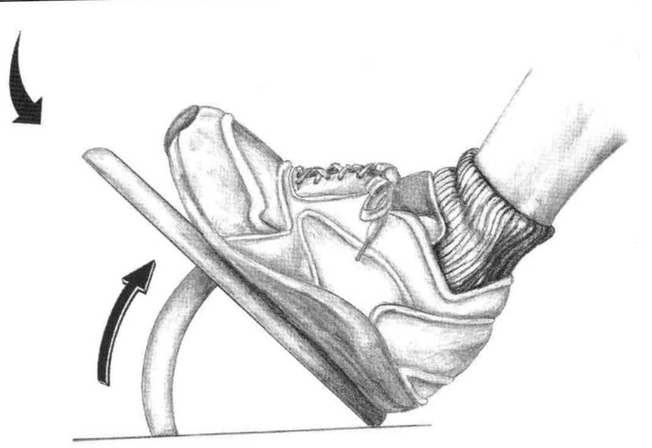

2) Action

____

3) Synergists

T ______________________________

E ______________________________

4) Antagonist

F ______________________________

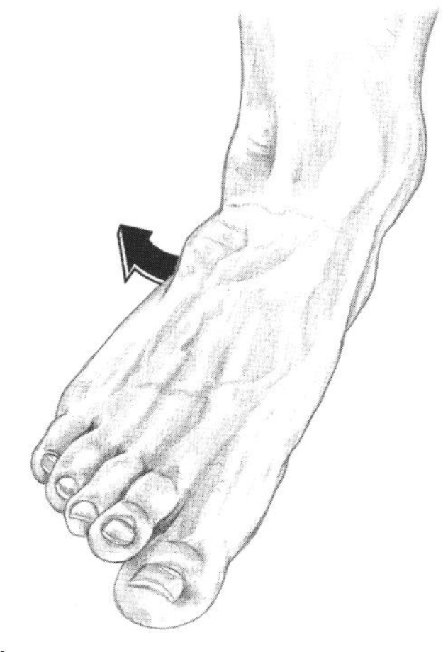

5) Action

____

6) Synergists

P ______________________________

P ______________________________

7) Antagonist

T ______________________________

8) Action

____

9) Synergists

F ______________________________

F ______________________________

10) Antagonist

L ______________________________

Leg and Foot
Muscles and Movements #2

Please list the action demonstrated, two synergists and one antagonist. The first letter of each muscle has been provided.

1) This action happens at which three joints?

____

____

____

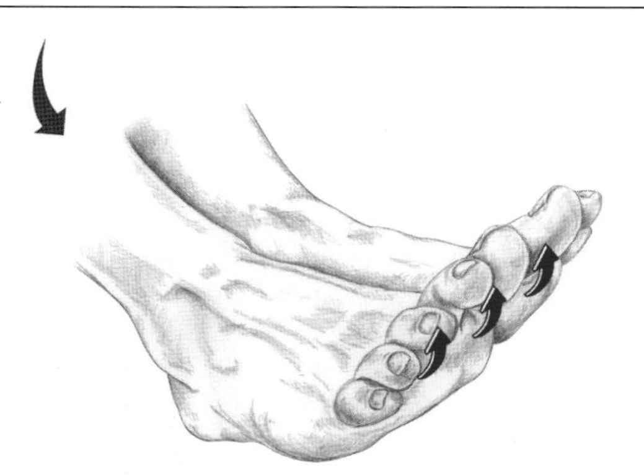

2) Action

____

3) Synergists

E ______________________

L ______________________

4) Antagonist

F ______________________

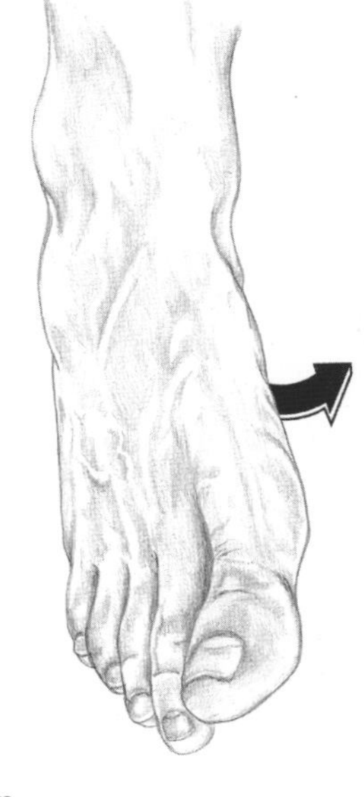

5) Action

____

6) Synergists

T ______________________

F ______________________

7) Antagonist

E ______________________

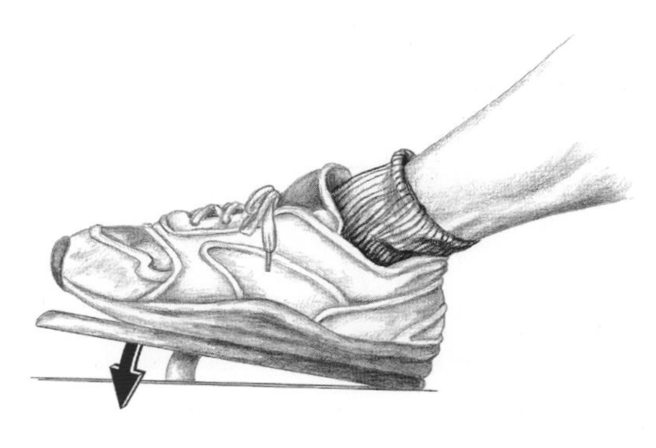

8) Action

____

9) Synergists

S ______________________

T ______________________

10) Antagonist

E ______________________

What's the Muscle? #1

Please identify the following muscles.

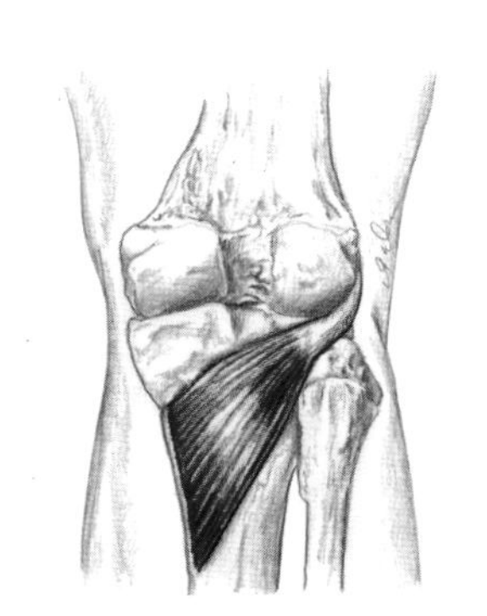

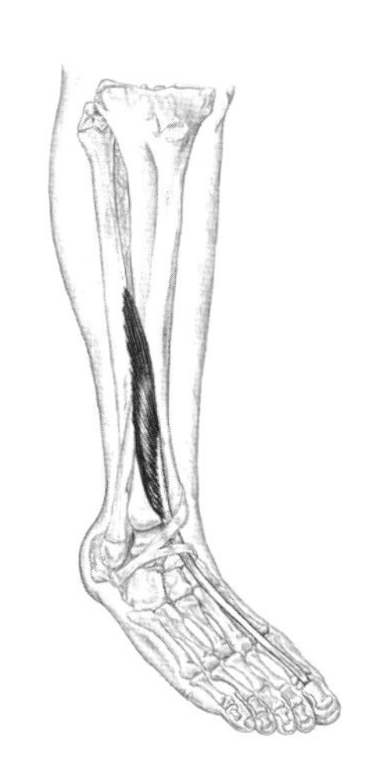

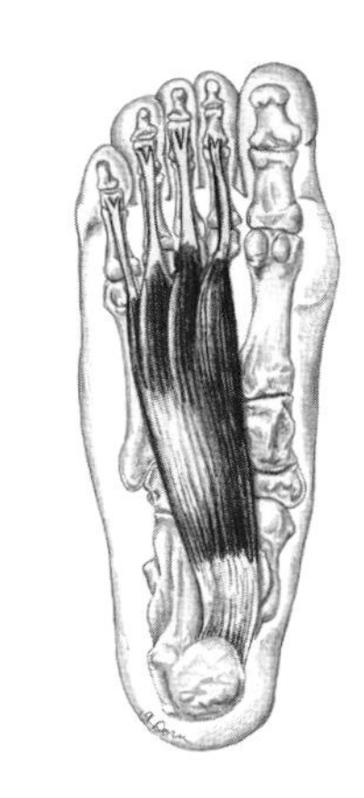

1) ____________________

2) ____________________

3) ____________________

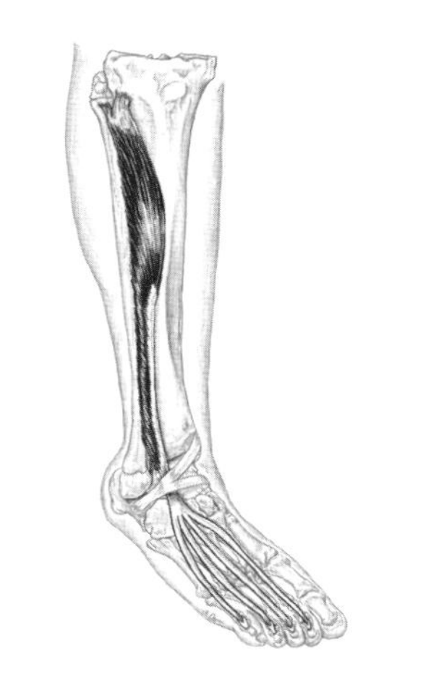

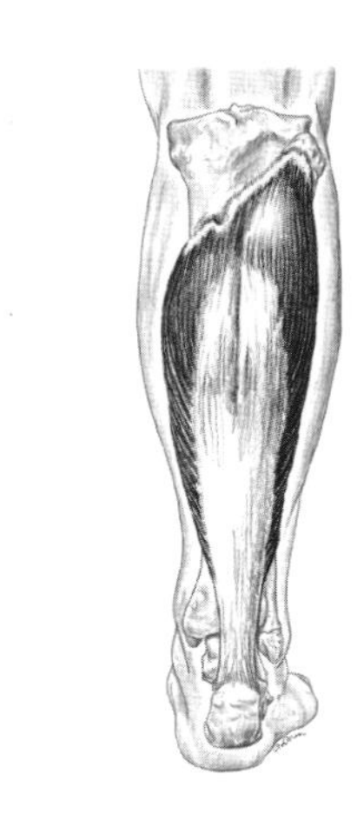

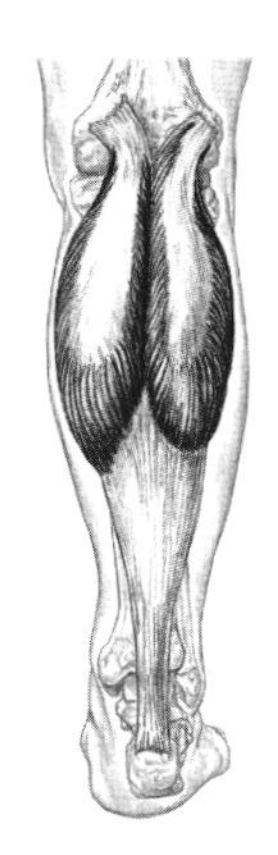

4) ____________________

5) ____________________

6) ____________________

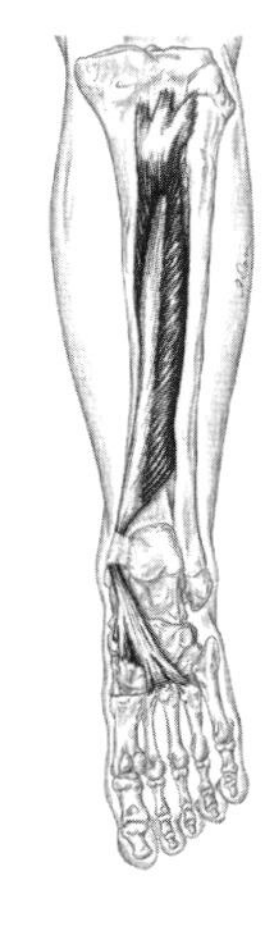

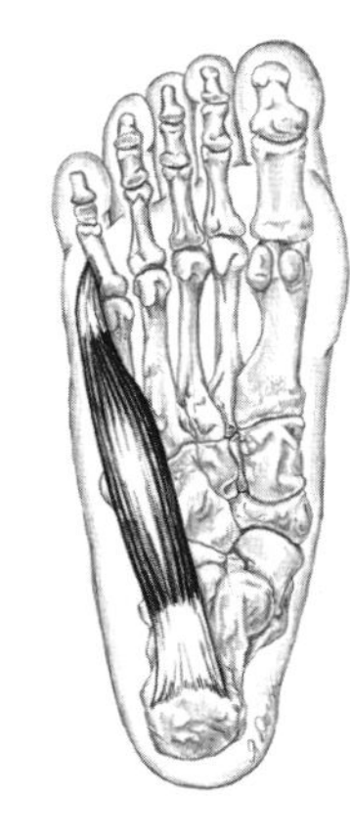

7) ____________________

8) ____________________

What's the Muscle? #2

Please identify the following muscles.

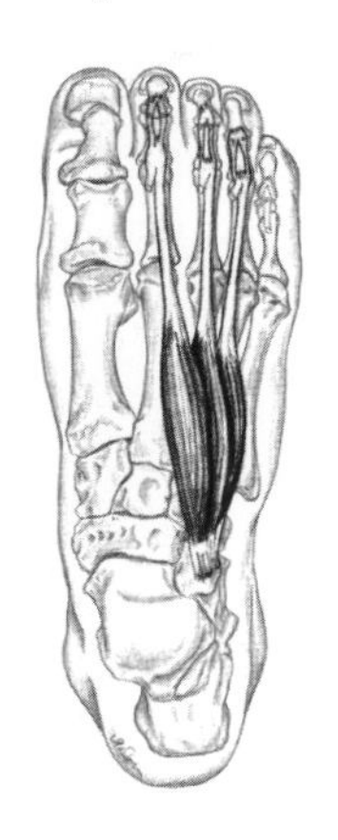

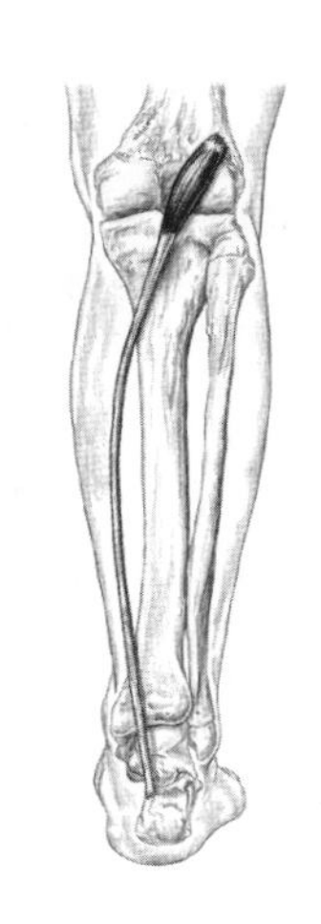

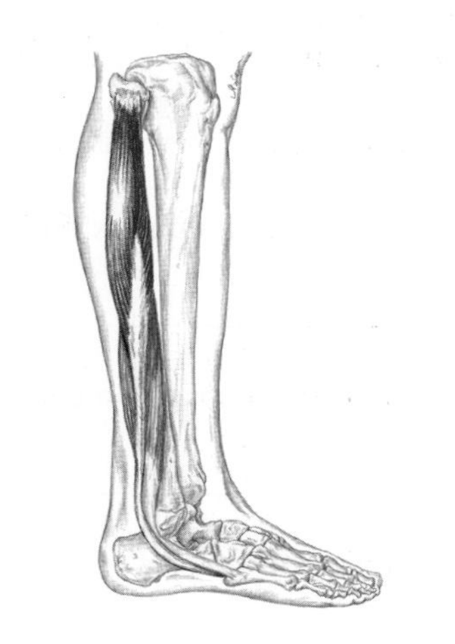

1) ____________________

2) ____________________

3) ____________________

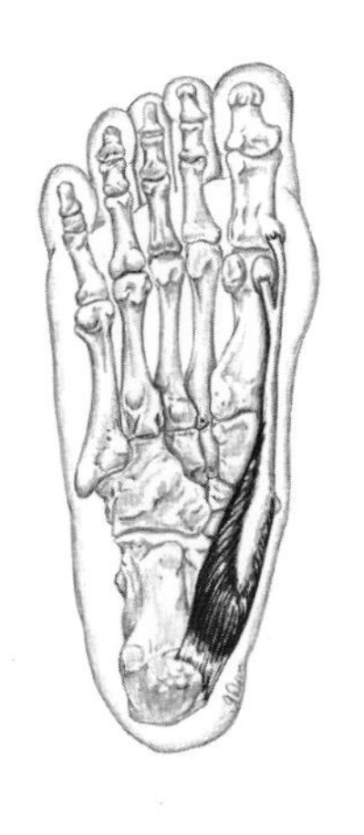

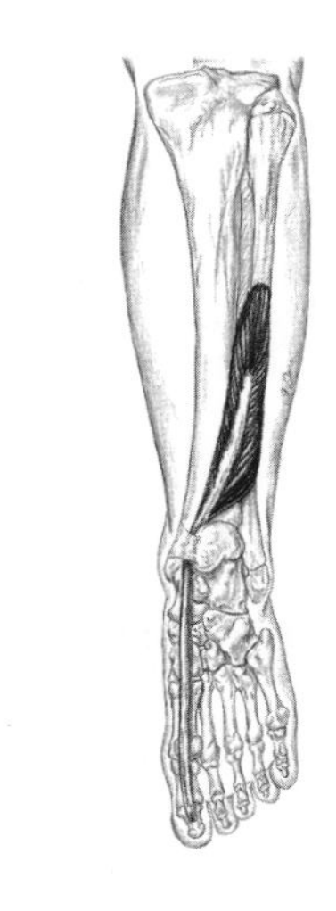

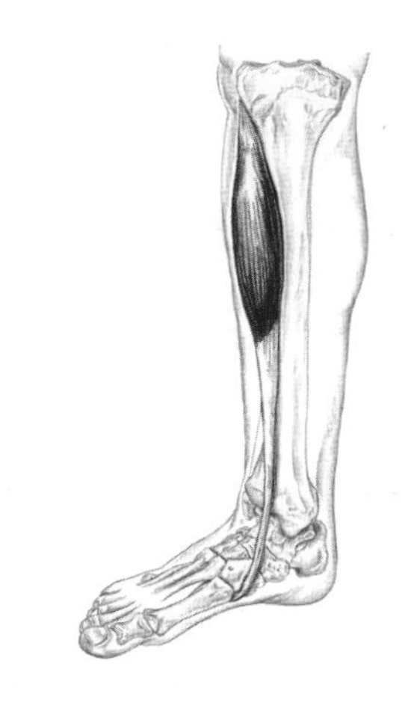

4) ____________________

5) ____________________

6) ____________________

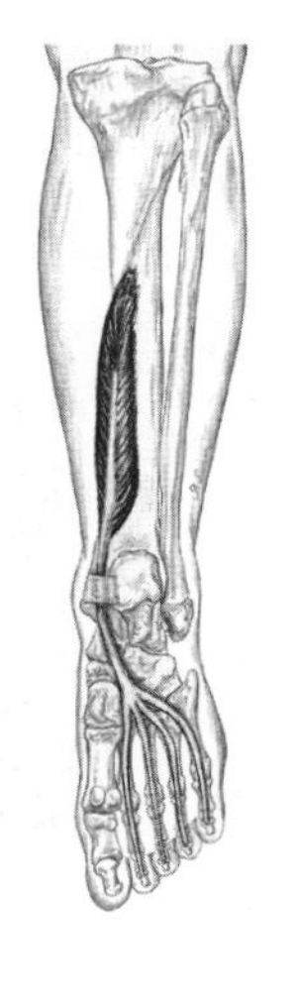

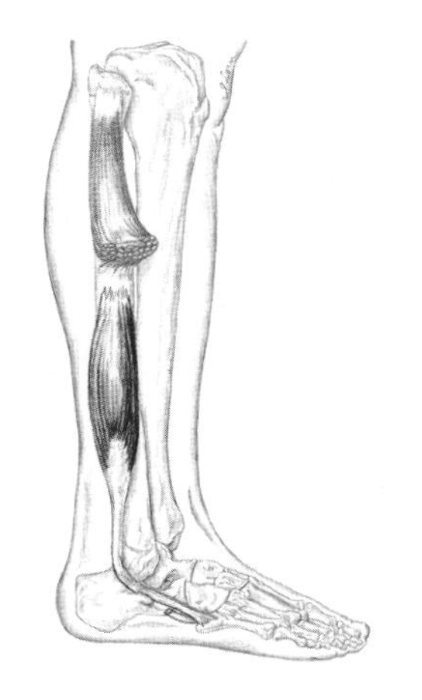

7) ____________________

8) ____________________

Leg and Foot, Muscle Group #1

Gastrocnemius, Soleus, Plantaris, Popliteus and Peroneals

p. 371-377

Please answer the following questions.

1) What are the two muscles that form the "triceps surae"? ______________________ ______________________

2) The gastrocnemius extends halfway down the leg before blending into which band of connective tissue?

____

3) What common action could you ask your partner to do to contract his gastrocnemius and soleus?

____

4) Palpate both bellies of the gastrocnemius. Which head extends further distally?

____

5) Although the gastrocnemius and soleus are located on the posterior leg, you can palpate them on the leg's anterior surface by sliding _________________________ off the _________________________.

6) The muscle belly accessed between the gastrocnemius heads, in the popliteal space, is the ____________________.

7) The belly of the plantaris can be distinguished by its _________________________-wide belly that runs at an _________________________ angle.

8) Which muscle is the deepest of the popliteal space? _________________________

9) When the knee is extended, the popliteus plays what important role?

____

10) To access the popliteus' tendinous attachment, you need to push the overlying edges of which muscles to the side?

____ ____

11) The peroneal muscles are located on the _________________________ side of the leg and lie between which two muscles? _________________________ _________________________

12) What are two bony landmarks that can help you isolate the peroneal bellies?

____ ____

13) What action could you ask your partner to perform to feel the peroneals tighten?

____

p. 371-377

Leg and Foot, Muscle Group #1

Gastrocnemius, Soleus, Plantaris, Popliteus and Peroneals

Matching

Match the origin and insertion to the correct muscle.

Origins

1) Distal two-thirds of lateral fibula
2) Lateral condyle of the femur
3) Lateral supracondylar line of femur
4) Condyles of the femur, posterior surfaces
5) Head of fibula and proximal two-thirds of lateral fibula
6) Soleal line; proximal, posterior surface of tibia and posteror aspect of head of fibula

Insertions

7) Base of the first metatarsal and medial cuneiform
8) Calcaneus via calcaneal tendon (3)
9) Proximal, posterior aspect of tibia
10) Tuberosity of fifth metatarsal

Muscle	O	I
Gastrocnemius	______	______
Peroneus brevis	______	______
Peroneus longus	______	______
Plantaris	______	______
Popliteus	______	______
Soleus	______	______

Shorten or Lengthen?

11) Passive dorsiflexion of the ankle would __________________ the soleus.
12) Passive lateral rotation of the knee would __________________ the popliteus.
13) Passive inversion of the foot would __________________ the peroneus longus.
14) Passive flexion of the knee would __________________ the gastrocnemius.

Let's Palpate!

Remember—there are no right or wrong answers here

Locate and explore the **gastrocnemius and soleus** on three individuals. Then write three words that describe what you feel. (See p. 371-373 in *Trail Guide*.)

Person #1 ____________ Person #2 ____________ Person #3 ____________

________________________ ________________________ ________________________

________________________ ________________________ ________________________

________________________ ________________________ ________________________

Leg and Foot, Muscle Group #2

Extensors, Flexors and Muscles of the Foot

Please answer the following questions.

1) The tibialis anterior belly can be easily located lateral to which bony landmark? ______________________

2) To feel the tibialis anterior belly contract, you could ask your partner to perform which action?

____

3) Along the ankle's dorsal surface, the extensor hallucis longus and extensor digitorum longus both pass underneath which band of connective tissue? ______________________

4) The flexors of the ankle and toes are virtually inaccessible, except on the medial side of the leg between which two structures?

______________________ ______________________

5) Please complete the following for the mnemonic device "**T**om, **D**ick **AN**' **H**arry".

T______________ ______________

______________ **D**______________ ______________

______________ **A**______________

______________ **N**______________

______________ **H**______________ ______________

6) What action at the toes could you ask your partner to perform to feel contraction of all the flexor bellies?

____

7) The dorsal surface of the foot is home to which muscle that extends down to the second, third and fourth toes?

____

8) The first layer of muscles on the foot's plantar surface is deep to which connective tissue structure?

____

9) Which tendons do you need to palpate beneath to locate the belly of the extensor digitorum brevis?

____

10) What two structures are helpful in isolating the flexor digitorum brevis?

______________________ ______________________

11) Not everyone has the coordination to abduct their first toe. What is another action your partner could do to contract the abductor hallucis? ______________________

12) Which three points of contact form a triangle with the three arches of the foot?

______________ ______________ ______________

p. 378-391

Leg and Foot, Muscle Group #2

Extensors, Flexors and Muscles of the Foot

Matching

Match the origin and insertion to the correct muscle.

Origins

1) Lateral condyle of tibia; proximal, anterior shaft of fibula and interosseous membrane

2) Lateral condyle of tibia; proximal, lateral surface of tibia and interosseous membrane

3) Middle anterior surface of fibula and interosseous membrane

4) Middle half of posterior fibula

5) Middle, posterior surface of tibia

6) Proximal, posterior shafts of tibia and fibula; and interosseous membrane

Muscle	O	I
Extensor digitorum longus	_______	_______
Extensor hallucis longus	_______	_______
Flexor digitorum longus	_______	_______
Flexor hallucis longus	_______	_______
Tibialis anterior	_______	_______
Tibialis posterior	_______	_______

Insertions

7) Distal phalanx of first toe (2)

8) Distal phalanges of second through fifth toes

9) Medial cuneiform and base of the first metatarsal

10) Middle and distal phalanges of second through fifth toes

11) All five tarsal bones and bases of second through fourth metatarsals

Shorten or Lengthen?

12) Passive flexion of the fifth toe would ____________________ the abductor digiti minimi.

13) Passive eversion of the foot would ____________________ the tibialis posterior.

14) Passive dorsiflexion of the ankle would ____________________ the extensor digitorum longus.

15) Passive eversion of the foot would ____________________ the tibialis anterior.

Let's Palpate!

Remember—there are no right or wrong answers here

Locate and explore the **tibialis anterior** on three individuals. Then write three words that describe what you feel. (See p. 378-379 in *Trail Guide.*)

Person #1 ______________ Person #2 ______________ Person #3 ______________

____________________ ____________________ ____________________

____________________ ____________________ ____________________

____________________ ____________________ ____________________

Muscles of the Foot #1

p. 384

Please identify the following structures.

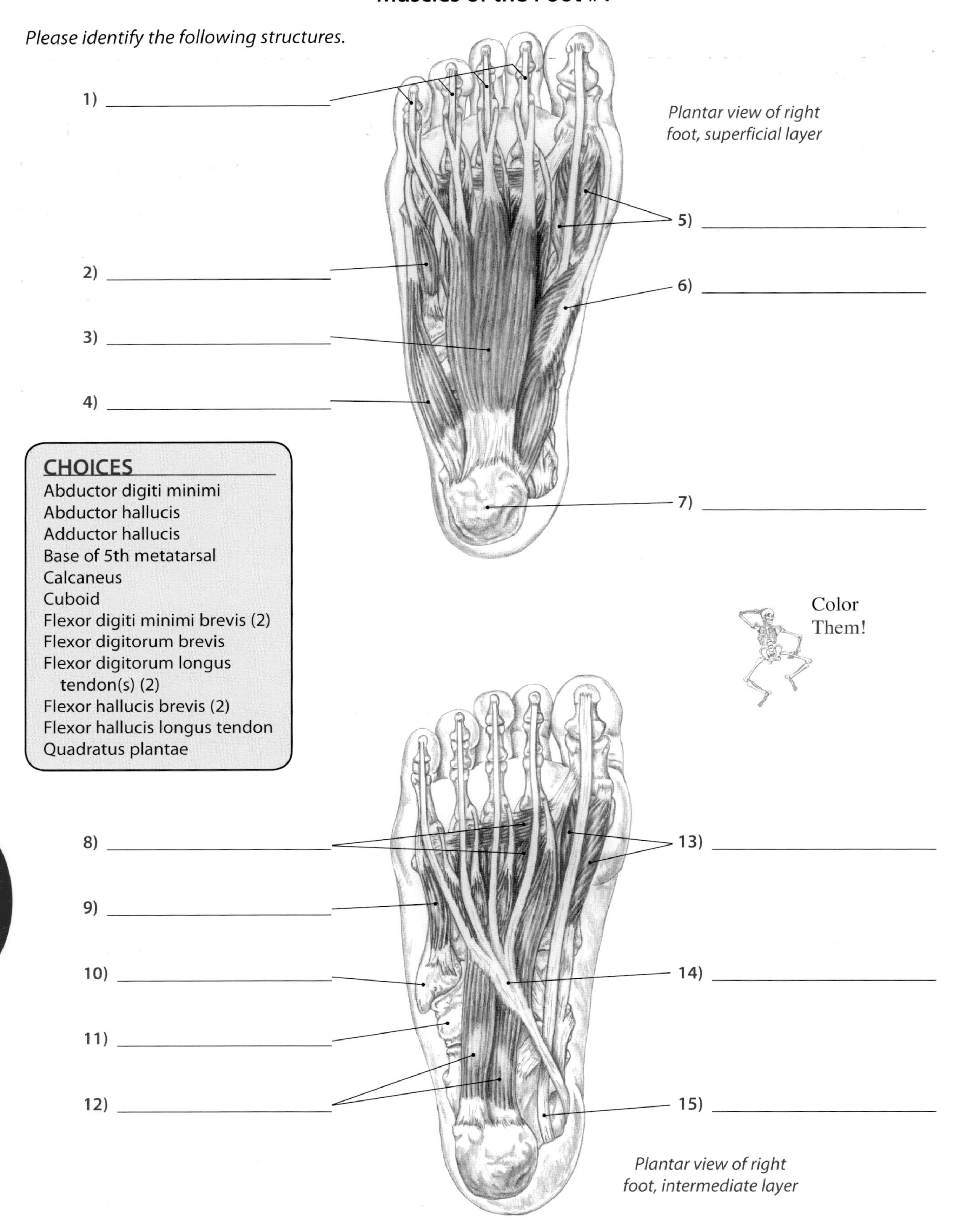

p. 385-386

Leg and Foot

Muscles of the Foot #2

Please identify the following structures.

1) ______________________

2) ______________________

3) ______________________

4) ______________________

5) ______________________

6) ______________________

7) ______________________

8) ______________________

9) ______________________

Dorsal view of right foot

CHOICES

- Abductor digiti minimi
- Calcaneal tendon
- Calcaneus
- Extensor digitorum brevis
- Extensor digitorum longus
- Extensor digitorum longus and peroneus tertius
- Extensor hallucis brevis
- Extensor hallucis longus
- Inferior extensor retinaculum (2)
- Inferior peroneal retinaculum (2)
- Peroneus brevis
- Peroneus longus
- Peroneus longus and brevis
- Soleus
- Superior extensor retinaculum (2)
- Superior peroneal retinaculum
- Tibialis anterior (2)

10) ______________________

11) ______________________

12) ______________________

13) ______________________

14) ______________________

15) ______________________

16) ______________________

17) ______________________

18) ______________________

19) ______________________

20) ______________________

21) ______________________

Lateral view of ankle and foot

Tibiofemoral Joint

Please identify the following structures.

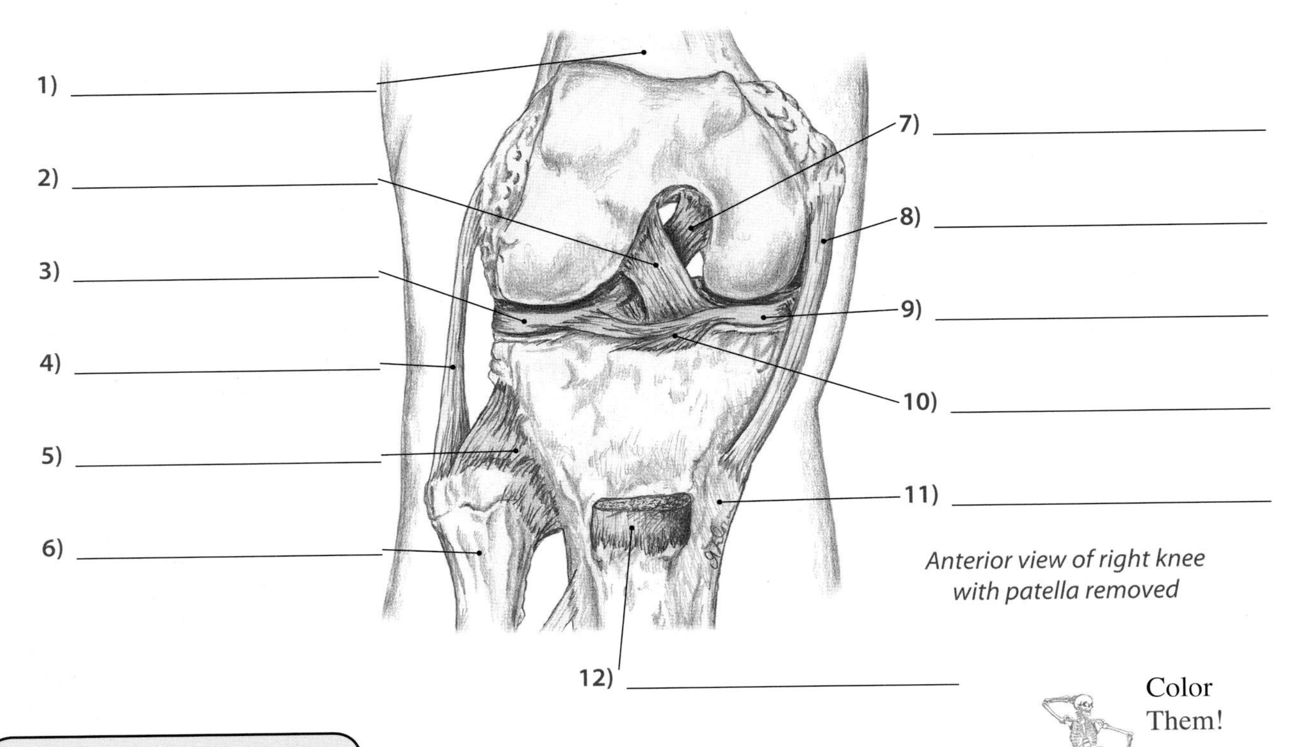

Anterior view of right knee with patella removed

Color Them!

CHOICES

Anterior cruciate ligament (2)
Anterior ligament of head of the fibula
Femur
Fibula
Fibular collateral ligament (2)
Lateral meniscus (2)
Medial meniscus (2)
Patellar ligament (cut)
Popliteus tendon (cut)
Posterior cruciate ligament (2)
Posterior ligament of head of the fibula
Posterior meniscofemoral ligament
Tibia
Tibial collateral ligament (2)
Transverse ligament of knee

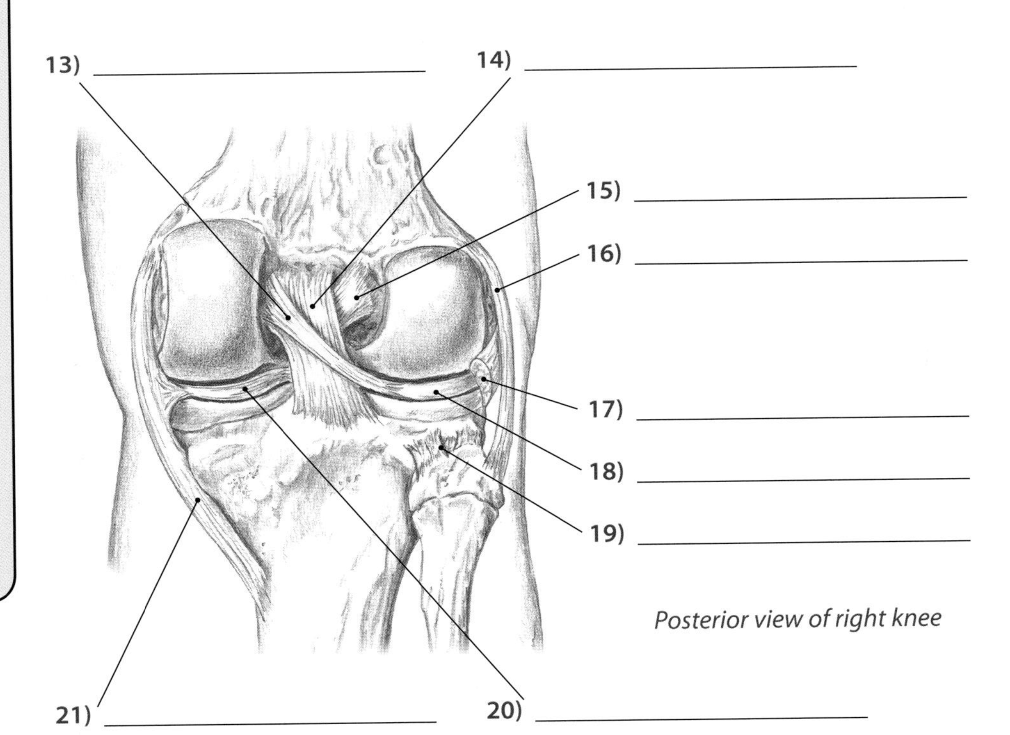

Posterior view of right knee

Leg and Foot

Tibiofemoral and Tibiofibular Joints

Please identify the following structures.

CHOICES

Anterior cruciate ligament (cut)
Anterior ligament of head of the fibula
Anterior talofibular ligament (cut)
Anterior tibiofibular ligament
Biceps femoris tendon (cut)
Cruciate ligaments (cut)
Fibula
Fibular collateral ligament (cut)
Iliotibial tract (cut)
Interosseous membrane
Lateral meniscus
Medial meniscus
Patellar ligament (cut)
Posterior cruciate ligament (cut)
Posterior meniscofemoral ligament (cut)
Tibia
Tibial collateral ligament (cut)

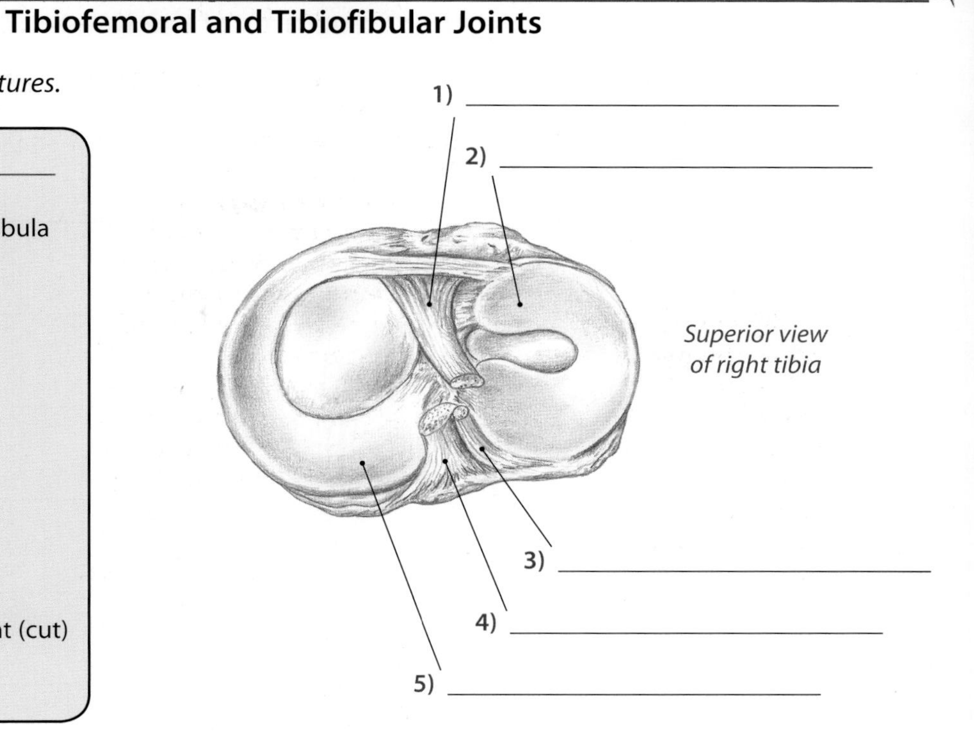

Superior view of right tibia

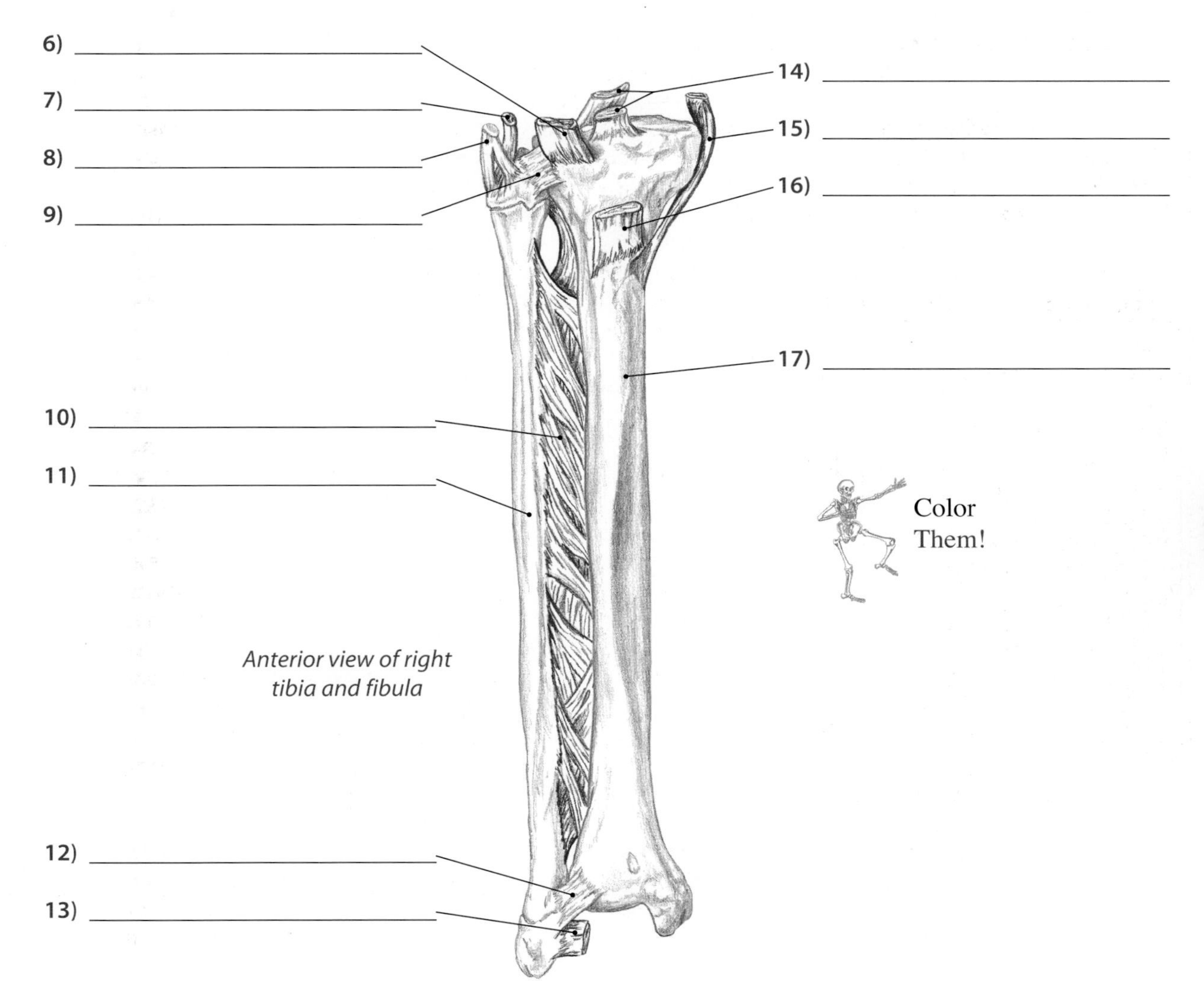

Anterior view of right tibia and fibula

Other Structures of the Knee

Please identify the following structures.

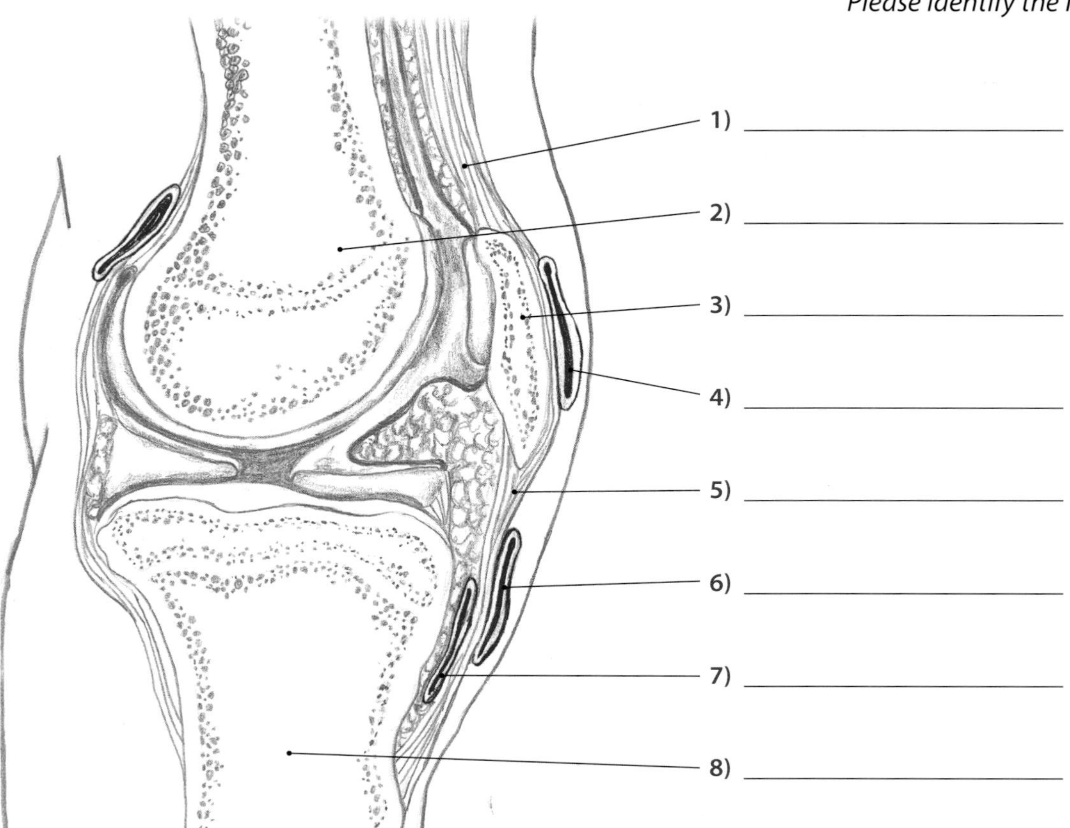

Lateral cross section of knee

CHOICES

- Common peroneal nerve
- Deep infrapatellar bursa
- Femur
- Gastrocnemius
- Hamstrings
- Lesser saphenous vein
- Patella
- Patellar ligament
- Popliteal artery and vein
- Prepatellar bursa
- Quadriceps femoris tendon
- Subcutaneous infrapatellar bursa
- Tibia
- Tibial nerve

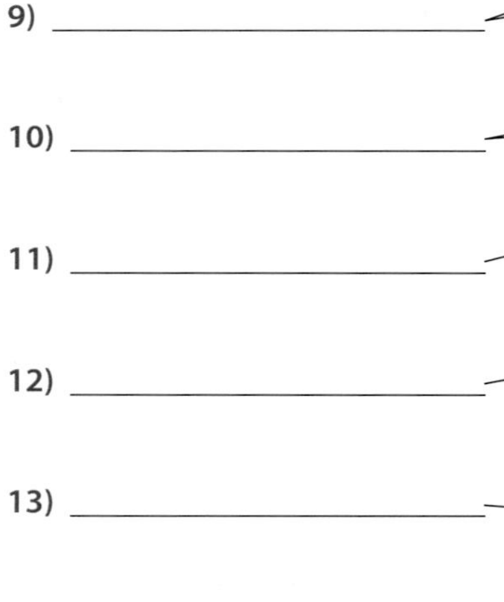

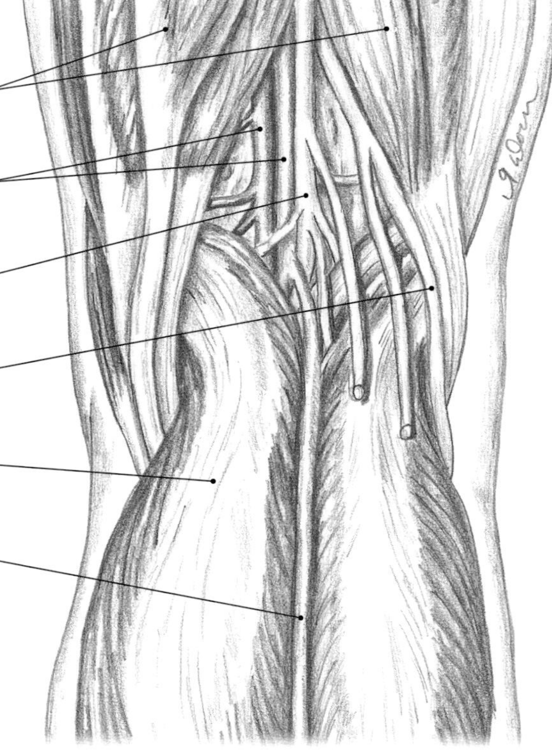

Posterior view of right knee

Color It!

Leg and Foot
Talocrural Joint

Please identify the following structures.

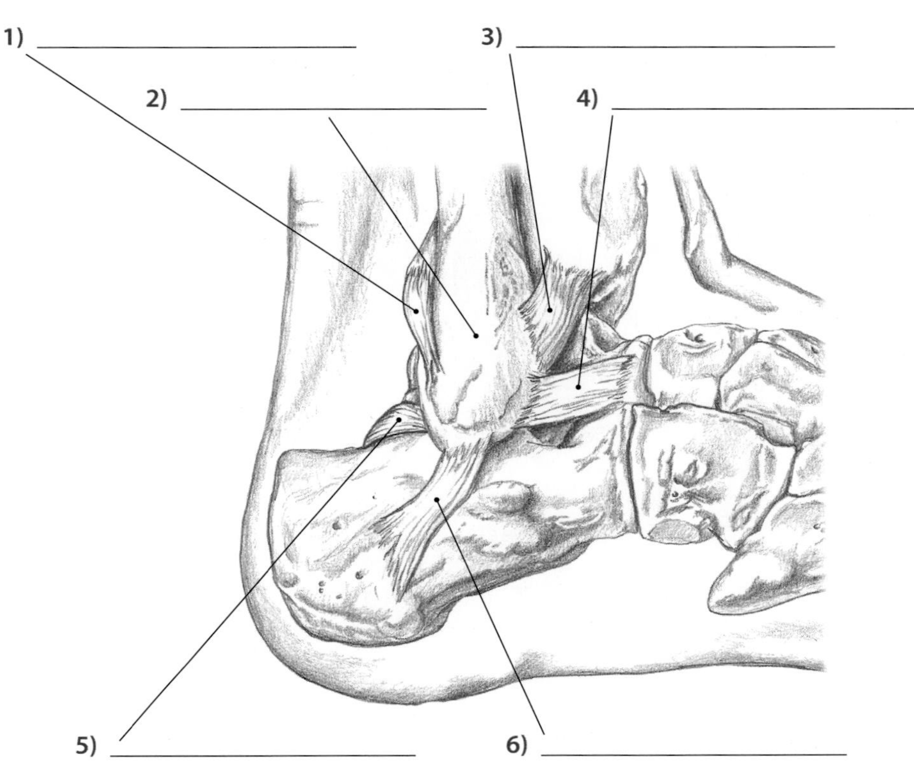

Lateral view of right ankle

CHOICES

- Anterior talofibular ligament
- Anterior tibiofibular ligament
- Anterior tibiotalar ligament
- Calcaneofibular ligament
- Deltoid ligament
- Lateral malleolus
- Medial malleolus
- Navicular
- Posterior talofibular ligament
- Posterior tibiofibular ligament
- Posterior tibiotalar ligament
- Sustentaculum tali
- Tibiocalcaneal ligament
- Tibionavicular ligament

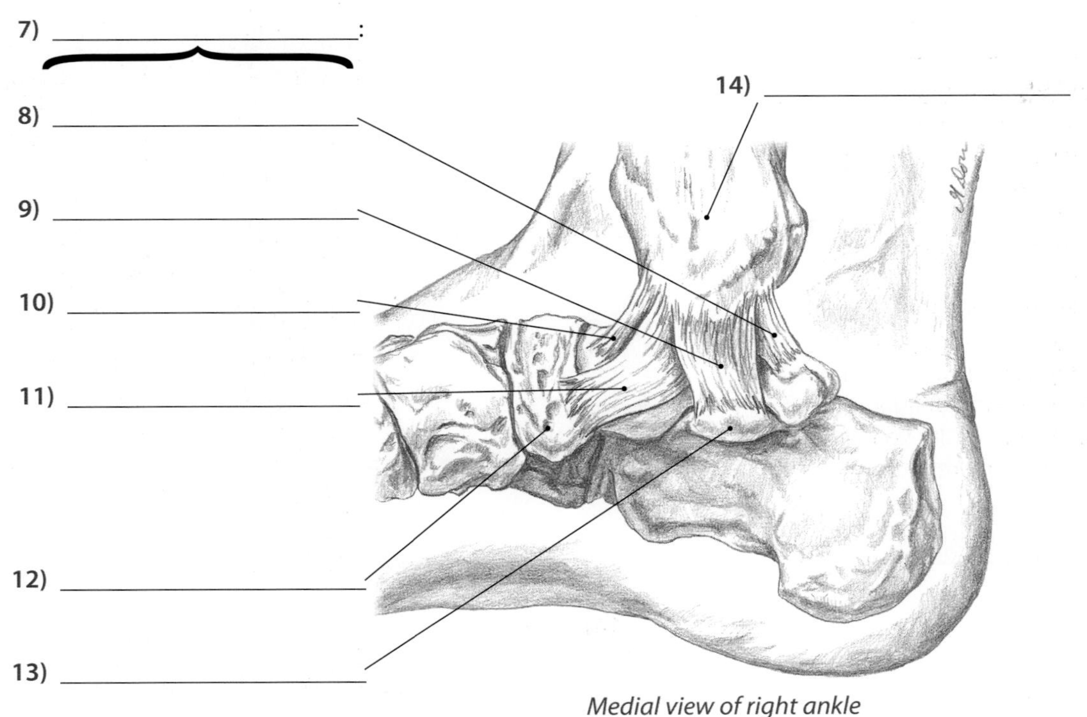

Medial view of right ankle

Leg and Foot
Talocrural and Talotarsal Joints

Please identify the following structures.

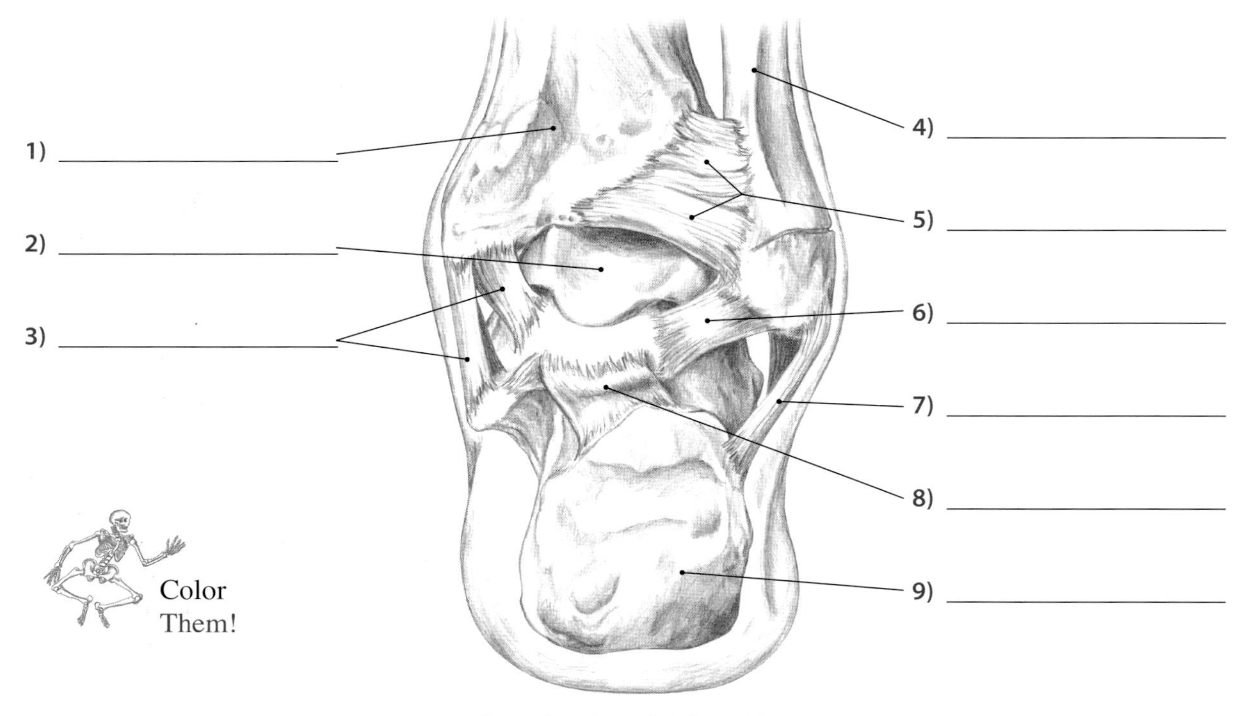

Posterior view of right ankle

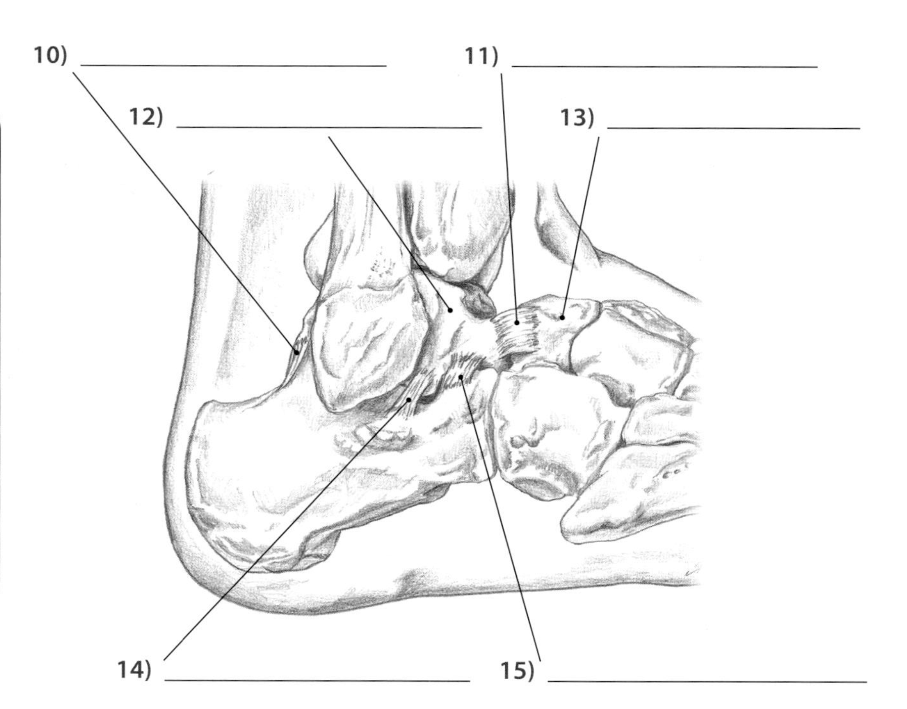

Lateral view of right ankle

CHOICES

- Calcaneofibular ligament
- Calcaneus
- Deltoid ligament
- Fibula
- Interosseous talocalcaneal ligament
- Lateral talocalcaneal ligament
- Navicular
- Posterior talocalcaneal ligament (2)
- Posterior talofibular ligament
- Posterior tibiofibular ligament
- Talonavicular ligament
- Talus (2)
- Tibia

Leg and Foot

Ligaments of the Foot

Please identify the following structures.

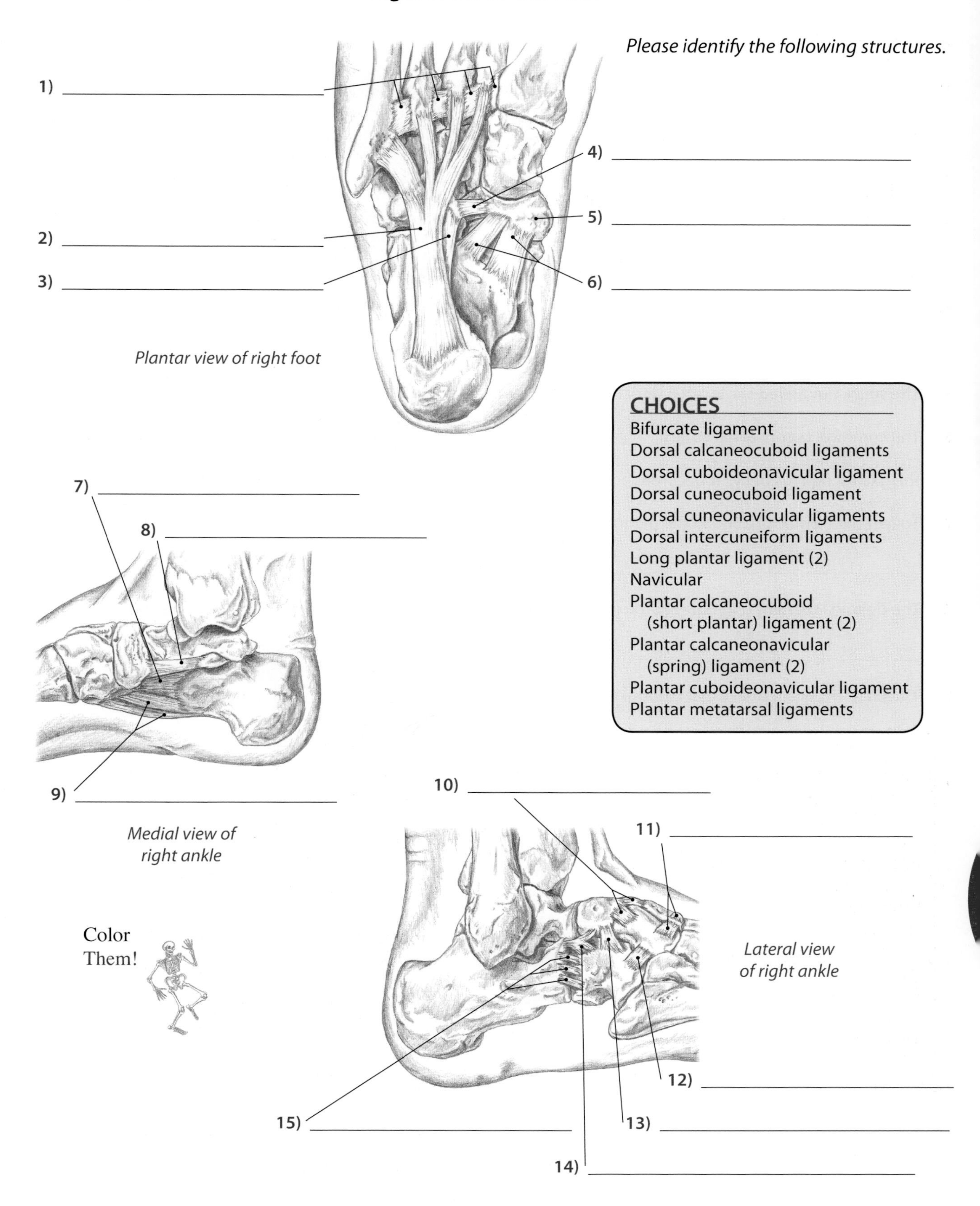

CHOICES

- Bifurcate ligament
- Dorsal calcaneocuboid ligaments
- Dorsal cuboideonavicular ligament
- Dorsal cuneocuboid ligament
- Dorsal cuneonavicular ligaments
- Dorsal intercuneiform ligaments
- Long plantar ligament (2)
- Navicular
- Plantar calcaneocuboid (short plantar) ligament (2)
- Plantar calcaneonavicular (spring) ligament (2)
- Plantar cuboideonavicular ligament
- Plantar metatarsal ligaments

Other Structures of the Knee, Leg and Foot

p. 392-405

Please answer the following questions.

1) The fibular collateral ligament spans between which two bony landmarks?

_______________ _______________

2) Both collateral ligaments resist which rotation of the tibia? _______________

3) Aside from helping the femoral condyles sit upon the tibial plateaus, the menisci of the knee are also important for _______________ and _______________.

4) To access the edge of the medial meniscus, you would slowly rotate the knee in which direction?

____

5) The small, fluid-filled sac located directly superficial to the patella is the _______________.

6) The common peroneal nerve lies _______________ to the biceps femoris tendon and _______________ to the gastrocnemius belly. It becomes accessible along the _______________ surface of the head of the fibula.

7) To feel the tension change in the plantar aponeurosis, what action could you passively perform at the foot?

____

8) The deltoid ligament originates at the medial malleolus and fans distally out to which three bones/bony landmarks?

_______________ _______________ _______________

9) The plantar calcaneonavicular (spring) ligament stretches from the _______________ to the _______________ and may be deep to the _______________.

10) The fibers of the extensor retinacula have two palpable distinctions from the extensor tendons. What are they?

_______________ _______________

11) The flexor retinaculum can be located between which two bony landmarks?

_______________ _______________

12) The posterior tibial artery can be located just _______________ and _______________ to the medial malleolus.

13) Between the first and second metatarsals of the foot you can feel the pulse of which artery?

____

14) The calcaneal bursa is located between the attachment of the _______________ and the _______________.

Answer Pages

Italicized page numbers after the answers indicate where the information can be found in *Trail Guide*.

Introduction

Tour Guide Tips #1, p. 1

1) bony landmarks—*p. 2*
2) Even though the topography, shape and proportion are unique, the body's composition and structures are virtually identical on all individuals.—*p. 2*
3) To examine or explore by touching (an organ or area of the body), usually as a diagnostic aid—*p. 4*
4) locating, aware, assessing—*p. 4*
5) directs movement, depth.—*p. 4*
6) • read the information
   • visualize what you are trying to access
   • verbalize to your partner what you feel
   • locate the structure first on yourself
   • read the text aloud
   • be patient—*p. 5*
7) across, along—*p. 6*
8) stay still—*p. 6*
9) active, passive—*p. 7*
10) lengths, shapes, edges—*p. 7*
11) • move slowly
    • avoid using excessive pressure
    • focus your awareness—be present—*p. 8*
12) muscle cells, layers of connective tissue—*p. 11*
13) tendon—*p. 11*

Tour Guide Tips #2, p. 2

1) agonist, antagonist—*p. 11*
2) striated texture, direction of the muscle fibers, it can be in contracted or relaxed state—*p. 11, 12*
3) attachments, variable tension —*p. 13*
4) tendon, ligament, fascia, periosteum, retinaculum, aponeurosis, adipose—*p. 13-17*
5) fibrous membrane, skin—*p. 14*
6) compression or impingement of a nerve—*p. 17*

Matching

1) N adipose—*p. 17*
2) F aponeurosis—*p. 13*
3) D artery—*p. 16*
4) H bone—*p. 10*
5) E bursa—*p. 16*
6) B fascia—*p. 14*
7) G ligament—*p. 13*
8) I lymph node—*p. 17*
9) A muscle—*p. 11*
10) J nerve—*p. 17*
11) K retinaculum—*p. 15*
12) L skin—*p. 10*
13) M tendon—*p. 13*
14) C vein—*p. 16*

Exploring Textures #1, p. 3

1) epidermis
2) dermis
3) arrector pili muscle
4) sweat gland
5) hair follicle
6) blood vessels
7) muscle fibers
8) endomysium
9) perimysium
10) epimysium
11) bone
12) blood vessels
13) neurovascular bundle
14) tendon
15) periosteum

Types of Muscle Bellies & Joints, p. 4

1) fusiform
2) multipennate
3) sphincter
4) bipennate
5) triangular
6) unipennate
7) gliding
8) hinge
9) ellipsoid
10) pivot
11) ball-and-socket
12) saddle

Exploring Textures #2, p. 5

1) muscle tissue
2) bone
3) periosteum
4) interosseous membrane
5) deep fascia
6) adipose (fatty) tissue
7) superficial fascia
8) skin
9) deep fascia

Navigating

Regions of the Body, p. 6

1) pectoral
2) axillary
3) brachial
4) cubital
5) abdominal
6) inguinal
7) pubic
8) femoral
9) facial
10) mandibular
11) supraclavicular
12) antecubital
13) patellar
14) crural
15) cranial
16) cervical
17) scapular
18) thoracic
19) lumbar
20) pelvic
21) sural
22) gluteal
23) popliteal

Planes, Directions, Positions & Movements #1, p. 7

1) frontal
2) sagittal
3) transverse
4) superior *or* cranial
5) inferior *or* caudal
6) posterior
7) anterior
8) proximal
9) distal
10) medial
11) lateral
12) superficial
13) deep

Planes, Directions, Positions & Movements #2, p. 8

1) C anterior
2) J deep
3) B distal

Answer Pages

4) E inferior
5) I lateral
6) H medial
7) A posterior
8) G proximal
9) F superficial
10) D superior
11) K abduction
12) V adduction
13) P circumduction
14) T dorsiflexion
15) U extension
16) S flexion
17) Q lateral flexion
18) M lateral rotation
19) L medial rotation
20) O plantar flexion
21) W pronation
22) R rotation
23) N supination

Movements of the Body #1-5, p. 9-13

1) supination of forearm
2) depression of scapula
3) depression of mandible
4) abduction of hip
5) adduction of shoulder
6) flexion of wrist
7) flexion of thumb
8) inversion of foot
9) rotation of spine
10) upward rotation of scapula
11) posterior tilt (upward rotation) of pelvis
12) lateral deviation of mandible
13) adduction (ulnar deviation) of wrist
14) extension of fingers
15) anterior tilt (downward rotation) of pelvis
16) extension of elbow
17) extension of wrist
18) elevation of scapula
19) elevation of mandible
20) lateral flexion of spine
21) adduction of hip
22) extension of hip
23) abduction of shoulder
24) extension of neck
25) adduction (or retraction) of scapula
26) lateral flexion of neck
27) adduction of fingers
28) flexion of fingers
29) lateral tilt (elevation) of pelvis
30) abduction (or protraction) of scapula
31) lateral rotation of hip
32) flexion of hip
33) flexion of elbow
34) pronation of forearm
35) abduction of fingers
36) lateral rotation of shoulder
37) extension of spine
38) flexion of knee
39) flexion of neck
40) abduction of thumb
41) dorsiflexion of ankle
42) medial rotation of shoulder
43) extension of knee
44) horizontal adduction of shoulder
45) abduction (radial deviation) of wrist
46) eversion of foot
47) medial rotation of hip
48) protraction of mandible
49) opposition of thumb
50) retraction of mandible
51) extension of thumb
52) plantar flexion of ankle
53) flexion of spine
54) downward rotation of scapula
55) extension of shoulder
56) adduction of thumb
57) rotation of neck
58) flexion of shoulder
59) horizontal abduction of shoulder
60) elevation/expansion of ribs (inhalation)

Skeletal System #1, p. 14

1) axial
2) skull
3) cranial portion
4) facial portion
5) mandible
6) cervical vertebra
7) clavicle
8) sternum
9) humerus
10) ulna
11) radius
12) carpals
13) metacarpals
14) phalanges
15) scapula
16) ribs
17) lumbar vertebra
18) pelvis
19) sacrum
20) coccyx
21) femur
22) patella
23) tibia
24) fibula
25) tarsals
26) metatarsals
27) phalanges

Skeletal System #2, p. 15

1) appendicular
2) cranium
3) mandible
4) scapula
5) thoracic vertebra
6) ribs
7) lumbar vertebra
8) pelvis
9) sacrum
10) coccyx
11) cervical vertebra
12) clavicle
13) humerus
14) ulna
15) radius
16) carpals
17) metacarpals
18) phalanges
19) femur
20) tibia
21) fibula
22) talus
23) calcaneus

Muscular System #1, p. 16

1) orbicularis oculi
2) omohyoid
3) pectoralis minor
4) coracobrachialis
5) deltoid (cut)
6) brachialis
7) rectus abdominis
8) internal oblique
9) flexor digitorum profundus
10) adductor longus
11) vastus intermedius
12) temporalis
13) masseter
14) sternocleidomastoid
15) trapezius
16) deltoid
17) pectoralis major
18) serratus anterior
19) biceps brachii
20) external oblique

21) brachioradialis
22) pectineus
23) sartorius
24) gracilis
25) rectus femoris
26) vastus lateralis
27) vastus medialis
28) gastrocnemius
29) peroneus longus
30) tibialis anterior

Muscular System #2, p. 17

1) supraspinatus
2) rhomboids
3) infraspinatus
4) teres major
5) triceps brachii
6) brachioradialis
7) internal oblique
8) gluteus minimus
9) piriformis
10) quadratus femoris
11) vastus lateralis
12) semimembranosus
13) plantaris
14) occipitalis
15) splenius capitis
16) trapezius
17) deltoid
18) latissimus dorsi
19) flexor carpi ulnaris
20) extensor digitorum
21) gluteus maximus
22) gracilis
23) biceps femoris (long head)
24) gastrocnemius
25) peroneus longus

Muscular System #3, p. 18

1) sternocleidomastoid
2) trapezius
3) pectoralis major
4) deltoid
5) latissimus dorsi
6) gluteus maximus
7) vastus lateralis
8) iliotibial tract
9) biceps femoris
10) gastrocnemius
11) peroneus longus
12) tibialis anterior
13) orbicularis oculi
14) platysma
15) flexors of the wrist and fingers
16) biceps brachii
17) triceps brachii
18) pectoralis major
19) serratus anterior
20) rectus abdominis
21) external oblique
22) tensor fasciae latae
23) sartorius
24) vastus medialis
25) adductor magnus

Fascial System #1, p. 19

1) brachial fascia
2) biceps brachii
3) humerus
4) lateral intermuscular septum
5) triceps brachii
6) medial intermuscular septum
7) antebrachial fascia
8) flexor muscles
9) radius
10) ulna
11) interosseous membrane
12) extensor muscles

Fascial System #2, p. 20

1) lateral intermuscular septum
2) quadriceps
3) iliotibial tract
4) femur
5) fascia lata
6) medial intermuscular septum
7) adductors
8) hamstrings
9) tibia
10) interosseous membrane
11) deep crural fascia
12) crural fascia
13) fibula

Cardiovascular System—Arteries, p. 21

1) right internal carotid
2) right vertebral
3) ascending aorta
4) heart
5) thoracic aorta
6) abdominal aorta
7) femoral
8) popliteal
9) anterior tibial
10) posterior tibial
11) dorsal artery of foot (dorsalis pedis)
12) arcuate
13) right common carotid
14) subclavian
15) brachiocephalic trunk
16) axillary
17) brachial
18) common iliac
19) radial
20) ulnar

Cardiovascular System—Veins, p. 22

1) right brachiocephalic
2) superior vena cava
3) coronary sinus
4) inferior vena cava
5) common iliac
6) femoral
7) great saphenous
8) right internal jugular
9) right external jugular
10) left subclavian
11) axillary
12) great cardiac
13) cephalic
14) brachial
15) splenic
16) left renal
17) popliteal
18) small saphenous
19) posterior tibial
20) anterior tibial

Nervous System, p. 23

1) brachial plexus
2) musculocutaneous
3) axillary
4) median
5) radial
6) ulnar
7) lumbar plexus
8) iliohypogastric
9) ilioinguinal
10) genitofemoral
11) lateral femoral cutaneous
12) femoral
13) obturator
14) sacral plexus
15) superior gluteal
16) inferior gluteal
17) tibial
18) common peroneal
19) saphenous
20) cervical plexus
21) lesser occipital
22) ansa cervicalis
23) transverse cervical

Answer Pages

24) supraclavicular
25) phrenic
26) thoracic (12 pairs)
27) median
28) ulnar
29) radial
30) sciatic

Lymphatic System, p. 24
1) cervical lymph nodes
2) thymus gland
3) axillary lymph nodes
4) lymphatic vessels
5) cisterna chyli
6) iliac lymph nodes
7) inguinal lymph nodes
8) lymphatic vessels
9) tonsils
10) internal jugular vein
11) subclavian vein
12) thoracic duct
13) spleen
14) aggregated lymphatic follicle
15) bone marrow

Shoulder & Arm

Topographical Views, p. 25
1) superior nuchal line of the occiput
2) trapezius
3) spine of the scapula
4) inferior angle of the scapula
5) triceps brachii
6) latissimus dorsi
7) triceps brachii
8) deltoid
9) axilla
10) latissimus dorsi
11) serratus anterior
12) acromion
13) deltoid
14) pectoralis major
15) biceps brachii
16) trapezius
17) clavicle

Bones & Bony Landmarks #1, p. 26
1) clavicle, scapula, humerus—*p. 48*
2) synovial—*p. 48*
3) sternoclavicular—*p. 48*
4) glenohumeral—*p. 48*
5) spine of the scapula—*p. 52*
6) in the small of the back—*p. 53*
7) serratus anterior—*p. 53*
8) levator scapula, trapezius—*p. 53*
9) teres major and minor—*p. 54*
10) use your broad thumbpad—*p. 54*

Bones & Bony Landmarks #2, p. 27
1) infraspinatus, supraspinatus, subscapularis—*p. 55, 56*
2) spine of the scapula, medial border, lateral border—*p. 55*
3) acromion, clavicle—*p. 55*
4) maneuver partner's arm and scapula in a way which allows thumb to sink in further—*p. 56*
5) sidelying with your partner's arm lying against his side—*p. 56*
6) trapezius, deltoid—*p. 57*
7) acromial, sternal—*p. 57*
8) elevation, depression—*p. 58*
9) deltopectoral—*p. 59*
10) shape, size—*p. 59*
11) supraspinatus, infraspinatus and teres minor—*p. 60*
12) long head of the biceps brachii—*p. 60*

Extra Credit: **16**
biceps brachii
coracobrachialis
deltoid
infraspinatus
levator scapula
omohyoid
pectoralis minor
rhomboid major
rhomboid minor
serratus anterior
subscapularis
supraspinatus
teres major
teres minor
trapezius
triceps brachii

Bones of Shoulder & Arm #1, p. 28
1) sternoclavicular (S/C) joint
2) clavicle
3) acromioclavicular (A/C) joint
4) glenohumeral joint
5) scapula
6) humerus
7) acromion
8) superior notch
9) coracoid process
10) superior angle
11) supraglenoid tubercle
12) glenoid cavity
13) infraglenoid tubercle
14) subscapular fossa
15) lateral border
16) medial border
17) inferior angle

Bones of Shoulder & Arm #2, p. 29
1) greater tubercle
2) deltoid tuberosity
3) lesser tubercle
4) intertubercular groove
5) head of humerus
6) greater tubercle
7) deltoid tuberosity
8) superior angle
9) supraspinous fossa
10) acromion
11) acromial angle
12) spine of the scapula
13) infraspinous fossa
14) lateral border
15) medial border
16) inferior angle

Muscles of Shoulder & Arm #1, p. 30
1) trapezius
2) deltoid
3) levator scapula
4) rhomboid minor
5) rhomboid major
6) supraspinatus
7) infraspinatus
8) teres minor
9) teres major
10) triceps brachii
11) erector spinae group
12) serratus posterior inferior
13) latissimus dorsi
14) thoracolumbar aponeurosis

Muscles of Shoulder & Arm #2, p. 31
1) levator scapula
2) trapezius
3) deltoid
4) infraspinatus
5) teres minor
6) teres major
7) latissimus dorsi
8) biceps brachii
9) brachialis
10) triceps brachii

11) serratus anterior
12) external oblique
13) trapezius
14) deltoid
15) pectoralis major
16) serratus anterior
17) biceps brachii
18) levator scapula
19) pectoralis minor
20) coracobrachialis

Color the Muscles #1 & 2, p. 32 & 33

Muscles and Movements #1, p. 34

1) glenohumeral
2) horizontal adduction of shoulder
3) deltoid (anterior fibers)
   pectoralis major (upper fibers)
4) deltoid (posterior fibers)
5) adduction (retraction) of scapula
6) trapezius (middle fibers)
   rhomboid major
   rhomboid minor
7) pectoralis minor
8) lateral (external) rotation of the shoulder
9) deltoid (posterior fibers)
   teres minor
10) subscapularis

Muscles and Movements #2, p. 35

1) glenohumeral
2) adduction of the shoulder
3) infraspinatus
   teres major
   teres minor
   triceps brachii (long head)
4) supraspinatus
5) downward rotation of the scapula
6) rhomboid major
   rhomboid minor
   levator scapula
7) trapezius (upper and lower fibers)
8) abduction (protraction) of the scapula
9) serratus anterior (with origin fixed)
   pectoralis minor
10) rhomboid major
    rhomboid minor
11) horizontal abduction of the shoulder
12) deltoid (posterior fibers)
13) pectoralis major (upper fibers)
    deltoid (anterior fibers)

Muscles and Movements #3, p. 36

1) scapulothoracic
2) depression of the scapula
3) serratus anterior (with origin fixed)
   pectoralis minor
4) rhomboid major
   rhomboid minor
5) extension of the shoulder
6) deltoid (posterior fibers)
   latissimus dorsi
7) biceps brachii
8) abduction of the shoulder
9) deltoid (all fibers)
   supraspinatus
10) pectoralis major (all fibers)

Muscles and Movements #4, p. 37

1) glenohumeral
2) medial (internal) rotation of the shoulder
3) subscapularis
   pectoralis major (all fibers)
4) infraspinatus
5) elevation of scapula
6) rhomboid major
   rhomboid minor
   levator scapula
7) serratus anterior (with origin fixed)
8) flexion of the shoulder
9) biceps brachii
   coracobrachialis
10) latissimus dorsi
11) upward rotation of scapula
12) trapezius (upper and lower fibers)
    serratus anterior (with origin fixed)
13) rhomboid major
    rhomboid minor

What's the Muscle? p. 38

1) pectoralis major
2) teres major
3) rhomboid minor
4) coracobrachialis
5) infraspinatus
6) serratus anterior
7) levator scapula
8) trapezius
9) triceps brachii
10) supraspinatus
11) rhomboid major
12) teres minor
13) subscapularis
14) deltoid
15) pectoralis minor
16) latissimus dorsi

Muscle Group #1, p. 39

1) trapezius—*p. 67*
2) abduct the shoulder—*p. 67*
3) antagonist—*p. 67-68*
4) depress—*p. 68*
5) extension—*p. 69*
6) adduction (retraction) of scapula or "bring your shoulder up off the table"—*p. 70*
7) middle portion—*p. 71*
8) grasp tissue and let it slip through your fingers; feel for the muscle's fibrous texture—*p. 72*
9) lateral border—*p. 73*

10) lengthen
11) shorten, lengthen
12) lengthen
13) shorten
14) shorten
15) lengthen
16) shorten
17) shorten

Muscle Group #1, p. 40

Muscle	O	I
deltoid	2	6
latissimus dorsi	4	8
teres major	3	5
trapezius	1	7

Muscle Group #2, p. 41

1) glenohumeral—*p. 74*
2) trapezius (upper fibers)—*p. 74*
3) supraspinatus—*p. 74*
4) thick, superficial fascia—*p. 74*
5) subscapular fossa, serratus anterior—*p. 74*
6) abduction of the shoulder—*p. 76*
7) spine of the scapula, medial border, lateral border—*p. 77*
8) deltoid—*p. 77*
9) teres minor is smaller, teres major medially rotates the shoulder, teres minor laterally rotates the shoulder—*p. 71, 74*
10) latissimus dorsi and teres major—*p. 78*
11) medially rotate the shoulder—*p. 78*
12) deltoid—*p. 79*

Answer Pages

13) flex the shoulder to 90°, then horizontally adduct and laterally rotate the shoulder 10°—*p. 81*
14) biceps brachii—*p. 81*
15) inferior and lateral—*p. 81*

Muscle Group #2, p. 42

Muscle	O	I
infraspinatus	1	5
subscapularis	2	6
supraspinatus	4	5
teres minor	3	5

7) lengthen
8) shorten
9) lengthen
10) lengthen

Muscle Group #3, p. 43

1) trapezius, erector spinae muscles—*p. 82*
2) adduct and elevate the scapula are two synergistic actions; the trapezius upwardly rotates the scapula, while the rhomboids downwardly rotate it—*p. 82, 68-69*
3) splenius capitis, posterior scalene—*p. 84, 85*
4) elevation of the scapula—*p. 85*
5) shifts the cervical TVPs further anterior, gives the levator more palpable tension and shortens and softens the overlying trapezius—*p. 85*
6) rhomboids—*p. 86*
7) latissimus dorsi and pectoralis major—*p. 86*
8) trapezius and rhomboids—*p. 88*
9) clavicular, sternal and costal—*p. 89*
10) *examples: giving a hug, doing a push-up, lifting a stack of heavy anatomy books*
11) communicating your intentions to your partner—*p. 90*
12) deltoid—*p. 90*
13) brings the pectoralis major off the chest wall, allows breast tissue to fall away—*p. 91*
14) brachial plexus, axillary artery and axillary vein—*p. 92*

Muscle Group #3, p. 44

Muscle	O	I
levator scapula	7	12
pectoralis major	2	10
pectoralis minor	6	9
rhomboid major	4	13
rhomboid minor	3	14
serratus anterior	5	8
subclavius	1	11

Muscle Group #4, p. 45

1) long head—*p. 95*
2) *examples: turning a doorknob, tightening your gasoline cap, digging in the sand*
3) deltoid (anterior fibers)—*p. 95*
4) bicipital aponeurosis—*p. 96*
5) teres major and minor—*p. 97*
6) olecranon process—*p. 98*
7) extend his elbow (against your resistance)—*p. 98*
8) pectoralis major and anterior deltoid—*p. 99*
9) laterally rotate and abduct the shoulder to 45°—*p. 99*
10) pectoralis major—*p. 99*

11) lengthen
12) lengthen
13) lengthen
14) lengthen

Muscle Group #4, p. 46

Muscle	O	I
biceps brachii	2	6
coracobrachialis	1	4
triceps brachii	3	5

Other Structures, p. 47

1) acromioclavicular ligament
2) acromion
3) coracoacromial ligament
4) coracohumeral ligament
5) biceps brachii tendon (cut)
6) coracoclavicular ligament
7) trapezoid
8) conoid
9) coracoid process
10) glenohumeral joint capsule

Fill In

11) immediately release and adjust your position posteriorly—*p. 101*
12) clavicle, coracoid process—*p. 104*
13) coracoacromial—*p. 104*
14) extend—*p. 104*
15) extend the shoulder—*p. 105*
16) biceps brachii and triceps brachii—*p. 106*

Glenohumeral Joint #1, p. 48

1) acromioclavicular joint and ligament
2) supraspinatus tendon
3) acromion
4) subacromial bursa
5) deltoid
6) capsular ligament
7) synovial membrane
8) glenoid labrum
9) cartilage of glenoid cavity
10) articular capsule

Glenohumeral Joint #2, p. 49

1) acromion
2) supraspinatus tendon
3) subacromial bursa
4) infraspinatus tendon
5) glenoid cavity
6) teres minor tendon
7) synovial membrane
8) coracoid process
9) superior glenohumeral ligament
10) biceps brachii tendon (long head)
11) subscapularis tendon
12) middle glenohumeral ligament
13) inferior glenohumeral ligament

Sternoclavicular Joint, p. 50

1) clavicle
2) anterior sternoclavicular ligament
3) first rib
4) costal cartilages
5) second rib
6) radiate sternocostal ligament
7) interclavicular ligament
8) articular disc
9) joint cavity
10) costoclavicular ligament
11) sternocostal synchondrosis
12) manubrium
13) sternocostal joints

Forearm & Hand

Topographical Views, p. 52

1) brachioradialis
2) extensor crease of the wrist
3) metacarpophalangeal joints
4) lateral epicondyle
5) olecranon process
6) extensor bellies
7) shaft of the ulna
8) head of the ulna
9) extensor digitorum tendons
10) biceps brachii tendon
11) brachioradialis
12) thenar eminence
13) medial epicondyle
14) flexor bellies
15) flexor carpi radialis tendon
16) palmaris longus tendon
17) flexor carpi ulnaris tendon
18) flexor crease of the wrist
19) hypothenar eminence

Bones & Bony Landmarks, p. 53

1) ulna—*p. 110*
2) pronation, supination—*p. 110*
3) humeroulnar, humeroradial —*p. 110*
4) "flexor crease"—*p. 110*
5) triceps brachii—*p. 114*
6) lateral epicondyle—*p. 114*
7) head of the ulna—*p. 116*
8) annular—*p. 117*
9) styloid process—*p. 118*
10) head of the ulna—*p. 118*
11) carpals—*p. 118*

Humerus, p. 54

1) greater tubercle
2) deltoid tuberosity
3) lateral supracondylar ridge
4) lateral epicondyle
5) head of humerus
6) lesser tubercle
7) intertubercular groove
8) medial supracondylar ridge
9) medial epicondyle
10) head of humerus
11) medial supracondylar ridge
12) medial epicondyle
13) greater tubercle
14) deltoid tuberosity
15) lateral supracondylar ridge
16) olecranon fossa
17) lateral epicondyle

Ulna & Radius, p. 55

1) carpals
2) metacarpals
3) ulna
4) radius
5) phalanges
6) head of the radius
7) radial tuberosity
8) styloid process of the radius
9) trochlear notch
10) coronoid process
11) shaft of the ulna
12) styloid process of the ulna
13) olecranon process
14) head of the radius
15) shaft of the radius
16) lister's tubercle
17) styloid process of the radius

Carpals, p. 56

1) scaphoid
2) lunate
3) triquetrum
4) pisiform
5) scaphoid tubercle
6) trapezium tubercle
7) trapezium
8) trapezoid
9) capitate
10) hook of the hamate
11) hamate
12) lunate
13) scaphoid
14) pisiform
15) triquetrum
16) hamate
17) capitate
18) trapezoid
19) trapezium

Bones & Bony Landmarks of Wrist & Hand #1, p. 57

1) palmar, dorsal, radial, ulnar —*p. 120*
2) flexor crease of the wrist—*p. 120*
3) pisiform—*p. 121*
4) flexor carpi ulnaris—*p. 121*
5) triquetrum—*p. 121*
6) hamate—*p. 122*
7) flexor retinaculum—*p. 122*
8) ulnar artery, ulnar nerve—*p. 122*
9) pisiform, (hook of the) hamate, scaphoid (tubercle), trapezium (tubercle)—*p. 122, 124*
10) scaphoid—*p. 123*
11) trapezium—*p. 123*
12) scaphoid—*p. 123*
13) lunate, capitate—*p. 125*
14) metacarpophalangeal—*p. 126*

Extra Credit—*p. 127*

scaphoid
lunate
triquetrum
pisiform
trapezium
trapezoid
capitate
hamate

(mnemonic example beginning with proximal row of carpals: Some Lovers Try Positions That They Can't Handle)

Bones & Bony Landmarks of Wrist & Hand #2, p. 58

1) metacarpals
2) base
3) shaft
4) head
5) base
6) shaft
7) head
8) phalanges
9) proximal phalange
10) middle phalange
11) distal phalange

Muscles of Forearm #1, p. 59

1) brachioradialis
2) flexor pollicis longus
3) biceps brachii
4) brachialis
5) pronator teres
6) bicipital aponeurosis
7) flexor carpi radialis
8) palmaris longus
9) flexor carpi ulnaris
10) flexor digitorum superficialis
11) antebrachial fascia
12) palmar aponeurosis

Muscles of Forearm #2, p. 60

1) anconeus
2) flexor carpi ulnaris
3) extensor carpi ulnaris
4) extensor digiti minimi

5) extensor digitorum
6) brachioradialis
7) extensor carpi radialis longus
8) extensor carpi radialis brevis
9) abductor pollicis longus
10) extensor pollicis brevis
11) extensor pollicis longus

Muscles of Forearm #3, p. 61

1) extensor carpi ulnaris
2) extensor digiti minimi
3) extensor digitorum
4) biceps brachii
5) brachialis
6) brachioradialis
7) extensor carpi radialis longus
8) extensor carpi radialis brevis
9) abductor pollicis longus
10) extensor pollicis brevis
11) extensor pollicis longus
12) brachioradialis
13) supinator
14) extensor carpi radialis brevis (cut)
15) extensor carpi radialis longus (cut)
16) abductor pollicis longus
17) extensor pollicis brevis
18) extensor pollicis longus
19) extensor indicis

Color the Muscles #1-3, p. 62—64

Muscles and Movements #1, p. 65

1) proximal and distal radioulnar
2) supination of forearm
3) biceps brachii
   brachioradialis (assists)
   supinator
4) pronator teres
   pronator quadratus
5) adduction of fingers
6) palmar interossei
7) abduction of thumb
8) abductor pollicis longus
   abductor pollicis brevis
9) adductor pollicis
10) flexion of elbow
11) flexor carpi radialis
   flexor carpi ulnaris (assists)
   palmaris longus (assists)
   pronator teres (assists)
12) triceps brachii (all heads)

Muscles and Movements #2, p. 66

1) radiocarpal
2) abduction (radial deviation) of wrist
3) extensor carpi radialis longus
   extensor carpi radialis brevis
   extensor pollicis longus
   extensor pollicis brevis
   flexor carpi radialis
4) extensor carpi ulnaris
5) extension of thumb
6) extensor pollicis longus
   extensor pollicis brevis
   abductor pollicis longus
7) flexor pollicis longus
   flexor pollicis brevis
8) pronation of the forearm
9) pronator teres
   pronator quadratus
   brachioradialis (assists)
10) biceps brachii
   brachioradialis (assists)
11) abduction of fingers (2nd-5th)
12) dorsal interossei (2nd-4th)

Muscles and Movements #3, p. 67

1) first carpometacarpal
2) adduction of thumb
3) adductor pollicis
   palmar interossei (1st)
4) abductor pollicis longus
   abductor pollicis brevis
5) opposition of thumb
6) opponens pollicis
   flexor pollicis brevis (assists)
7) flexion of wrist
8) flexor carpi radialis
   flexor carpi ulnaris
   flexor digitorum superficialis
   flexor digitorum profundus (assists)
   flexor pollicis longus (assists)
9) extensor carpi radialis longus
   extensor carpi radialis brevis
   extensor carpi ulnaris
   extensor digitorum (assists)
   extensor indicis (assists)
10) adduction (ulnar deviation) of wrist
11) extensor carpi ulnaris
   flexor carpi ulnaris
12) flexor carpi radialis

Muscles and Movements #4, p. 68

1) humeroulnar, humeroradial
2) extension of elbow
3) triceps brachii (all heads)
   anconeus
4) biceps brachii
   brachialis
   brachioradialis
5) flexion of thumb
6) flexor pollicis longus
   flexor pollicis brevis
   adductor pollicis (assists)
7) abductor pollicis longus
8) extension of wrist
9) extensor carpi radialis longus
   extensor carpi radialis brevis
   extensor carpi ulnaris
   extensor digitorum (assists)
   extensor indicis (assists)
10) palmaris longus

What's the Muscle? #1, p. 69

1) flexor carpi radialis
2) extensor carpi ulnaris
3) extensor carpi radialis longus
4) brachioradialis
5) supinator
6) palmaris longus
7) flexor digitorum profundus
8) opponens pollicis

What's the Muscle? #2, p. 70

1) flexor digitorum superficialis
2) adductor pollicis
3) extensor carpi radialis brevis
4) pronator teres
5) flexor carpi ulnaris
6) brachialis
7) extensor digitorum

Muscle Group #1, p. 71

1) brachialis—*p. 132*
2) flexors, extensors—*p. 133*
3) brachioradialis—*p. 133*
4) flexor, lateral—*p. 147*
5) pronator teres—*p. 146*
6) biceps brachii—*p. 146*
7) extensors—*p. 147*
8) lengthen
9) shorten
10) lengthen
11) shorten

Answer Pages

Fill In

12) pronator teres
13) brachialis
14) pronator quadratus
15) brachioradialis
16) supinator

Muscle Group #1, p. 72

Muscle	O	I
brachialis	1	10
brachioradialis	2	9
pronator quadratus	3	6
pronator teres	4	8
supinator	5	7

Muscle Group #2, p. 73

1) extensor, flexor—*p. 134*
2) shaft of the ulna—*p. 134*
3) • it's a **flexor**
   • there must be an **extensor** carpi radialis
   • it extends the **carpals** (crosses the wrist joint)
   • there's a muscle that flexes the **digits**
   • on the **radial** side of forearm
   • must be a flexor carpi **ulnaris** —*p. 129*
4) extensor carpi ulnaris—*p. 135*
5) second through fifth—*p. 136*
6) extensor, flexor—*p. 135*
7) brachioradialis, extensor carpi radialis longus and brevis—*p. 138*
8) abduct—*p. 137*
9) flexor carpi radialis, palmaris longus and flexor carpi ulnaris —*p. 140*
10) four, carpal tunnel—*p. 140*
11) palmaris longus—*p. 143*
12) flexor carpi ulnaris—*p. 144*
13) ulnar shaft—*p. 145*

Muscle Group #2, p. 74

Muscle	O	I
extensor carpi radialis brevis	2	10
extensor carpi radialis longus	5	9
extensor carpi ulnaris	2	7
extensor digitorum	2	14
flexor carpi radialis	3	8
flexor carpi ulnaris	6	15
flexor digitorum profundus	1	11
flexor digitorum superficialis	4	12
palmaris longus	3	13

16) shorten
17) lengthen
18) shorten
19) lengthen
20) lengthen

Muscles of the Hand #1, p. 75

1) flexor digitorum superficialis
2) pronator quadratus
3) flexor retinaculum
4) abductor digiti minimi
5) flexor digiti minimi brevis
6) lumbricals
7) flexor digitorum profundus
8) flexor digitorum profundus
9) flexor pollicis longus
10) opponens pollicis
11) opponens digiti minimi
12) adductor pollicis

Muscles of the Hand #2, p. 76

1) radius
2) opponens pollicis
3) adductor pollicis
4) ulna
5) flexor retinaculum (cut)
6) opponens digiti minimi
7) palmar interossei
8) extensor digitorum
9) extensor digiti minimi
10) extensor carpi ulnaris
11) abductor digiti minimi
12) 2nd-4th dorsal interosseous
13) abductor pollicis longus
14) extensor pollicis brevis
15) extensor pollicis longus
16) adductor pollicis
17) first dorsal interosseous

Muscle Group #3, p. 77

1) thenar, hypothenar—*p. 156, 159*
2) eight, four—*p. 149*
3) opponens pollicis—*p. 154*
4) abductor pollicis, extensor pollicis longus and brevis—*p. 151, 153*
5) lumbrical, metacarpal—*p. 157*
6) flexor digitorum profundus —*p. 157*
7) abductor digiti minimi—*p. 159*

What's the Muscle?

8) adductor pollicis
9) palmar interossei
10) lumbricals
11) dorsal interossei

Muscle Group #3, p. 78

Muscle	O	I
abductor pollicis longus	4	6
adductor pollicis	2	7
extensor pollicis longus	5	8
flexor pollicis longus	1	8
opponens pollicis	3	9

Other Structures, p. 79

1) radial and ulnar collateral ligaments—*p. 160, 161*
2) annular ligament—*p. 161*
3) medial epicondyle and olecranon process—*p. 162*
4) olecranon bursa—*p. 162*
5) flexor tendons, median—*p. 163*
6) flexor retinaculum—*p. 163*
7) palmar aponeurosis—*p. 163*
8) radial—*p. 164*

Fill In

9) flexor retinaculum
10) antebrachial fascia
11) median nerve
12) carpals
13) carpal tunnel

Humeroulnar & Proximal Radioulnar Joints, p. 80

1) humerus
2) head of radius (deep)
3) annular ligament
4) radius
5) ulna
6) radial collateral ligament
7) articular capsule
8) radius
9) annular ligament
10) articular capsule
11) humerus
12) medial epicondyle
13) ulnar collateral ligament
14) olecranon process
15) ulna

Answer Pages

Radiocarpal Joint, p. 81

1) palmar radiocarpal ligament
2) radioscapholunate part
3) radiotriquetral part
4) radiocapitate part
5) palmar radioulnar ligament
6) palmar ulnocarpal ligament
7) ulnolunate part
8) ulnotriquetral part
9) ulnar collateral ligament
10) dorsal radioulnar ligament
11) dorsal radiocarpal ligament
12) radial collateral ligament

Intercarpal Joints & More, p. 82

1) palmar intercarpal ligaments
2) radiate carpal ligaments
3) pisohamate ligament
4) dorsal intercarpal ligaments
5) distal intercarpal ligaments
6) palmar carpometacarpal ligaments
7) palmar metacarpal ligaments
8) pisometacarpal ligament
9) dorsal carpometacarpal ligaments
10) dorsal metacarpal ligaments

Spine & Thorax

Topographical Views, p. 84

1) jugular notch
2) sternum
3) ribs
4) edge of rib cage
5) rectus abdominis
6) external oblique
7) umbilicus
8) iliac crest
9) medial border of the scapula
10) erector spinae group
11) twelfth rib
12) iliac crest
13) spinous process of C-7
14) spinous processes of thoracic and lumbar vertebrae
15) posterior superior iliac spine (PSIS)
16) sacrum

Bones & Bony Landmarks #1, p. 85

1) cervical—*p. 170*
2) sternum and rib cage—*p. 170*
3) spinous processes—*p. 176*
4) D T-12
   C T-2
   A L-4
   B C-7
   E T-7 —*p. 174*
5) flexion, extension—*p. 176*
6) body type, muscular contraction —*p. 178*
7) C-2, C-7—*p. 178, 179*
8) ligamentum nuchae—*p. 176*
9) sternocleidomastoid—*p.180*
10) mastoid process, center of shaft of clavicle—*p. 180*
11) spinous and transverse processes—*p. 181*

Bones & Bony Landmarks #2, p. 86

1) erector spinae, connecting aspect of the ribs—*p. 182*
2) two inches—*p. 182*
3) second—*p. 184*
4) costal cartilage—*p. 185*
5) intercostals—*p. 185*
6) the sides of the trunk—*p. 185*
7) clavicle—*p. 186*
8) scalenes—*p. 186*
9) slow, deep breath into upper chest—*p. 186*
10) anterior/posterior, lateral, superior—*p. 187*
11) 45 degrees—*p. 187*
12) erector spinae group—*p. 187*

Bones of Spine & Thorax, p. 87

1) cervical vertebra
2) ribs
3) thoracic vertebra
4) lumbar vertebra
5) sacrum
6) coccyx
7) manubrium
8) sternum
9) costal cartilage
10) lumbar vertebra
11) cervical spine, seven, lordotic
12) thoracic spine, twelve, kyphotic
13) lumbar spine, five, lordotic

First & Second Cervical Vertebrae, p. 88

1) posterior tubercle
2) transverse process
3) lamina
4) superior facets
5) articular facet for odontoid process
6) transverse process
7) lamina
8) vertebral foramen
9) groove for vertebral artery
10) transverse foramen
11) atlas
12) spinous process
13) vertebral foramen
14) odontoid process
15) transverse process
16) transverse foramen
17) lamina
18) axis
19) superior facets
20) odontoid process
21) lamina
22) spinous process
23) vertebral foramen
24) transverse process

Cervical Vertebrae, p. 89

1) posterior tubercle
2) anterior tubercle
3) canal for spinal nerve
4) transverse process
5) spinous process
6) transverse foramen
7) body
8) anterior tubercle
9) canal for spinal nerve
10) posterior tubercle
11) lamina groove
12) spinous process
13) lamina
14) superior facet
15) transverse process

Thoracic & Lumbar Vertebrae, p. 90

1) transverse process
2) superior facet
3) body
4) costal facets
5) spinous process
6) vertebral foramen
7) transverse process
8) lamina
9) spinous process
10) body
11) superior facet
12) lamina groove
13) spinous process
14) transverse process
15) body
16) transverse process
17) vertebral foramen

18) spinous process
19) body
20) superior facet
21) lamina groove

Rib Cage & Sternum, p. 91
1) first rib
2) second rib
3) sternocostal joint
4) costochondral joint
5) body of sternum
6) costal cartilage
7) true ribs (1-7)
8) false ribs (8-12)
9) floating ribs (11-12)
10) jugular notch
11) manubrium
12) sternal angle
13) body of sternum
14) articulations with ribs
15) xiphoid process

Muscles of Spine & Thorax #1, p. 92
1) splenius capitis
2) sternocleidomastoid
3) trapezius
4) deltoid
5) triceps brachii
6) latissimus dorsi
7) external oblique
8) thoracolumbar aponeurosis
9) semispinalis capitis
10) splenius capitis
11) splenius cervicis
12) levator scapula
13) supraspinatus
14) rhomboids
15) infraspinatus
16) teres minor
17) teres major
18) erector spinae group
19) serratus posterior inferior
20) external oblique
21) internal oblique

Muscles of Spine & Thorax #2, p. 93
1) semispinalis capitis
2) splenius capitis
3) serratus posterior superior
4) iliocostalis
5) longissimus thoracis
6) spinalis thoracis
7) serratus posterior inferior
8) internal oblique
9) rectus capitis posterior minor
10) oblique capitis superior
11) rectus capitis posterior major
12) oblique capitis inferior
13) longissimus capitis
14) spinalis cervicis
15) iliocostalis
16) longissimus thoracis
17) spinalis thoracis
18) transverse abdominis
19) thoracolumbar aponeurosis

Muscles of Spine & Thorax #3, p. 94
1) semispinalis capitis
2) splenius capitis
3) levator scapula
4) ligamentum nuchae
5) splenius cervicis
6) longissimus capitis
7) semispinalis capitis
8) splenius cervicis
9) semispinalis capitis
10) multifidi
11) rotatores

Cross Section of the Neck, p. 95
1) sternocleidomastoid
2) anterior scalene
3) middle scalene
4) posterior scalene
5) levator scapula
6) trapezius
7) multifidi and spinalis cervicis
8) semispinalis cervicis
9) longissimus cervicis
10) longissimus capitis
11) splenius cervicis
12) semispinalis capitis
13) splenius capitis

Cross Section of the Thorax #1, p. 96
1) intercostals
2) lung
3) iliocostalis
4) longissimus
5) multifidi and rotatores
6) trapezius
7) abdominal aorta
8) heart

Cross Section of the Thorax #2, p. 97
1) rectus abdominis
2) external oblique
3) internal oblique
4) transverse abdominis
5) body of L-3
6) intestines
7) psoas minor
8) psoas major
9) quadratus lumborum
10) erector spinae group

Color the Muscles #1-4, p. 98-101

Muscles and Movements #1, p. 102
1) rotation of spine (vertebral column)
2) external oblique, his left
   internal oblique, his right
3) multifidi, his right
4) extension of spine (vertebral column)
5) semispinalis capitis
   spinalis (bilaterally)
   iliocostalis (bilaterally)
   interspinalis
   intertransversarii (bilaterally)
6) rectus abdominis
7) depression/collapse (exhalation) of ribs
8) internal intercostals (assists)
   serratus posterior inferior
9) external intercostals (assists)

Muscles and Movements #2, p. 103
1) flexion of spine (vertebral column)
2) external oblique (bilaterally)
   internal oblique (bilaterally)
   iliacus (with the origin fixed)
3) quadratus lumborum (assists)
4) elevation/expansion (inhalation) of ribs
5) scalenes—anterior, middle, posterior (bilaterally)
   sternocleidomastoid (assists)
   serratus posterior superior
   serratus anterior (if scapula is fixed)
   subclavius (first rib)
6) internal intercostals (assists)
7) lateral flexion of spine (vertebral column)
8) quadratus lumborum, his right
   external oblique, his right
9) spinalis, his left

What's the Muscle? #1, p. 104
1) rectus abdominis
2) rectus capitis posterior major
3) splenius cervicis
4) oblique capitis inferior

5) quadratus lumborum
6) longissimus
7) multifidi
8) splenius capitis

What's the Muscle? #2, p. 105

1) oblique capitis superior
2) iliocostalis
3) external oblique
4) semispinalis capitis
5) rectus capitis posterior minor
6) diaphragm
7) spinalis
8) rotatores

Muscle Group #1, p. 106

1) spinalis, iliocostalis—*p. 196, illustration 4.60*
2) thoracolumbar aponeurosis —*p. 196*
3) raise and lower his feet slightly —*p. 199*
4) trapezius, rhomboids or serratus posterior superior—*p. 199*
5) short, diagonal—*p. 200*
6) lamina grooves—*p. 200*

Fill In

7) longissimus capitis
8) longissimus cervicis
9) longissimus thoracis
10) iliocostalis cervicis
11) iliocostalis thoracis
12) iliocostalis lumborum
13) spinalis cervicis
14) spinalis thoracis

Muscle Group #1, p. 107

Muscle	O	I
iliocostalis	1	12
longissimus	2	7
multifidi	3	9
rotatores	5	8
semispinalis capitis	6	11
spinalis	4	10

13) lengthen
14) shorten
15) shorten
16) lengthen

Muscle Group #2, p. 108

1) left—*p. 203*
2) trapezius, sternocleidomastoid —*p. 203*
3) rotate his head slightly toward the side you are palpating—*p. 204*
4) the trap's lateral edge is the same width as the suboccipitals—*p. 205*
5) spinous process of C-2, TVPs of C-1, the space between the superior nuchal line of the occiput and C-2—*p. 205*

Draw the muscle

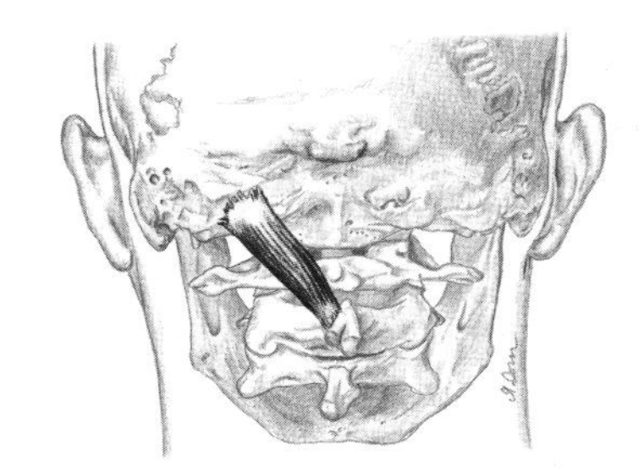

1) rectus capitis posterior major, *p. 205*

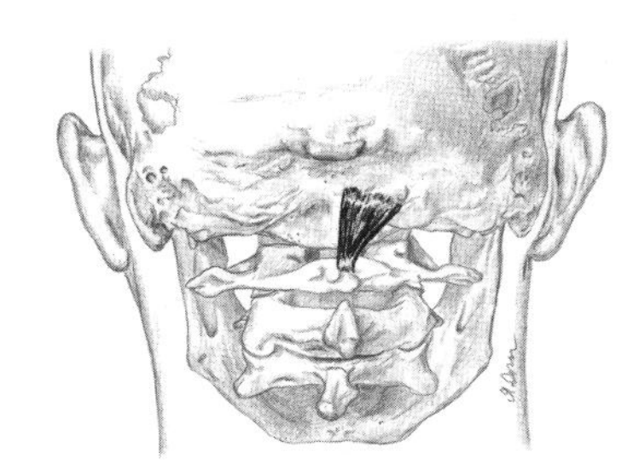

2) rectus capitis posterior minor, *p. 205*

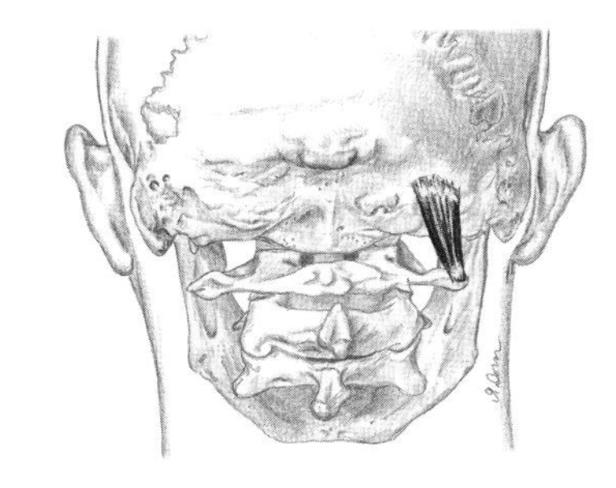

3) oblique capitis superior, *p. 205*

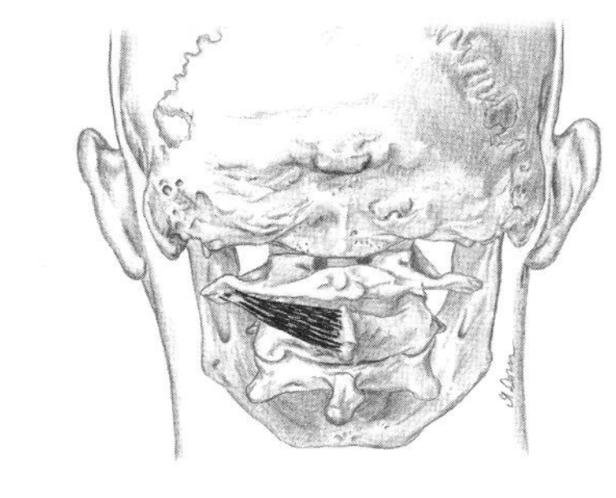

4) oblique capitis inferior, *p. 205*

Muscle Group #2, p. 109

Muscle	O	I
oblique capitis inferior	2	9
oblique capitis superior	4	6
rectus capitis post. major	2	7
rectus capitis post. minor	5	7
splenius capitis	1	8
splenius cervicis	3	10

11) shorten
12) shorten
13) shorten
14) lengthen
15) lengthen

Muscle Group #3, p. 110

1) lateral edge (side)—*p. 207*
2) twelfth rib, posterior iliac crest, transverse processes of lumbar vertebrae—*p. 208*
3) laterally tilt (elevate) the hip —*p. 208*
4) rectus abdominis—*p. 209, 210*
5) right—*p. 211*
6) external oblique—*p. 212*
7) diaphragm—*p. 213*
8) the central tendon—*p. 213*
9) only as your partner exhales —*p. 214*
10) latissimus dorsi, pectoralis major or external oblique—*p. 215*

Fill In

11) serratus anterior
12) external oblique
13) abdominal aponeurosis
14) inguinal ligament
15) rectus abdominis
16) linea alba
17) pubic crest
18) pubic symphysis

Muscle Group #3, p. 111

Muscle	O	I
diaphragm	1	10
external oblique	4	8
internal oblique	2	11
quadratus lumborum	5	12
rectus abdominis	6	9
transverse abdominis	3	7

13) shorten
14) lengthen
15) shorten
16) shorten
17) lengthen

Other Structures, p. 112
1) external occipital protuberance, spinous process of C-7—*p. 218*
2) flexion and extension—*p. 218*
3) supraspinous—*p. 219*
4) medial—*p. 219*
5) latissimus dorsi, any branches of the erector spinae group—*p. 220*

Craniovertebral Joints #1, p. 113
1) basilar portion of occiput
2) capsule of atlantooccipital joint
3) atlas (C-1)
4) capsule of lateral atlantoaxial joint
5) axis (C-2)
6) capsule of zygapophyseal (lateral) joint
7) anterior longitudinal ligament
8) alar ligaments
9) cruciform ligament
10) superior longitudinal fibers
11) transverse ligament of atlas
12) inferior longitudinal fibers
13) atlas (C-1)
14) axis (C-2)

Craniovertebral Joints #2, p. 114
1) ligamentum nuchae
2) posterior atlantooccipital membrane
3) superior longitudinal fibers of cruciform ligament
4) occiput
5) Posterior tubercle of atlas
6) posterior atlantoaxial membrane
7) posterior longitudinal ligament
8) apical ligament
9) odontoid process of axis
10) anterior tubercle of atlas
11) transverse ligament of atlas
12) anterior longitudinal ligament
13) odontoid process of axis
14) synovial cavities
15) alar ligament
16) atlas (C-1)
17) transverse ligament of atlas

Intervertebral Joints, p. 115
1) body of vertebra
2) pedicle (cut)
3) posterior longitudinal ligament
4) posterior surface of vertebral body
5) intervertebral disc
6) pedicle (cut)
7) transverse process
8) ligamentum flavum
9) superior articular process
10) lamina
11) inferior articular facet

Costovertebral & Intervertebral Joints, p. 116
1) intervertebral disc (cut)
2) body of vertebra
3) synovial cavities
4) interarticular ligament
5) radiate ligament
6) superior costotransverse ligament (cut)
7) costotransverse ligament
8) lateral costotransverse ligament
9) anterior longitudinal ligament
10) transverse process
11) supraspinous ligament
12) spinous process
13) interspinous ligament
14) intervertebral foramen
15) ligamentum flavum
16) intervertebral disc
17) body of vertebra
18) posterior longitudinal ligament

Costovertebral & Sternocostal Joints, p. 117
1) transverse process
2) ligamentum flavum
3) radiate ligaments
4) lateral costotransverse ligament
5) superior costotransverse ligament
6) rib (cut)
7) interclavicular ligament
8) articular disc
9) clavicle
10) sternomanubrial joint
11) ribs
12) radiate ligaments
13) costoxiphoid ligament
14) costoclavicular ligament
15) sternocostal joints
16) costal cartilages
17) xiphoid process

Head, Neck & Face

Topographical View, p. 119
1) temporalis
2) zygomatic arch
3) condyle of the mandible
4) masseter
5) sternocleidomastoid
6) trapezius
7) scalenes
8) clavicle
9) base of the mandible
10) hyoid bone
11) thyroid cartilage
12) jugular notch

Bones & Bony Landmarks, p. 120
1) sternocleidomastoid, base of the mandible, trachea—*p. 226*
2) posterior triangle—*p. 226*
3) twenty-two—*p. 228*
4) fibrous—*p. 228*
5) occiput—*p. 231*
6) external occipital protuberance —*p. 231*
7) superior nuchal line—*p. 231*
8) parietal—*p. 232*
9) mastoid process—*p. 233*
10) temporalis—*p. 233*
11) frontal—*p. 234*
12) submandibular fossa—*p. 235*
13) submandibular fossa—*p. 235*

Skull #1, p. 121
1) frontal
2) parietal
3) sphenoid
4) temporal
5) nasal
6) ethmoid
7) lacrimal
8) zygomatic
9) vomer
10) maxilla
11) mandible
12) sagittal suture
13) parietal
14) lambdoid suture
15) occiput
16) superior nuchal line
17) external occipital protuberance
18) mastoid process
19) maxilla
20) mandible

Answer Pages

Skull #2, p. 122
1) occiput
2) parietal
3) temporal
4) frontal
5) sphenoid
6) ethmoid
7) lacrimal
8) nasal
9) zygomatic
10) maxilla
11) mandible
12) external occipital protuberance
13) temporal lines
14) external auditory meatus
15) mastoid process
16) condyle of the mandible
17) styloid process
18) temporomandibular joint
19) zygomatic arch
20) coronoid process
21) occiput
22) temporal
23) sphenoid
24) zygomatic
25) maxilla
26) palatine
27) vomer
28) mastoid process
29) foramen magnum
30) inferior nuchal line
31) superior nuchal line
32) external occipital protuberance

Mandible & Hyoid, p. 123
1) coronoid process
2) head
3) pterygoid fossa
4) neck
5) condyle
6) ramus
7) angle
8) body
9) mental foramen
10) base
11) greater horn
12) lesser horn
13) body

Temporomandibular Joint, p. 124
1) joint capsule
2) sphenomandibular ligament
3) lateral temporomandibular ligament
4) zygomatic arch
5) external auditory meatus
6) mastoid process
7) styloid process
8) stylomandibular ligament
9) mandible
10) articular disc of temporomandibular joint
11) lateral pterygoid
12) joint capsule
13) condyle of mandible (cut)
14) mandible
15) sphenomandibular ligament

Muscles of the Head, Neck & Face #1, p. 125
1) galea aponeurotica
2) temporalis
3) occipitalis
4) digastric (posterior belly)
5) stylohyoid
6) splenius capitis
7) levator scapula
8) trapezius
9) posterior scalene
10) middle scalene
11) anterior scalene
12) omohyoid (inferior belly)
13) frontalis
14) masseter
15) digastric (anterior belly)
16) thyrohyoid
17) omohyoid (superior belly)
18) sternohyoid
19) sternothyroid
20) sternocleidomastoid

Muscles of the Head, Neck & Face #2, p. 126
1) stylohyoid
2) digastric
3) internal jugular vein
4) common carotid artery
5) thyroid cartilage
6) omohyoid (cut)
7) sternothyroid (cut)
8) sternohyoid (cut)
9) mylohyoid
10) submandibular gland
11) thyrohyoid
12) omohyoid (superior belly)
13) sternohyoid
14) scalenes
15) trapezius
16) omohyoid (inferior belly)
17) sternocleidomastoid

Muscles of Facial Expression, p. 127
1) frontalis
2) procerus
3) corrugator supercili
4) orbicularis oculi
5) nasalis
6) levator anguli oris
7) zygomaticus major
8) orbicularis oris
9) depressor anguli oris
10) mentalis
11) platysma

Color the Muscles #1-3, p. 128-130

Muscles and Movements #1, p. 131
1) protraction of mandible
2) lateral pterygoid (bilaterally)
   medial pterygoid (bilaterally)
   masseter (assists)
3) temporalis
   digastric
4) extension of neck (cervical spine)
5) splenius capitis (bilaterally)
   splenius cervicis (bilaterally)
   semispinalis capitis
6) sternocleidomastoid (bilaterally)
   scalene, anterior (bilaterally)
7) flexion of neck (cervical spine)
8) longus capitis (bilaterally)
   longus colli (bilaterally)
9) levator scapula (bilaterally)
   longissimus capitis (assists)
   longissimus cervicis (assists)

Muscles and Movements #2, p. 132
1) depression of mandible
2) geniohyoid
   digastric (with hyoid bone fixed)
3) temporalis
4) rotation of neck (cervical spine)
5) middle scalene, his left
   multifidi (his left)
   trapezius (upper fibers), his left
6) levator scapula, his left
   longus colli, his left
   longus capitis, his left
   longissimus capitis, his left (assists)
   longissimus cervicis, his left (assists)
7) retraction of mandible
8) temporalis
   digastric
9) lateral pterygoid (bilaterally)

Muscles and Movements #3, p. 133

1) temporomandibular
2) elevation of mandible
3) temporalis
masseter
medial pterygoid
4) geniohyoid
5) lateral flexion of neck (cervical spine)
6) sternocleidomastoid
scalenes—anterior, middle, posterior (with ribs fixed)
splenius capitis
splenius cervicis

What's the Muscle? #1, p. 134

1) temporalis
2) sternothyroid
3) sternocleidomastoid
4) anterior scalene
5) stylohyoid
6) longus colli
7) medial pterygoid
8) posterior scalene
9) occipitofrontalis

What's the Muscle? #2, p. 135

1) longus capitis
2) middle scalene
3) platysma
4) masseter
5) omohyoid
6) lateral pterygoid
7) mylohyoid
8) sternohyoid
9) digastric

Muscle Group #1, p. 136

1) top of manubrium, medial 1/3 of clavicle—*p. 244*
2) rotate head slightly to opposite side—*p. 245*
3) posterior—*p. 246*
4) scalenes—*p. 246*
5) anterior scalene, middle scalene —*p. 246*
6) scalenes—*p. 248*
7) sternocleidomastoid—*p. 248*
8) elevate the scapula—*p. 249*
9) masseter—*p. 250*
10) temporalis—*p. 251*
11) open your (mouth) jaw to access the coronoid process—*p. 252*
12) lengthen
13) shorten
14) shorten
15) shorten
16) lengthen
17) shorten

Muscle Group #1, p. 137

Muscle	O	I
anterior scalene	5	9
masseter	6	7
middle scalene	4	9
posterior scalene	3	11
sternocleidomastoid	2	10
temporalis	1	8

Muscle Group #2, p. 138

1) digastric, geniohyoid, mylohyoid, stylohyoid—*p. 253*
2) digastric—*p. 253*
3) press the tip of the tongue firmly against the roof of the mouth —*p. 254*
4) omohyoid—*p. 255*
5) platysma—*p. 257*
6) occipitalis, frontalis—*p. 258*
7) raise his eyebrows—*p. 258*
8) longus capitis, longus colli—*p. 260*

Muscle Group #2, p. 139

Muscle	O	I
digastric	2	8
geniohyoid	5	7
mylohyoid	5	7
omohyoid	4	7
sternohyoid	1	7
sternothyroid	1	9
stylohyoid	3	7
thyrohyoid	6	7

10) shorten
11) shorten
12) lengthen
13) shorten
14) lengthen

Muscles of Facial Expression #1, p. 140

1) corrugator supercili
2) depressor anguli oris
3) mentalis
4) levator labii superioris alaeque nasi
5) levator anguli oris
6) buccinator
7) frontalis (lateral portion)
8) risorius
9) zygomaticus major

Muscles of Facial Expression #2, p. 141

1) contempt
2) surprise
3) anger
4) sadness
5) disgust
6) happiness
7) fear

Other Structures, p. 142

1) superficial temporal artery
2) facial nerve
3) external auditory meatus
4) parotid gland
5) common carotid artery
6) parotid duct
7) facial artery
8) submandibular gland
9) thyroid cartilage
10) cricoid cartilage
11) thyroid gland
12) trachea

Fill In

13) common carotid artery—*p. 271*
14) in front of the ear along the zygomatic arch—*p. 271*
15) facial artery—*p. 271*
16) jugular notch, cricoid cartilage —*p. 273*

Pelvis & Thigh

Topographical Views, p. 143

1) rectus abdominis
2) iliac crest
3) anterior superior iliac spine (ASIS)
4) inguinal ligament
5) pubic crest
6) adductors
7) sartorius
8) rectus femoris
9) vastus medialis
10) patella
11) gluteus medius

12) greater trochanter
13) vastus lateralis
14) iliotibial tract
15) posterior superior iliac spine (PSIS)
16) sacrum
17) coccyx
18) gluteus maximus
19) gluteal cleft
20) gluteal fold
21) hamstrings
22) hamstring tendons
23) popliteal fossa

Bones & Bony Landmarks of the Pelvis #1, p. 144

1) ilium, ischium, pubis—*p. 278*
2) sacrum, coccyx—*p. 278*
3) female pelvis:
   - broader for childbearing
   - wider iliac crest
   - larger pelvic "bowl"
   - greater distance between ischial tuberosities—*p. 278*
4) iliac crest—*p. 285*
5) posterior superior iliac spine (PSIS)—*p. 286*
6) ischial tuberosities—*p. 287*
7) greater trochanter—*p. 287*
8) iliac fossa—*p. 289*

Bones & Bony Landmarks of the Pelvis #2, p. 145

1) hip
2) ilium
3) pubis
4) ischium
5) lumbar vertebra
6) sacroiliac joint
7) sacrum
8) sacrococcygeal joint
9) coccyx
10) coxal (hip) joint
11) femur
12) fifth lumbar vertebra
13) posterior superior iliac spine (PSIS)
14) median sacral crest
15) greater trochanter
16) lesser trochanter
17) gluteal tuberosity
18) ischial tuberosity
19) obturator foramen

Hip, p. 146

1) iliac crest
2) iliac fossa
3) anterior superior iliac spine (ASIS)
4) anterior inferior iliac spine (AIIS)
5) pectineal line
6) superior ramus of the pubis
7) pubic tubercle
8) symphyseal surface
9) inferior ramus of the pubis
10) posterior superior iliac spine (PSIS)
11) posterior inferior iliac spine (PIIS)
12) greater sciatic notch
13) ischial spine
14) lesser sciatic notch
15) obturator foramen
16) ischial tuberosity
17) ramus of the ischium
18) anterior gluteal line
19) posterior gluteal line
20) posterior superior iliac spine (PSIS)
21) inferior gluteal line
22) posterior inferior iliac spine (PIIS)
23) greater sciatic notch
24) ischial spine
25) lesser sciatic notch
26) obturator foramen
27) ischial tuberosity
28) iliac crest
29) iliac tubercle
30) anterior superior iliac spine (ASIS)
31) anterior inferior iliac spine (AIIS)
32) acetabulum
33) superior ramus of the pubis
34) pubic tubercle
35) inferior ramus of the pubis

Pelvis & Sacrum, p. 147

1) posterior superior iliac spine (PSIS)
2) posterior inferior iliac spine (PIIS)
3) ischial spine
4) ischial tuberosity
5) obturator foramen
6) ramus of ischium
7) inferior ramus of the pubis
8) pubic symphysis
9) sacrum
10) gluteal surface of ilium
11) coccyx
12) acetabulum

Femur, p. 148

1) greater trochanter
2) patellar surface
3) lateral epicondyle
4) lateral condyle
5) head
6) fovea of head
7) neck
8) lesser trochanter
9) intertrochanteric line
10) shaft
11) adductor tubercle
12) medial epicondyle
13) medial condyle
14) head
15) neck
16) intertrochanteric crest
17) lesser trochanter
18) pectineal line
19) medial lip of linea aspera
20) lateral lip of linea aspera
21) adductor tubercle
22) medial epicondyle
23) medial condyle
24) greater trochanter
25) trochanteric fossa
26) gluteal tuberosity
27) intercondylar fossa
28) lateral epicondyle
29) lateral condyle

Bones & Bony Landmarks of the Pelvis #3, p. 149

1) sacrum, coccyx—*p. 290, 291*
2) median sacral crest—*p. 290*
3) gluteal cleft—*p. 291*
4) sacroiliac—*p. 291*
5) flex partner's knee to 90 degrees and rotate the hip laterally and medially—*p. 291*
6) gluteal tuberosity—*p. 292*
7)
   - explain what you are doing
   - ask permission
   - use partner's hand to palpate with your hand guiding on top—*p. 293*
8) pubic tubercles—*p. 293*
9) pectineus—*p. 294*
10) pubic crest, ischial tuberosity —*p. 294*
11) supine, with your flexed knee under your partner's knee—*p. 294*
12) gluteal fold—*p. 295*

Muscles of Pelvis & Thigh #1, p. 150

1) psoas major
2) psoas minor
3) iliacus
4) inguinal ligament

5) tensor fasciae latae
6) sartorius
7) iliotibial tract
8) rectus femoris
9) vastus lateralis
10) vastus medialis
11) pectineus
12) adductor longus
13) gracilis

Muscles of Pelvis & Thigh #2, p. 151

1) gracilis
2) adductor magnus
3) semitendinosus
4) semimembranosus
5) gluteus medius
6) tensor fasciae latae
7) gluteus maximus
8) iliotibial tract
9) biceps femoris (long head)
10) biceps femoris (short head)

Muscles of Pelvis & Thigh #3, p. 152

1) gluteus maximus
2) gluteus medius
3) gluteal fascia
4) tensor fasciae latae
5) sartorius
6) rectus femoris
7) iliotibial tract
8) vastus lateralis
9) biceps femoris (long head)
10) biceps femoris (short head)

Muscles of Pelvis & Thigh #4, p. 153

1) ASIS
2) iliacus
3) psoas minor
4) pubic symphysis
5) vastus medialis
6) sartorius
7) gracilis
8) pes anserinus tendon
9) sacrum
10) piriformis
11) obturator internus
12) coccygeus
13) gluteus maximus
14) levator ani
15) adductor magnus
16) semimembranosus
17) semitendinosus

Muscles of Pelvis & Thigh #5, p. 154

1) psoas major
2) gluteus medius
3) piriformis
4) sciatic nerve (cut)
5) quadratus femoris
6) gluteus maximus (cut and reflected)
7) sacrotuberous ligament
8) adductor magnus
9) iliac crest
10) gluteus minimus
11) gemellus superior
12) obturator internus
13) gemellus inferior
14) sacrum
15) piriformis
16) gemellus superior
17) obturator internus
18) gemellus inferior
19) psoas major
20) iliac crest
21) ASIS
22) psoas major

Perineum and Pelvic Floor, p. 155

1) ischiocavernosus
2) deep transverse perineal
3) levator ani
4) anococcygeal ligament
5) coccyx
6) superficial transverse perineal
7) ischial tuberosity
8) sacrotuberous ligament
9) obturator canal
10) pubic symphysis
11) tendinous arch of levator ani
12) levator ani
13) puborectalis
14) pubococcygeus
15) iliococcygeus
16) coccygeus
17) piriformis
18) sacrum
19) coccyx

Color the Muscles #1-4, p. 156-159

Muscles and Movements #1, p. 160

1) coxal
2) extension of hip
3) biceps femoris (long head)
adductor magnus (posterior fibers)
4) rectus femoris
5) adduction of hip
6) gracilis
gluteus maximus (lower fibers)
pectineus
7) tensor fasciae latae
8) medial rotation of flexed knee
9) semitendinosus
semimembranosus
sartorius
popliteus
10) biceps femoris

Muscles and Movements #2, p. 161

1) tibiofemoral
2) extension of knee
3) vastus lateralis
vastus medialis
vastus intermedius
4) gracilis
gastrocnemius
5) abduction of hip
6) gluteus maximus (all fibers)
gluteus medius (all fibers)
gluteus minimus
7) pectineus
8) medial rotation of hip
9) gluteus medius (anterior fibers)
gluteus minimus
gracilis
pectineus
10) biceps femoris (long head, assists)

Muscles and Movements #3, p. 162

1) tibiofemoral
2) flexion of knee
3) biceps femoris
gracilis
gastrocnemius
4) rectus femoris
5) lateral rotation of flexed knee
6) biceps femoris
7) semitendinosus
semimembranosus
sartorius
8) lateral (external) rotation of hip
9) piriformis
psoas major
iliacus
10) adductor magnus
adductor longus
adductor brevis
11) flexion of hip
12) tensor fasciae latae
sartorius
13) gluteus maximus (all fibers)
gluteus medius (posterior fibers)

Answer Pages

What's the Muscle? #1, p. 163

1) gluteus maximus
2) adductor brevis
3) vastus intermedius
4) obturator externus
5) semitendinosus
6) iliacus
7) sartorius
8) gemellus superior
9) pectineus

What's the Muscle? #2, p. 164

1) piriformis
2) vastus intermedius
3) adductor magnus
4) adductor longus
5) psoas minor
6) tensor fasciae latae
7) quadratus femoris
8) gluteus medius
9) semimembranosus

What's the Muscle? #3, p. 165

1) obturator internus
2) vastus medialis
3) gracilis
4) biceps femoris (long head)
5) gluteus minimus
6) rectus femoris
7) psoas major
8) gemellus inferior

Muscle Group #1, p. 166

1) coxal (hip), tibiofemoral (knee) —*p. 296*
2) rectus femoris—*p. 306*
3) vastus lateralis—*p. 306*
4) anterior inferior iliac spine, patella—*p. 309*
5) vastus medialis—*p. 310*
6) ischial tuberosity—*p. 311*
7) vastus lateralis, adductor magnus—*p. 311*
8) laterally—*p. 312*
9) semitendinosus —*p. 311, diagram 6.68*

Muscle Group #1, p. 167

Muscle	O	I
biceps femoris	4	7
rectus femoris	2	10
semimembranosus	3	8
semitendinosus	3	9
vastus intermedius	1	10
vastus lateralis	5	10
vastus medialis	6	10

11) lengthen
12) lengthen
13) shorten
14) shorten
15) shorten
16) lengthen, shorten

Muscle Group #2, p. 168

1) gluteus maximus—*p. 315*
2) gluteus medius—*p. 315*
3) gluteus maximus—*p. 317*
4) gluteus medius—*p. 318*
5) "abduct your hip"—*p. 318*
6) superior ramus of the pubis, ischial tuberosity—*p. 319*
7) adductor magnus—*p. 319*
8) knee—*p. 319*
9) adduct the hip, medially rotate the hip—*p. 319*
10) pubic tubercle—*p. 322*
11) pectineus—*p. 323*
12) adductor magnus—*p. 323*

Muscle Group #2, p. 169

Muscle	O	I
adductor brevis	4	13
adductor longus	6	11
adductor magnus	5	12
gluteus maximus	1	9
gluteus medius	3	10
gluteus minimus	2	8
gracilis	4	15
pectineus	7	14

16) lengthen
17) shorten
18) shorten
19) lengthen
20) lengthen
21) lengthen
22) shorten
23) lengthen

Muscle Group #3, p. 170

1) tensor fasciae latae—*p. 324*
2) iliotibial tract—*p. 324*
3) "medially rotate your hip"—*p. 325*
4) sartorius—*p. 326*
5) femoral—*p. 326*
6) semitendinosus, gracilis, sartorius—*p. 327*
7) piriformis—*p. 328*
8) coccyx, posterior superior iliac spine, greater trochanter—*p. 330*
9) quadratus femoris—*p. 331*
10) psoas major—*p. 332*
11) anterior superior iliac spine, navel—*p. 334*
12) • explain what you are doing
 • communicate with your partner
 • slowly remove your hands if your partner feels unsafe or uncomfortable
 • have client take a deep breath and compress on their exhale
 • use small circles as you compress
 • be mindful of the pulse of the abdominal aorta and reposition laterally if you feel it—*p. 334*
13) "flex your hip ever so slightly" —*p. 334*

Muscle Group #3, p. 171

Muscle	O	I
gemellus inferior	8	15
gemellus superior	7	15
iliacus	6	14
obturator externus	11	19
obturator internus	10	15
piriformis	2	12
psoas major	4	14
psoas minor	3	18
quadratus femoris	9	16
sartorius	1	17
tensor fasciae latae	5	13

Muscle Group #3, p. 172

1) shorten
2) lengthen
3) lengthen
4) shorten
5) lengthen
6) shorten
7) shorten
8) shorten
9) lengthen
10) shorten

Other Structures #1, p. 173
1) inguinal ligament
2) sartorius
3) adductor longus
4) femoral nerve
5) femoral artery
6) femoral vein
7) inguinal ligament
8) adductor longus
9) inguinal lymph nodes
10) great saphenous vein
11) sartorius

Joints & Ligaments #1, p. 174
1) anterior longitudinal ligament
2) iliolumbar ligament
3) anterior sacroiliac ligament
4) sacrotuberous ligament
5) inguinal ligament
6) sacrospinous ligament
7) pubic symphysis
8) supraspinous ligament
9) iliolumbar ligament
10) posterior sacroiliac ligament
11) sacrotuberous ligament
12) hamstrings tendon
13) posterior sacrococcygeal ligaments
14) sacrospinous ligament

Joints & Ligaments #2, p. 175
1) posterior sacroiliac ligament
2) sacrotuberous ligament
3) sacrospinous ligament
4) articular capsule of coxal joint
5) tendon of rectus femoris (cut)
6) acetabulum
7) lunate surface of acetabulum
8) round ligament (ligamentum capitis femoris—cut)
9) obturator membrane
10) anterior sacroiliac ligament
11) sacrospinous ligament
12) sacrotuberous ligament
13) obturator membrane
14) pubic symphysis

Coxal Joint, p. 176
1) iliofemoral ligament
2) pubofemoral ligament
3) femur
4) iliofemoral ligament
5) ischiofemoral ligament
6) zona orbicularis
7) articular cartilage
8) lunate surface of acetabulum
9) acetabular labrum
10) transverse acetabular ligament
11) round ligament (ligamentum capitis femoris—cut)

Other Structures #2, p. 177
1) anterior superior iliac spine, pubic tubercle—*p. 339*
2) femoral artery, femoral nerve, femoral vein—*p. 339*
3) between ASIS and pubic tubercle, just distal to the inguinal ligament—*p. 339*
4) sacrotuberous ligament—*p. 340*
5) sacroiliac—*p. 340*
6) iliolumbar—*p. 341*
7) sciatic nerve—*p. 341*
8) trochanteric bursa—*p. 342*

Leg & Foot

Topographical Views, p. 179
1) popliteal fossa
2) patella
3) tibial tuberosity
4) pes anserinus attachment site
5) gastrocnemius
6) tibialis anterior
7) shaft of the tibia
8) calcaneal tendon
9) lateral malleolus
10) medial malleolus
11) tibialis anterior tendon
12) tibialis anterior tendon
13) extensor hallucis longus tendon
14) extensor digitorum longus tendons

Bones & Bony Landmarks of the Knee & Leg, p. 180
1) tibiofemoral—*p. 346*
2) flexed—*p. 346*
3) tibia, fibula—*p. 346*
4) proximal tibia, femoral condyles —*p. 350*
5) tibial tuberosity—*p. 350*
6) patellar ligament—*p. 307 (box), 350*
7) biceps femoris, soleus, fibular collateral ligament—*p. 351*
8) the edges—*p. 351*
9) sartorius, gracilis, semitendinosus—*p. 352*
10) edges of femoral condyles—*p. 352*
11) iliotibial tract—*p. 353*
12) adductor tubercle, adductor magnus—*p. 353*

Bones of the Knee, Leg & Foot, p. 181
1) femur
2) patella
3) tibia
4) fibula
5) talus
6) tarsals
7) metatarsals
8) phalanges
9) medial and lateral intercondylar tubercles
10) lateral condyle
11) head of the fibula
12) lateral malleolus
13) medial condyle
14) tibial tuberosity
15) soleal line
16) medial malleolus
17) medial malleolus
18) fossa of lateral malleolus
19) lateral condyle
20) lateral malleolus

Bony Landmarks of the Knee and Leg, p. 182
1) tibial tuberosity
2) adductor tubercle
3) medial epicondyle
4) medial condyle
5) tibial plateau
6) pes anserinus attachment site
7) lateral epicondyle
8) lateral condyle
9) tibial plateau
10) tibial tubercle
11) head of the fibula
12) tibial tuberosity
13) tibia

Bones of the Foot, p. 183
1) calcaneus
2) talus
3) cuboid
4) navicular
5) cuneiforms
6) metatarsals
7) phalanges
8) phalanges
9) sesamoid bones

10) metatarsals
11) cuneiforms
12) navicular
13) cuboid
14) talus
15) calcaneus

Bones & Bony Landmarks of the Foot #1, p. 184

1) lateral and middle cuneiforms
2) metatarsals
3) base
4) shaft
5) head
6) tuberosity of fifth metatarsal
7) cuboid
8) peroneal trochlea
9) calcaneus
10) talus
11) navicular
12) base of first metatarsal
13) trochlea of the talus
14) talus
15) medial tubercle of talus
16) sustentaculum tali
17) calcaneus
18) tuberosity of calcaneus
19) head of the talus
20) navicular tubercle
21) medial cuneiform
22) head
23) shaft
24) base
25) phalanges

Calcaneus & Talus, p. 185

1) body
2) peroneal trochlea
3) groove for peroneus longus tendon
4) articular surfaces for talus
5) tuberosity
6) sustentaculum tali
7) groove for flexor hallucis longus tendon
8) trochlea
9) neck
10) head
11) lateral tubercle
12) lateral process
13) tarsal sinus
14) trochlea
15) neck
16) head
17) medial tubercle

Bones & Bony Landmarks of the Foot #2, p. 186

1) calcaneus, talus—*p. 354*
2) dorsal—*p. 354*
3) lateral malleolus—*p. 357*
4) invert the foot—*p. 357*
5) distal, one inch—*p. 359*
6) medial malleolus, navicular tubercle—*p. 360*
7) invert and plantar flex—*p. 361*
8) medial cuneiform—*p. 362*
9) dorsal, medial—*p. 362*
10) proximal interphalangeal, distal interphalangeal—*p. 362*
11) peroneus brevis—*p. 363*
12) tibialis anterior—*p. 364*
13) tuberosity of fifth metatarsal —*p. 365*
14) tuberosity of fifth metatarsal, lateral malleolus—*p. 365*

Muscles of the Leg & Foot #1, p. 187

1) plantaris
2) gastrocnemius
3) soleus
4) calcaneal tendon
5) tendons of flexors of ankle and toes
6) peroneal tendons
7) superior peroneal retinaculum
8) flexor retinaculum
9) gastrocnemius (cut)
10) plantaris
11) popliteus
12) soleus
13) gastrocnemius (cut)
14) calcaneal tendon

Muscles of Leg & Foot #2, p. 188

1) peroneus longus
2) tibialis anterior
3) gastrocnemius
4) soleus
5) extensor digitorum longus
6) peroneus brevis
7) peroneus longus
8) tibialis anterior
9) gastrocnemius
10) soleus
11) peroneus brevis
12) extensor digitorum longus
13) extensor hallucis longus

Muscles of Leg & Foot #3, p. 189

1) tibialis anterior
2) extensor hallucis longus
3) extensor digitorum longus
4) peroneus longus
5) peroneus brevis
6) flexor hallucis longus
7) tibialis posterior
8) flexor digitorum longus
9) soleus
10) calcaneal tendon

Color the Muscles #1-2, p. 190 & 191

Muscles and Movements #1, p. 192

1) talocrural
2) dorsiflexion of ankle
3) tibialis anterior
extensor digitorum longus
extensor hallucis longus
4) flexor digitorum longus (weak)
flexor hallucis longus (weak)
5) eversion of foot
6) peroneus longus
peroneus brevis
7) tibialis anterior
tibialis posterior
8) flexion of toes
9) flexor digitorum longus
flexor digitorum brevis
flexor digiti minimi brevis (5th)
10) lumbricals

Muscles and Movements #2, p. 193

1) metatarsophalangeal, proximal interphalangeal, distal interphalangeal
2) extension of toes
3) extensor digitorum longus
extensor digitorum brevis (2nd—4th)
lumbricals
4) flexor digitorum longus
flexor digitorum brevis
flexor digiti minimi brevis (5th)
5) inversion of foot
6) tibialis anterior
tibialis posterior
flexor digitorum longus
flexor hallucis longus
7) extensor digitorum longus
8) plantar flexion of ankle
9) soleus
tibialis posterior
10) extensor digitorum longus
extensor hallucis longus

Answer Pages

What's the Muscle? #1, p. 194

1) popliteus
2) extensor hallucis longus
3) flexor digitorum brevis
4) extensor digitorum longus
5) soleus
6) gastrocnemius
7) tibialis posterior
8) abductor digiti minimi

What's the Muscle? #2, p. 195

1) extensor digitorum brevis
2) plantaris
3) peroneus longus
4) abductor hallucis
5) flexor hallucis longus
6) tibialis anterior
7) flexor digitorum longus
8) peroneus brevis

Muscle Group #1, p. 196

1) gastrocnemius, soleus—*p. 371*
2) calcaneal tendon—*p. 371*
3) stand on his toes—*p. 372*
4) medial—*p. 372*
5) medially, shaft of the tibia—*p. 373*
6) plantaris—*p. 374*
7) inch, oblique—*p. 374*
8) popliteus—*p. 375*
9) unlocking the joint—"the key which unlocks the knee"—*p. 375*
10) soleus, gastrocnemius—*p. 375*
11) lateral, extensor digitorum longus, soleus—*p. 376*
12) head of the fibula, lateral malleolus—*p. 377*
13) evert the foot—*p. 377*

Muscle Group #1, p. 197

Muscle	O	I
gastrocnemius	4	8
peroneus brevis	1	10
peroneus longus	5	7
plantaris	3	8
popliteus	2	9
soleus	6	8

11) lengthen
12) lengthen
13) lengthen
14) shorten

Muscle Group #2, p. 198

1) tibial shaft—*p. 378*
2) dorsiflex or invert the foot—*p. 379*
3) extensor retinacula—*p. 378, 380*
4) tibial shaft, edge of the soleus/calcaneal tendon—*p. 381*
5) tibialis posterior
   flexor digitorum longus
   tibial artery
   tibial nerve
   flexor hallucis longus—*p. 383*
6) wiggle all his toes—*p. 383*
7) extensor digitorum brevis—*p. 387*
8) plantar aponeurosis—*p. 387*
9) extensor digitorum longus—*p. 387*
10) plantar surface of the heel, second through fifth toes—*p. 389*
11) flex the first toe—*p. 389*
12) calcaneus, head of the first metatarsal, head of the fifth metatarsal—*p. 385*

Muscle Group #2, p. 199

Muscle	O	I
extensor digitorum longus	1	10
extensor hallucis longus	3	7
flexor digitorum longus	5	8
flexor hallucis longus	4	7
tibialis anterior	2	9
tibialis posterior	6	11

12) shorten
13) lengthen
14) shorten
15) lengthen

Muscles of the Foot #1, p. 200

1) flexor digitorum longus tendons
2) flexor digiti minimi brevis
3) flexor digitorum brevis
4) abductor digiti minimi
5) flexor hallucis brevis
6) abductor hallucis
7) calcaneus
8) adductor hallucis
9) flexor digiti minimi brevis
10) base of 5th metatarsal
11) cuboid
12) quadratus plantae
13) flexor hallucis brevis
14) flexor digitorum longus tendon
15) flexor hallucis longus tendon

Muscles of the Foot #2, p. 201

1) peroneus longus and brevis
2) extensor digitorum longus and peroneus tertius
3) superior extensor retinaculum
4) inferior peroneal retinaculum
5) extensor digitorum brevis
6) tibialis anterior
7) extensor hallucis longus
8) inferior extensor retinaculum
9) extensor hallucis brevis
10) soleus
11) peroneus longus
12) peroneus brevis
13) calcaneal tendon
14) superior peroneal retinaculum
15) inferior peroneal retinaculum
16) calcaneus
17) abductor digiti minimi
18) tibialis anterior
19) extensor digitorum longus
20) superior extensor retinaculum
21) inferior extensor retinaculum

Tibiofemoral Joint, p. 202

1) femur
2) anterior cruciate ligament
3) lateral meniscus
4) fibular collateral ligament
5) anterior ligament of head of the fibula
6) fibula
7) posterior cruciate ligament
8) tibial collateral ligament
9) medial meniscus
10) transverse ligament of knee
11) tibia
12) patellar ligament (cut)
13) posterior meniscofemoral ligament
14) posterior cruciate ligament
15) anterior cruciate ligament
16) fibular collateral ligament
17) popliteus tendon (cut)
18) lateral meniscus
19) posterior ligament of head of the fibula
20) medial meniscus
21) tibial collateral ligament

Answer Pages

Tibiofemoral & Tibiofibular Joints, p. 203

1) anterior cruciate ligament (cut)
2) lateral meniscus
3) posterior meniscofemoral ligament (cut)
4) posterior cruciate ligament (cut)
5) medial meniscus
6) iliotibial tract (cut)
7) fibular collateral ligament (cut)
8) biceps femoris tendon (cut)
9) anterior ligament of head of fibula
10) interosseous membrane
11) fibula
12) anterior tibiofibular ligament
13) anterior talofibular ligament (cut)
14) cruciate ligaments (cut)
15) tibial collateral ligament (cut)
16) patellar ligament (cut)
17) tibia

Other Structures of the Knee, p. 204

1) quadriceps femoris tendon
2) femur
3) patella
4) prepatellar bursa
5) patellar ligament
6) subcutaneous infrapatellar bursa
7) deep infrapatellar bursa
8) tibia
9) hamstrings
10) popliteal artery and vein
11) tibial nerve
12) common peroneal nerve
13) gastrocnemius
14) lesser saphenous vein

Talocrural Joint, p. 205

1) posterior tibiofibular ligament
2) lateral malleolus
3) anterior tibiofibular ligament
4) anterior talofibular ligament
5) posterior talofibular ligament
6) calcaneofibular ligament
7) deltoid ligament
8) posterior tibiotalar ligament
9) tibiocalcaneal ligament
10) anterior tibiotalar ligament
11) tibionavicular ligament
12) navicular
13) sustentaculum tali
14) medial malleolus

Talocrural & Talotarsal Joints, p. 206

1) tibia
2) talus
3) deltoid ligament
4) fibula
5) posterior tibiofibular ligament
6) posterior talofibular ligament
7) calcaneofibular ligament
8) posterior talocalcaneal ligament
9) calcaneus
10) posterior talocalcaneal ligament
11) talonavicular ligament
12) talus
13) navicular
14) lateral talocalcaneal ligament
15) interosseous talocalcaneal ligament

Ligaments of the Foot, p. 207

1) plantar metatarsal ligaments
2) long plantar ligament
3) plantar calcaneocuboid (short plantar) ligament
4) plantar cuboideonavicular ligament
5) navicular
6) plantar calcaneonavicular (spring) ligament
7) plantar calcaneocuboid (short plantar) ligament
8) plantar calcaneonavicular (spring) ligament
9) long plantar ligament
10) dorsal cuneonavicular ligaments
11) dorsal intercuneiform ligaments
12) dorsal cuneocuboid ligament
13) dorsal cuboideonavicular ligament
14) bifurcate ligament
15) dorsal calcaneocuboid ligaments

Other Structures of the Knee, Leg & Foot, p. 208

1) lateral epicondyle of femur, head of the fibula—*p. 394*
2) medial—*p. 394*
3) weight distribution, friction reduction—*p. 395*
4) medially—*p. 395*
5) prepatellar bursa—*p. 396*
6) medial, lateral, posterior—*p. 397*
7) flex and extend the toes—*p. 404*
8) talus, sustentaculum tali, navicular—*p. 401*
9) sustentaculum tali, navicular tubercle, tibialis posterior tendon—*p. 401*
10) fibers are superficial and perpendicular—*p. 403*
11) medial malleolus, medial calcaneus—*p. 403*
12) inferior, posterior—*p. 404*
13) dorsalis pedis—*p. 405*
14) calcaneal tendon, overlying skin—*p. 405*